Lecture Notes in Artificial Intelligence   10022

Subseries of Lecture Notes in Computer Science

More information about this series at http://www.springer.com/series/1244

Manuel Montes-y-Gómez · Hugo Jair Escalante
Alberto Segura · Juan de Dios Murillo (Eds.)

Advances in Artificial Intelligence – IBERAMIA 2016

15th Ibero-American Conference on AI
San José, Costa Rica, November 23–25, 2016
Proceedings

 Springer

Editors
Manuel Montes-y-Gómez
INAOE
Tonantzintla
Mexico

Hugo Jair Escalante
INAOE
Puebla
Mexico

Alberto Segura
Universidad Nacional de Costa Rica
Heredia
Costa Rica

Juan de Dios Murillo
Universidad Nacional de Costa Rica
Heredia
Costa Rica

ISSN 0302-9743   ISSN 1611-3349   (electronic)
Lecture Notes in Artificial Intelligence
ISBN 978-3-319-47954-5   ISBN 978-3-319-47955-2   (eBook)
DOI 10.1007/978-3-319-47955-2

Library of Congress Control Number: 2016938377

LNCS Sublibrary: SL7 – Artificial Intelligence

Printed on acid-free paper

This Springer imprint is published by Springer Nature
The registered company is Springer International Publishing AG
The registered company address is: Gewerbestrasse 11, 6330 Cham, Switzerland

Preface

This volume of *Lecture Notes in Artificial Intelligence* contains the papers presented at the 15th Ibero-American Conference on Artificial Intelligence (IBERAMIA 2016), held at San José, Costa Rica, during November 23–25, 2016.

IBERAMIA is a biennial international conference supported by the main Ibero-American societies of artificial intelligence. It is the leading symposium where the Ibero-American AI community comes together to share research results and experiences with researchers in Artificial Intelligence from all over the world. It has been held uninterruptedly since 1988, and with the years, it has expanded its scope to become a well-recognized international conference in the AI field, with papers published in English by Springer in the LNCS/LNAI series since the sixth edition (Lisbon 1998).

The technical structure of this year's conference was composed of the main technical sessions, keynote talks, and workshops. The organizational structure of the scientific program of IBERAMIA 2016 was as follows: the conference was organized along several areas of AI, each coordinated by one or two area chairs (ACs) who are recognized experts in their fields of expertise. ACs were responsible for selecting the Program Committee (PC) members. In total, IBERAMIA 2016 involved 18 ACs, more than 150 PC members (from 25 countries, including 15 that were outside the Ibero-American geographical area), and a considerable number of additional reviewers.

IBERAMIA 2016 received 75 papers from 31 different countries, attesting to the truly international nature of the conference. After review by the international PC, 34 papers were accepted for presentation and publication in this volume. These papers are structured into the following nine sections:

- Knowledge Engineering, Knowledge Representation and Probabilistic Reasoning
- Agent Technology and Multi-agent Systems
- Planning and Scheduling
- Natural Language Processing
- Machine Learning
- Big Data, Knowledge discovery and Data Mining
- Computer Vision and Pattern Recognition
- Computational Intelligence and Soft Computing
- AI in Education, Affective Computing, and Human–Computer Interaction

IBERAMIA 2016 was honored by the participation of three outstanding women as keynote speakers: Emilia Mendes from the Blekinge Institute of Technology, Sweden, Rada Mihalcea from the University of Michigan, USA, and Rosa Maria Vicari from the Federal University of Rio Grande do Sul, Brazil. We thank them for giving us such excellent keynote lectures.

We would like to thank: the authors for submitting their work to IBERAMIA; the ACs, PC members, and additional reviewers for their hard work and valuable time; the sponsors; and Springer for agreeing to print this volume. Also, we thank Francisco

Garijo and Federico Barber from IBERAMIA's Executive Board for their continuous support in administrative matters, as well as for supporting the website of the conference. We also want to acknowledge EasyChair for the facilities provided to support the submission and review of the papers, as well as for the preparation of the proceedings.

Finally, it is important to mention that nothing would have been possible without the initiative and dedication of the Organizing Committee from the Universidad Nacional de Costa Rica. We are very grateful to all the people who helped in the large variety of organizing tasks.

November 2016

Manuel Montes-y-Gómez
Hugo Jair Escalante
Alberto Segura Gutiérrez
Juan de Dios Murillo

Organization

Program Chair

Manuel Montes-y-Gómez   Instituto Nacional de Astrofísica, Óptica y Electrónica,
Mexico

Organization Chair

Alberto Segura Gutiérrez   Universidad Nacional de Costa Rica, Costa Rica

Area Chairs

Knowledge Engineering and Representation, Probabilistic Reasoning

Renata Wassermann   Universidade de So Paulo, Brazil
Eduardo Fermé   Universidad de Madeira, Portugal

Agent Technology and Multi-agent Systems

Juan Antonio   IIIA-CSIC, Spain
    Rodríguez-Aguilar
Juan Carlos Burguillo   Universidad de Vigo, Spain

Planning and Scheduling

Miguel Salido   Universidad Politécnica de Valencia, Spain

Robotics

Luis Chaimowicz   Universidade Federal de Minas Gerais, Brazil

Natural Language Processing

Thamar Solorio   University of Houston, USA
Laura Alonso Alemany   Universidad de Córdoba, Argentina

Machine Learning

Fabio González   Universidad Nacional de Colombia, Colombia
Eduardo Morales   Instituto Nacional de Astrofísica, Óptica y Electrónica,
Mexico

Big Data, Knowledge Discovery and Data Mining

João Gama   Universidade do Porto, Portugal
Ángel Kuri   Instituto Tecnológico Autónomo de México, Mexico

Computer Vision and Pattern Recognition

Juan Carlos Niebles   Universidad del Norte, Colombia; Stanford, USA
Alvaro Soto   Universidad Católica de Chile, Chile

Computational Intelligence and Soft Computing

Leticia Cagnina   Universidad de San Luis, Argentina
Luis Correia   Universidade de Lisboa, Portugal

AI in Education, Affective Computing, and Human–Computer Interaction

Rosa Vicari   Universidade Federal do Rio Grande do Sul, Brazil
Silvia Schiaffino   Universidad Nacional del Centro de la Provincia de
                           Buenos Aires, Argentina

Program Committee

Silvana Aciar   Instituto de Informática, Universidad Nacional de San
                           Juan, Argentina
Iñaki Alegria   Euskal Herriko Unibertsitatea, Spain
Laura Alonso Alemany   Universidad Nacional de Córdoba, Argentina
Matias Alvarado   Centro de Investigación y de Estudios Avanzados del
                           IPN, Mexico
Javier Apolloni   Universidad Nacional de San Luis, Argentina
Carlos Argueta   National Tsing Hua University, Taiwan
Marcelo Gabriel Armentano Instituto Superior de Ingeniería de Software Tandil
                           (CONICET - UNCPBA), Argentina
Paulo Azevedo   Universidade do Minho, Portugal
Silvia Margarita Baldiris Universitat de Girona, Spain
  Navarro
Federico Barber   Universidad Politécnica de Valencia, Spain
Raúl Cruz Barbosa   Universidad Tecnológica de la Mixteca, Mexico
Roman Barták   Charles University in Prague, Czech Republic
Núria Bel   Universitat Pompeu Fabra, Spain
Albert Bifet   University of Waikato, New Zealand
Blai Bonet   Universidad Simon Bolivar, Venezuela
Silvia Botelho   Fundação Universidade Federal do Rio Grande, Brazil
Ramón F. Brena   Tecnológico de Monterrey, Mexico
Facundo Bromberg   Universidad Tecnológica Nacional, Argentina
Juan Carlos Burguillo   Universidad de Vigo, Spain
Benjamin Bustos   University of Chile, Chile
Aleksander Byrski   AGH University Science and Technology, Poland
Pedro Cabalar   University of Corunna, Spain
Daniel Cabrera-Paniagua   Universidad de Valparaiso, Chile
Leticia Cagnina   Universidad Nacional de San Luis, Argentina

Jorge E. Camargo	Universidad Antonio Nariño, Colombia
Javier Carbo	Universidad Carlos III de Madrid, Spain
Douglas Cardoso	Universidade Federal do Rio de Janeiro, Brazil
Jesús Ariel Carrasco-Ochoa	Instituto Nacional de Astrofísica, Óptica y Electrónica, Mexico
Henry Carrillo	Pontificia Universidad Javeriana, Colombia
Andre Carvalho	Universidade de Sao Paolo, Brazil
Ana Casali	Universidad Nacional de Rosario, Centro Internacional Franco Argentino de Ciencias de la Información y de Sistemas, Argentina
Jose Cascalho	Universidade dos Azores, Portugal
José Castaño	Universidad de Buenos Aires, Argentina
Irene Castellón	Universitat de Barcelona, Spain
Luis Fernando Castillo	Universidad de Caldas, Colombia
Amedeo Cesta	National Research Council of Italy, Italy
Luiz Chaimowicz	Universidade Federal de Minas Gerais, Brazil
Carlos Chesñevar	Universidad Nacional del Sur, Argentina
Luis Correia	Universidade de Lisboa, Portugal
Paulo Cortez	Universidade do Minho, Portugal
Anna Helena Reali Costa	Universidade de São Paulo, Brazil
Mayela Coto	Universidad Nacional, Costa Rica
Marco Cristo	UFAM, Brazil
Claudio Cubillos	Pontificia Universidad Católica de Valparaíso, Chile
Agostinho Da Rosa	Universidade de Lisboa, Portugal
José Del Campo-Ávila	Universidad de Málaga, Spain
Jorge Dias	Institute of Systems and Robotics, Portugal
Juan J. Durillo	University of Innsbruck, Austria
Marcelo Errecalde	Universidad Nacional de San Luis, Argentina
Hugo Jair Escalante	Instituto Nacional de Astrofísica, Óptica y Electrónica, Mexico
Victor Escorcia	King Abdullah University of Science and Technology, Saudi Arabia
Anna I Esparcia Alcazar	Universitat Politècnica de València, Spain
Paula Estrella	Universidad Nacional de Córdoba, Argentina
Ramon Fabregat	Universitat de Girona, Spain
Elaine Faria	Federal University of Uberlandia, Brazil
Alessandro Farinelli	Verona University, Italy
Eduardo Fermé	Universidade da Madeira, Portugal
Jose Luis Fernandez-Marquez	University of Geneva, Switzerland
Carlos Ferreira	LIAAD INESC Porto LA, Portugal
James Foster	University of Idaho, USA
Victor Fragoso	West Virginia University, USA
Rubén Fuentes-Fernández	Universidad Complutense de Madrid, Spain
João Gama	Universidade do Porto, Portugal
Roberto García	Universitat de Lleida, Spain

Enrico Gerding	University of Southampton, UK
Daniela Godoy	Instituto Superior de Ingeniería de Software Tandil (CONICET - UNCPBA), Argentina
Sergio Alejandro Gómez	Universidad Nacional del Sur, Argentina
Luiz Goncalves	Universidade Federal do Rio Grande do Norte, Brazil
Carina González	Universidad de La Laguna, Spain
Enrique González	Pontificia Universidad Javeriana, Colombia
Fabio González	Universidad Nacional de Colombia, Colombia
Juan Carlos González Moreno	Universidad de Vigo, Spain
Nathan Griffiths	University of Warwick, UK
Miguel Ángel Guevara López	University of Porto, Portugal
Francisco Luis Gutiérrez Vela	Universidad de Granada, Spain
Carlos Hernandez	Universidad Andres Bello, Chile
Ángel Kuri	Instituto Tecnológico Autónomo de México, Mexico
Nuno Lau	University of Aveiro, Portugal
Weidong Li	Conventry University, UK
Patricio Loncomilla	Universidad de Chile, Chile
Magalí Teresinha Longhi	Federal University of Rio Grande do Sul, Brazil
Antonio López	Universitat Autònoma de Barcelona, Spain
Douglas Macharet	Universidade Federal de Minas Gerais, Brazil
Samhar Mahmoud	King's College London, UK
Benedita Malheiro	Instituto Superior de Engenharia do Porto, Portugal
Enrico Marchioni	University of Oxford, UK
Joao Marques-Silva	University of Lisbon, Portugal
María Vanina Martínez	Universidad Nacional del Sur in Bahia Blanca and CONICET, Argentina
José Fco. Martínez-Trinidad	Instituto Nacional de Astrofísica, Óptica y Electrónica, Mexico
Vicente Matellan	University of Leon, Spain
Denis Maua	Universidade de São Paulo, Brazil
João Mendes-Moreira	FEUP/DEI, Portugal
Domingo Mery	Pontificia Universidad Católica de Chile, Chile
Diego Milone	Universidad Nacional del Litoral, Argentina
José M. Molina	Universidad Carlos III de Madrid, Spain
Manuel Montes-y-Gómez	Instituto Nacional de Astrofísica, Óptica y Electrónica, Mexico
Eduardo Morales	Instituto Nacional de Astrofísica, Optica y Electrónica, Mexico
Julian Moreno	Universidad Nacional de Colombia, Colombia
Enrique Muñoz De Cote	National Institute of Astrophysics, Optics and Electronics, Mexico
Olfa Nasraoui	University of Louisville, USA

Rogelio Soto	Tecnológico de Monterrey, Mexico
Álvaro Soto	Universidad Católica de Chile, Chile
Armando Sousa	Institute for Systems and Computer Engineering, University of Porto, Portugal
Elaine Sousa	University of Sao Paulo, Brazil
Jorge Sánchez	Universidad Nacional de Córdoba, Argentina
Dunbing Tang	Nanjing University of Aeronautics and Astronautics, China
Vicente R. Tomás López	Universitat Jaume I, Spain
Flavio Tonidandel	Centro Universitario da FEI, Brazil
Juan-Manuel Torres-Moreno	Université d'Avignon et des Pays de Vaucluse, France
Tomas Trescak	University of Western Sydney, Australia
Leonardo Trujillo	Instituto Tecnológico de Tijuana, Mexico
Abril Uriarte	Instituto Politécnico Nacional, Mexico
Leonardo Vanneschi	Universidade Nova de Lisboa, Portugal
Ramiro Varela	University of Oviedo, Spain
Ivan Varzinczak	Universidade Federal do Rio de Janeiro, Brazil
Wamberto Vasconcelos	University of Aberdeen, UK
Rodrigo Verschae	Universidad de Chile, Chile
Rosa Vicari	Universidade Federal do Rio Grande do Sul, Brazil
Esaú Villatoro-Tello	Universidad Autónoma Metropolitana, Mexico
Aline Villavicencio	Universidade Federal do Rio Grande do Sul, Brazil
Meritxell Vinyals	Commissariat à l'énergie atomique et aux énergies alternatives, France
Renata Wassermann	Universidade de São Paulo, Brazil
Denis Wolf	University of Sao Paulo, Brazil
Dina Wonsever	Universidad de la República, Uruguay
Neil Yorke-Smith	American University of Beirut, Lebanon
Pablo Zegers	Universidad de los Andes, Chile

Additional Reviewers

Avila Garzon, Cecilia	Etcheverry, Mathias	Richard, Gilles
Azevedo, Paulo	Gruska, Damas	Rubiolo, Mariano
Britz, Arina	Lloberes, Marina	Scaliante Wiese, Igor
Bula, Gustavo	Longhi Rossi, Luiz Henrique	Shrestha, Prasha
Carvalho, Andre		Simari, Gerardo
Chiruzzo, Luis	Longhi, Magalí	Teyseyre, Alfredo
Deagustini, Cristhian Ariel David	Rens, Gavin	Wilkens, Rodrigo
	Rey-Villamizar, Nicolas	Zilio, Leonardo

Contents

Natural Language Processing

Machine Learning

Big Data, Knowledge Discovery and Data Mining

Computer Vision and Pattern Recognition

Computational Intelligence Soft Computing

AI in Education, Affective Computing, and Human-Computer Interaction

Knowledge Engineering, Knowledge Representation and Probabilistic Reasoning

Towards an Integration of Workflows and Clinical Guidelines: A Case Study

Paolo Terenziani[1](✉) and Salvatore Femiano[2]

[1] DISIT – Universita' del Piemonte Orientale "Amedeo Avogadro",
Viale Teresa Michel 11, 15121 Alessandria, Italy
terenz@di.unito.it
[2] Dipartimento di Scienze della Sanità Pubblica e Pediatriche,
Univ. di Torino, Turin, Italy
fcl-med-inftol@unito.it

Abstract. The integration of workflows and guidelines modeling healthcare processes is a hot topic of research in Artificial Intelligence in Medicine, and is likely to provide a major advance in the IT support to healthcare [1]. In this position paper, we use a case study in order to identify commonalities and differences between workflows and guidelines. As a result of the analysis, we argue in favor of an integrated architecture in which workflow and guideline models are independently managed and supported, while integration is obtained through a mapping onto a system-internal format, where traditional AI-style inferential capabilities are supported.

Keywords: Knowledge representation · Clinical practice guidelines · Workflows · Integration · System architecture

1 Introduction

Workflows (WFs) are large pieces of knowledge used in order to describe healthcare processes within organizations (e.g., hospitals, hospital departments). They usually describe all the activities carried on by the healthcare practitioners (e.g., physicians, nurses) within a specific context (e.g., an hospital), often detailing actors, resources, and times needed in order to accomplish each activity. WFs mostly focus on the pattern of operations that describe the organizational structure, and\or on resources. In many cases, the goal is to model the organization of activities, to analyse and improve them (e.g., to discover and remove bottlenecks). Thus, a WF model of hospital activities can be seen as the representation of such activities from the viewpoint of a manager, or of an analyst. Most computerized approaches to WF provide simulation facilities that are used by analysts and managers in order to coordinate processes and to optimize costs and time of healthcare services.

Clinical practice guidelines (CPGs) are increasingly used in the healthcare context, in order to grant the quality of healthcare, putting evidence based medicine (EBM) into practice. CPGs are, in the definition of the MeSH dictionary, *"work consisting of a set of directions or principles to assist the health care practitioners with patient care decisions about appropriate diagnostic, therapeutic, or other clinical procedures for*

M. Montes-y-Gómez et al. (Eds.): IBERAMIA 2016, LNAI 10022, pp. 3–13, 2016.
DOI: 10.1007/978-3-319-47955-2_1

specific clinical circumstances". Thousands of CPGs have been devised in the last years. For instance, the Guideline International Network (http://www.g-i-n.net) groups 77 organizations of 4 continents, and provides a library of more than 5000 CPGs. CPGs are mostly complex knowledge sources aimed at supporting healthcare practitioners on therapeutic and diagnostic decisions, by providing them evidence-based recommendations. CPGs are built in order to support physicians in the treatment of "typical" patients, so that they describe clinical processes according to such a task. The adoption of AI-style computerized approaches to acquire, represent, execute and reason with CPGs can further increase the advantages of CPGs, by making them widely distributed and easily consultable (e.g. on the Web), by facilitating the contextualization of CPGs to the specific patients, and by supporting physicians in their decision-making activity [1–3].

The development of an advanced AI architecture to acquire, represent, manage and reason on WFs and CPGs is a hot topic within the Artificial Intelligence in medicine community, and an open and challenging issue is whether and how to **combine** such two knowledge sources. The idea underlying our approach is that WFs and CPGs, when applied to the healthcare domain, act as **complementary** knowledge-based frameworks looking at the same reality (the clinical treatment of patients) from different *viewpoints*, and with different *focuses* and *goals*. Although both WFs and CPGs deal with the healthcare processes, in most cases they have been coped with independently of each other, using different approaches, and leading to different and independent models of reality.

Both WFs and CPGs describe processes on patients, so that several healthcare tasks demands for an integrated treatment of WFs and CPGs [1]. For instance,

- as regards the general viewpoint (of healthcare organizations), integration is needed to have a model of activities, context, resources and times that focus accurately on activities executed on patients, and that is sensitive to the different patients (patient-centered)
- as regards the specific viewpoint (of healthcare practitioners coping with a specific patient), integration is needed to have an accurate schedule of activities, in which the decision logic of CPGs is merged with the usual pattern of activities in the organization.

Such a need of integration is widely recognized within the Artificial Intelligence in Medicine community. Thus, in the last years, several computerized approaches have started to define suitable representation formalisms and frameworks for integration [4, 5]. The main trend of such a research area is to extend current WF or CPG frameworks to represent and manage integrated models which merge the content of both CPGs and WFs. For instance, in the area of CPGs, it has been shown that the PROForma approach to CPGs [6] can model and deal with hospital WFs, and the GUIDE approach adopts a workflow-based formalism to cope with an integration of CPGs and (part of) workflow activities [7]. Within the WF community, Van der Aalst et al. [8] have shown that the primitives provided by WF formalisms to cope with patterns of activities are expressive enough to cope also with the patterns in CPGs. Additionally, several WF approaches have extended their formalisms and execution engine to provide the flexibility needed in order to cope with CPGs and their execution (consider, e.g., [9]).

This is a position paper, in which we argue in favor of a different way of integrating CPGs and WFs. While no actual implementation has been developed yet, this paper provides three main types of AI contributions:

(1) a *knowledge representation* contribution: we propose a representation of both the CPG and the WF of the vinorelbine treatment for advanced non-small cell lung cancer

(2) a *knowledge analysis* contribution: starting from the case study in (1), we generalize and analyze the differences between CPGs and WFs as regards contents, focus, goals, users and editors. Such an analysis leads us to conclude that building a unique model and execution engine integrating both a CPG and its related WFs (as many approaches in the literature suggest – see the Introduction above) is not only very difficult both from the theoretical and practical point of view, but, even more important, it is not likely to provide user-friendly results.

(3) an *architectural* contribution: as a consequence of the analysis in (2), we propose an *abstract* architecture for an alternative approach, in which different models and tools are provided to cope with CPGs and WFs separately, while the integration of them is obtained at a system-internal level, where general inferential mechanisms operate.

A final step would be needed: implementation. It may take advantage of GLARE (GuideLine Acquisition, Representation and Execution) [10, 11], but it is just started, and is out of the scope of this position paper (see however Sect. 4, where we also discuss open issues).

2   Comparing WFs and CPGs: A Case Study

We are well aware that proposing a general comparison between "typical" CPGs and "typical" healthcare WFs is a very difficult task, given the great variety of CPGs and WFs in the healthcare literature. Instead of providing a general discussion, we thus present a specific case study, that we consider quite significant and representative.

Specifically, we consider the case of vinorelbine treatment for advanced non-small cell lung cancer. As regards WFs, Femiano [12] has formalized the WF of activities regarding the vinorelbine treatment in the day hospital of the Azienda Ospedaliera San Giovanni Battista, Turin, Italy, one of the major hospitals in Italy (see Fig. 1). As regards CPGs, we consider the guideline described in [13], which is one of the reference CPGs considered in the Azienda (see Fig. 2 in the following). In both cases, for the sake of clarity and brevity, the description in this paper is simplified and not formally rigorous. Actually, the WF has been modeled using iGrafXProcess [14], and the CPG has been acquired using GLARE [10, 11].

2.1   Workflow for the Vinorelbine Treatment (in Day Hospital)

In Fig. 1, rectangles represent standard actions, big diamonds represent decisions, small diamonds (containing "+") represent the fork and join of patterns of activities.

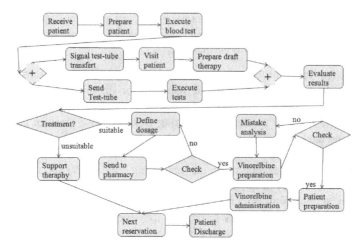

Fig. 1. Workflow about the day hospital vinorelbine treatment

The WF starts and ends with administrative actions: patient reception\discharge, and reservation of the next treatment. After reception in day hospital, the patient is seated on a chair\bed in case s\he has walking problems ("Prepare patient" node in Fig. 1) and a blood test is executed. After that, for the sake of optimization, two patterns of actions are executed in parallel. On one side, the test-tube is sent to the laboratory, and laboratory tests are executed; on the other side, the transfer of the test-tube is signaled to the information system. Then, the oncologist visits the patient and defines a draft (provisional) treatment to optimize waiting times. When the laboratory results become available, the physician can evaluate results. Then, the physician checks whether the vinorelbine treatment (as defined in the draft treatment) is suitable for the patient (decision node "treatment"). If it is not, a support therapy is applied (out of the scope of this WF); Otherwise, the proper dosage is defined, and the request is sent to the pharmacy. Then, a double loop of checks involves the chemist at the pharmacy and the physician (in Italy they are both legally responsible of the treatment), in order to grant the correctness of the transmission of the treatment to the chemist (first loop), and of the drug back to the physician (second loop). After that, the patient is prepared for the administration and, finally, vinorelbine is administered to the patient.

2.2 Guideline for the Vinorelbine Treatment of Advanced Non-small Cell Lung Cancer

In Fig. 2, the guideline for vinorelbine treatment of advanced non-small cell lung cancer is shown. In the figure parallelograms represents data requests (and laboratory tests), diamonds represent decisions, oval nodes represent "basic" actions, and rectangles represent composite actions.

The CPG starts with a data request ("Clinical evaluation"), to get the patient data needed to check patient eligibility for the CPG itself. Specifically, a blood test is

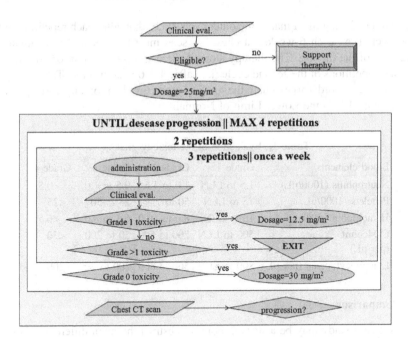

Fig. 2. Guideline for the vinorelbine treatment of advanced non-small cell lung cancer

required to get white blood cell and platelets counts, as well as a visit to the patient, to evaluate her\his Karnofsky performance status and ask her\his consent to the treatment. Eligibility is evaluated by checking the following condition:

$$White\ blood\ cell\ counts\ >\ 4000\ mm^3$$

$$AND\ platelets\ >\ 120000\ mm^3$$

$$AND\ Karnofsky\ performance\ status\ >\ 40$$

$$AND\ Consent$$

If the patient is not eligible, then a support therapy is needed (out of the scope of this CPG). Otherwise, the "basic" dosage of 30 mg/m^2 is fixed. After that, the CPG consists of three nested cycles of actions, to be repeated periodically. The inner cycle consists of at most three repetitions (once each week) of the following pattern of actions. An administration of vinorelbine (according to the chosen dosage) is performed; after that a "Clinical evaluation" (as above) is executed, followed by two checks: if the grade of toxicity for the patient is more than 1, than the therapy must be discontinued ("Exit" from the CPG), while if toxicity is grade 1, then the dosage must be reduced to 12.5 mg/m^2 (in case of grade 0 toxicity, the dosage is maintained unchanged and the treatment goes on). The second (intermediate) cycle specifies that, after each repetition of the inner cycle (i.e., each 3 weeks), if the grade of toxicity is still 0, than the dosage can be fixed to 30 mg/m^2. Finally, the outer cycle states that, after each repetition of the second cycle (i.e., each 6 weeks), a chest CT scan must be executed to evaluate the progression of the disease. The therapy stops if there is a progression, or after a maximum of 4 repetitions of the second cycle (i.e., after 4 * 6 = 24 weeks). Table 1 reports the criteria used in order to evaluate the toxicity grade (grade 5 toxicity corresponds to patient death; LLN is the Lower Limit of Normality).

Table 1. Evaluation of the toxicity grade.

Blood elements	Grade 1	Grade 2	Grade 3	Grade 4
Neutrophilis (1000/μL)	1.5 to LLN	1.0 to 1.5	0.5 to 1.0	<0.5
Platelets (1000/μL)	75 to LLN	50 to 75	25 to 50	<25
Hemoglobin (gr/dL)	10 to LLN	8.0 to 10.0	6.5 to 8.0	<6.5
CD4count (per μL)	500 to LLN	200 to 500	50 to 200	<50
Lymphocytopenia (per μL)	800 to LLN	500 to 800	200 to 500	<200

2.3   Comparisons

The above case study may be a starting point to abstract the main differences between "typical" WFs and CPGs, considering content, focus, goals, users, and editors.

Content and Focus. WFs and CPGs usually focus on different aspects of healthcare processes. Considering the case study, the WF contains a description of administrative actions, and of the interaction between the physician and the chemist at the pharmacy, which are not present in the CPG. The optimization of performing the patient visit and defining a draft treatment while the laboratory test is being executed is not considered in the CPG (indeed, the CPG recommends that blood test results must be available to define the treatment for the patient!). Such aspects are "out of the scope" of the CPG, which mostly focuses on the decision logic used to define the proper therapy.

On the other hand, the CPG contains a detailed description of the different decisions that have to be taken to define the proper treatment and dosage. The WF just contains three nodes ("Evaluate results", "Treatment", and "define dosage") to describe the fact that, given the laboratory test results and the visit of the patient, the physician has to check whether the therapy is suitable, and to define the dosage. The criteria used to perform the check and to define the dosage are out of the scope of the WF, which focuses on the implementation and organization of processes, rather than on the decision logic. On the other hand, the decision logic is the focus of the CPG, which details the different decisions to be taken at different times during the execution of the CPG (thus, the description of the periodicity at which actions have to be executed is critical in the CPG), and the decision criteria. Notably, the "chest CT scan" action is not mentioned in the WF, since it just focuses on the operations to be performed in day-hospital on a patient who has to receive a vinorelbine administration (the "chest CT scan" is part of a different Hospital WF, concerning patient follow-up).

Goals and Users. In general, WFs are used by managers and\or analysts to simulate and analyze the flow of healthcare activities, with the goal of increasing service efficiency through a proper allocation of resources and\or re-organization of processes. On the other hand, CPGs are mostly consulted by physicians to be up-to-date as regards the EBM- clinical procedures, and are used as a decision-support tool to cope with specific patients. As a matter of fact, such differences about goals are strictly related to the differences about focus (process flow in WFs vs. decision logic in CPGs). In principle, the content of both WFs and CPGs (concerning the same topic) might be merged into a unique model. However, given the different users and goals, such a unifying model would not be optimal either for analysts\managers (focusing on the organization issues) or for physicians (focusing on the decision logic).

Editors. Last, but not least, building such a unifying model would be very difficult in practice. CPGs are usually "edited" by national or international agencies. They are a way of putting EBM into practice, and are mostly context independent. On the other hand, WFs are usually defined by teams in local contexts (e.g., in a specific hospital), often as a result of direct observation of the activities. It seems to us very unlikely that editors of CPGs and WFs can work together to build a "unifying" model coping with both general CPG principles and local organization issues.

3   Towards an Integrated Architecture

The above analysis substantiate our general criticisms with respect to many recent approaches trying to provide a unifying treatment of WFs and CPGs (see the Introduction). Thus, we propose a different approach to integration, based on three basic considerations:

(i)   Each kind of user wants to focus only on the part of the healthcare process s\he is interested in. Therefore, the models of CPGs (for physicians) and WFs (for analysts and managers) should be kept separate.

(ii) Each kind of user wants a user-friendly framework representing knowledge as closely as possible to the way s\he is used to look at it. In the last twenty years, (computerized) WF tools have been devised in order to provide a user-friendly treatment of WF for managers\analysts, and CPGs tools have been developed to present clinical activities in a physician-oriented way. Therefore, we advocate the use of different tools for WFs and CPGs.

(iii) Last, but not least, integration is needed to accomplish many important tasks. Such an integration, when needed, can be obtained by mapping both WFs and CPGs onto a unifying internal system representation, on which the inferential operations needed to solve such tasks can be executed.

These considerations motivate the system architecture we propose in Fig. 3.

In the abstract system architecture in Fig. 3, a WF tool is used to model and cope with the "process-organization" aspects of the problem; a CPG approach is used to cope with clinical guidelines; both models are mapped onto a common system-internal representation, and inferences requiring integration are performed on it.

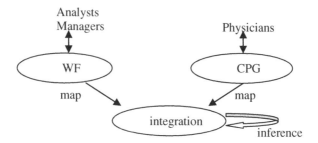

Fig. 3. An architecture for integration.

4   Ongoing and Future Work

We are starting to devise an implementation of the general architecture shown in Fig. 3, taking advantage of

(1) iGrafXProcess [14] to cope with WFs,
(2) GLARE [10, 11] to cope with CPGs, and
(3) Petri Nets to deal with integration.
(1) Concerning WFs, iGrafXProcess [14] is a standard commercial tool.
(2) Concerning CPGs, we chose to adopt GLARE (Guideline Acquisition, Representation and Execution) [10, 11], a user-friendly and domain-independent system which has been built since 1997 in a long-term cooperation between the Department of Computer Science of the University of Eastern Piedmont Alessandria, Italy, and the Azienda Ospedaliera San Giovanni Battista in Turin (the third largest hospital in Italy). GLARE is based on a "physician-friendly" representation formalism to model CPGs, and provides a tool to facilitate expert-physicians and knowledge engineers in the acquisition of a new CPG, and a tool to support user-physicians in the application of a given (acquired) CPG to a specific patient. GLARE has been successfully tested in many different domains, including bladder cancer, reflux esophagitis, heart failure, alcohol-related problems, emergency treatment of polytrauma, and ischemic stroke. Besides supporting CIG acquisition, representation, storage and execution, GLARE is characterized by the adoption of advanced Artificial Intelligence formal techniques to provide advanced supports for different tasks [15], including reasoning about temporal constraints [16, 17], model-checking verification [18], decision support (based on Decision Theory) [19].
(3) Concerning integration, we think that Petri Nets (and their extensions) are good candidates to cope with both WF and CPG formalisms, and to provide suitable and efficient inferential mechanisms. Petri Nets constitute the underlying semantics of several WF formalisms (consider, e.g., [20]), and they have been also used in order to model CPGs [21, 22]. Different families of Petri Nets are available. In particular, we think that Well-formed Net [23] are well suited to model the WF and CPG integration, since they can provide a more compact and readable representation of the knowledge, and the possibility of using efficient

solution techniques. Additionally, Well-formed Net provide a suite of *consolidated inferential capabilities* (see [23]), which can be easily exploited in this context. As regards the mapping from GLARE CPGs into Well-formed Nets, we exploit the approach described in [22].

4.1 Open Issues

However, several difficult problems have still to be faced in order to make our approach fully operative. First of all, the mapping of WF and (above all) CPG formalisms onto Petri Nets must be further studied, especially as concerns the treatment of complex issues such as, e.g., temporal aspects.

Second, the task of integrating the (translation into Petri Nets of the) model of the CPG with the (translation into Petri Nets of the) model of the WF is quite complex. Common places\transitions must be identified, as well as overlaps, and refinements. The adoption of a common *vocabulary\ontology* is important in this context. Indeed, it might facilitate the integration into a unique model, and it can make the CPG model, the WF model, and the internal integrated model more understandable and user-friendly.

Third, in real hospitals, hundreds of CPGs can be concurrently executed on patients. The dimension of the resulting Petri Net, modeling the hospital daily activities, is likely to rise temporal complexity problems for the underlying inferential mechanism.

Fourth, the architecture sketched in Fig. 3 is aimed to allow each user (physician or manager\analyst) to look at the processes through her\his preferred "glasses". However, in order to consistently support such a goal, one should also reconvert the output of the inferential mechanism (which is provided at the internal leyer, i.e., at the level of Petri Nets) into the proper users' interface formalisms (CPG for physicians, Workflow for analysts\managers). This is, once again, a challenging task.

Finally, Petri Nets are quite rigid for Knowledge Representation. In particular, their elaboration tolerance (in McCarthy's terms) is limited: each time a slight variation of the original problem is made, one must reconsider the whole set of places and transitions. Other formalisms (for instance, those based on *non-monotonic reasoning*; see, e.g., [24, 25]) are worth to be investigated for knowledge integration.

5 Conclusion

Integrating WFs and CPGs is a demanding task in many healthcare applications [1]. In the last years, several computerized approaches have been devised to achieve integration. Most of such approaches advocate the adoption of a unique (WF or CPG) framework to cope with both WFs and CPGs, and on the development of integrated models (merging WFs and CPGs) for users. In this paper, we analyze a case study in order to show the limitations of such a way of facing integration. The case study highlights typical differences between WFs and CPGs as regards content, focus, users, goals and editors, and motivates our new approach towards integration, aiming at providing each kind of user (manager\analysts vs. physician) with a user-oriented computerized framework to cope with healthcare processes.

Our proposal advances the state of the art, since it is the only approach that reconciles (i) the need of performing inferences on the integration of WF and CPG knowledge, with the users' requirements (ii) to focus only with the aspects of knowledge (decision logic vs. process organization) they are interested in, and (iii) to have a user-friendly representation of such a knowledge.

References

1. Lenz, R., Reicher, M.: IT support for healthcare processes – premises, challenges, perspectives. Data Knowl. Eng. **61**, 39–58 (2007)
2. Ten Teije, A., Miksch, S., Lucas, P. (Eds.): Computer-based medical guidelines and protocols: a primer and current trends. In: Studies in Health Technology and Informatics, vol. 139 (2008)
3. Peleg, M.: Computer-interpretable clinical guidelines: a methodological review. J. Biomed. Inf. **46**(4), 744–763 (2013)
4. PROHealth 2009 Workshop program. http://mis.hevra.haifa.ac.il/~morpeleg/events/prohealth09/PROHealthProgram.html
5. Fridsma, D.B.: Special issue on workflow management and clinical guidelines. JAMIA **22** (1), 1–80 (2001)
6. Fox, J., Johns, N., Rahmanzadeh, A., Thomson, R.: Disseminating medical knowledge: the PROforma approach. Artif. Intell. Med. **14**, 157–181 (1998)
7. Quaglini, S., Stefanelli, M., Cavallini, A., Micieli, G., Fassino, C., Mossa, C.: GL-based careflow systems. Artif. Intell. Med. **20**(1), 5–22 (2000)
8. Mulyar, N., Pesic, M., van der Aalst, W.M.P., Peleg, M.: Towards the flexibility in clinical guideline modelling languages. BPM Center report BPM-07-04 (2007). BPMcenter.org
9. Thao, L.L., Reichert, M., Dadam, P.: Integration and verification of semantic constraints in adaptive process management systems. Data Knowl. Eng. **64**, 3–23 (2008)
10. Terenziani, P., Molino, G., Torchio, M.: A modular approach for representing and executing clinical guidelines. Artif. Intell. Med. **23**, 249–276 (2001)
11. Terenziani, P., Montani, S., Bottrighi, A., Molino, G., Torchio, M.: Applying artificial intelligence to clinical guidelines: the GLARE approach. In: Computer-Based Medical Guidelines and Protocols: A Primer and Current Trends, in Studies in Health Technology and Informatics, vol. 139, pp. 273–282 (2008)
12. Femiano, S.: Analysis and restructuring processes of patient care in complex health systems. University of Turin, Ph.D. dissertation (2009)
13. Veronesi, A., Crivellari, D., Magri, M.D., Cartei, G., Mansutti, M., Foladore, S., Monfardini, S.: Vinorelbine treatment of advanced non-small cell lung cancer with special emphasis on elderly patients. Eur. J. Cancer **32**(10), 1809–1811 (1996)
14. iGrafx.com: Unlocking the Potential of Business Process Management. http://portal.igrafx.com/downloads/documents/bpm_whitepaper.pdf
15. Terenziani, P., Montani, S., Bottrighi, A., Torchio, M., Molino, G., Anselma, L., Correndo, G.: Applying artificial intelligence to clinical guidelines: the GLARE approach. In: Cappelli, A., Turini, F. (eds.) AI*IA 2003. LNCS, vol. 2829. Springer, Heidelberg (2003)

16. Anselma, L., Terenziani, P., Montani, S., Bottrighi, A.: Towards a comprehensive treatment of repetitions, periodicity and temporal constraints in clinical guidelines. Artif. Intell. Med. **38**(2), 171–195 (2006)
17. Console, L., Terenziani, P.: Efficient processing of queries and assertions about qualitative and quantitative temporal constraints. Comput. Intell. **15**(4), 442–465 (1999)
18. Bottrighi, A., Giordano, L., Molino, G., Montani, S., Terenziani, P., Torchio, M.: Adopting model checking techniques for clinical guidelines verification. Artif. Intell. Med. **48**(1), 1–19 (2010)
19. Anselma, L., Bottrighi, A., Molino, G., Montani, S., Terenziani, P., Torchio, M.: Supporting knowledge-based decision making in the medical context: the GLARE approach. Int. J. Knowl.-Based Organ. **1**(1), 42–60 (2011)
20. van der Aalst, W.M.P.: The application of petri nets to workflow management. J. Circuits Syst. Comput. **8**(1), 21–66 (1998)
21. Grando, M.A., Glasspool, D.W., Fox, J.: Petri nets as a formalism for comparing expressiveness of workflow-based clinical guideline languages. In: Ardagna, D., Mecella, M., Yang, J. (eds.) Business Process Management Workshops. LNBIP, vol. 17, pp. 348–360. Springer, Heidelberg (2009)
22. Beccuti, M., Bottrighi, A., Franceschinis, G., Montani, S., Terenziani, P.: Modeling clinical guidelines through petri nets. In: Combi, C., Shahar, Y., Abu-Hanna, A. (eds.) AIME 2009. LNCS, vol. 5651, pp. 61–70. Springer, Heidelberg (2009)
23. Chiola, G., Dutheillet, C., Franceschinis, G., Haddad, S.: Stochastic well-formed colored nets for symmetric modelling applications. IEEE Trans. Comput. **42**(11), 1343–1360 (1993)
24. Leone, N., Eiter, T., Faber, W., Fink, M., Gottlob, G., Greco, G.: Boosting information integration: the INFOMIX system. In: SEBD 2005, pp. 55–66
25. Leone, N., Greco, G., Ianni, G., Lio, V., Terracina, G., Eiter, T., Faber, W., Fink, M., Gottlob, G., Rosati, R., Lembo, D., Lenzerini, M., Ruzzi, M., Kalka, E., Nowicki, B., Staniszkis, W.: The INFOMIX system for advanced integration of incomplete and inconsistent data. In: SIGMOD Conference 2005, pp. 915–917

Anomalies Detection in the Behavior of Processes Using the Sensor Validation Theory

Pablo H. Ibargüengoytia[(✉)], Uriel A. García, Alberto Reyes,
and Mónica Borunda

Instituto de Investigaciones Eléctricas, Cuernavaca, Morelos, Mexico
{pibar,uriel.garcia,areyes,monica.borunda}@iie.org.mx

Abstract. Behavior can be defined as combination of variable's values according to external inputs or environmental changes. This definition can be applied to persons, equipment, social systems or industrial processes. This paper proposes a probabilistic mechanism to represent the behavior of industrial equipment and an algorithm to identify deviations to this behavior. The anomaly detection mechanisms, together with the sensor validation theory are combined to propose an efficient manner to diagnose industrial equipment. A case study is presented with the failure identification of a wind turbine. The diagnosis is conducted when detecting deviations to the turbine normal behavior.

Keywords: Anomaly detection · Model of behavior · Bayesian networks · Wind turbines

1 Introduction

Anomaly detection refers to the problem of finding patterns in data that do not conform with the expected behavior [3].

In this sentence, several issues are identified and tackled in this paper. These are:

Behavior: a probabilistic model for the behavior of a process is proposed. Specifically, Bayesian networks are used to represent the relation that some variables have with others in a process. This is precisely the function of Bayesian networks.

Patterns in data: related variables maintain combinations of values according to the dynamic of the process. Some combinations respond to some contexts while different combinations or patterns corresponds to other contexts. Some patterns represent normal behavior while others represent failures.

Expected behavior: when a process changes due to external inputs or context changes, the process behavior can be recognized or expected using knowledge on the process dynamics.

This paper proposes a probabilistic model for representing the behavior of a process and a mechanism to detect patterns that reflects deviations of the

© Springer International Publishing AG 2016
M. Montes-y-Gómez et al. (Eds.): IBERAMIA 2016, LNAI 10022, pp. 14–24, 2016.
DOI: 10.1007/978-3-319-47955-2_2

expected behavior. The idea is to collect historical data from a process when it is behaving properly according to experts. The historical data is formed by time series of several variables. The data forms a matrix where columns are the variables and the rows are the samples, measurements or instances. In this paper, the terms variable, sensor and node are used interchangeable.

This project develops the approach of anomaly detection in the behavior for the diagnosis of wind turbines. Historical data is collected and used to identify and model the normal behavior of the turbine. Related work has been revised for the diagnosis of wind turbines using advanced computational algorithms and artificial intelligence. Most of the consulted work is subscribed to the Condition Monitory (CM) community. The survey in [8] include the most common methods for fault detection. The main methods are vibration analysis, acoustic emission, ultrasonic testing techniques, oil analysis, strain measurement, thermography, shock pulse methods, radiographic inspections and others. However, those are traditional methods that usually require experts in the field and complex models difficult to construct and maintain. Computational methods include the work by [11], based on ontology and Failure Mode, Effects and Criticality Analysis (FEMCA). With that analysis, the method creates ontology and a knowledge base that is used on a expert system shell. However, uncertainty management is not considered. Other computational method is the work reported in [10] that also considers behavioral models obtained with SCADA historical data. However, their approach follows an adapted neuro-fuzzy interference system (ANFIS), but they make no distinction between the different operating modes of the turbine.

This paper models the behavior of the wind turbine and uses anomaly detection to find deviations to its normal behavior.

This paper is organized as follows. The next section introduces anomaly detection techniques and situates the proposal of this paper. Section 3 briefly explains the developed sensor validation theory. Next, Sect. 4 develops the behavior analysis of processes in order to detect deviations. Section 5 introduces the case study, namely the diagnosis of wind turbines with the detection of deviations in its normal behavior. Initial experiments are presented and discussed. Finally, Sect. 6 concludes the paper and indicates the future work in this project.

2   Anomaly Detection

Anomaly detection is an active area of research that is being used in several fields [3]. In finances, illegal transactions are identified in credit cards purchases. In a computer network, undesirable participants can be discovered. Anomaly detection is the identification of unexpected patterns when dealing with data collected from some process. The data recompilation can be conducted using sensors, measure instruments or even human sources like social networks or conversations.

Different types of anomalies can be detected in data representing the behavior of a process:

point anomalies- a single point is deviated in the observed process.

contextual anomalies- a data instance can be anomaly in some cases and normal in others. These cases are called *contexts*.

collective anomalies- a group of data instances is anomalous with respect to the entire data set.

In this project, the focus is in the contextual anomalies. They are data combinations that can be correct on one context, but invalid behavior in other contexts. For example, the behavior of the electric generator in a wind turbine is diverse when there is low wind (low speed) that when high winds blow in the park (high speed in the generator). For this detection, two kind of attributes are differentiated. First, **contextual attributes** are those that determine the neighborhood or context where the process is developed. Second, **behavioral attributes** are those that characterize the behavior in a specific context. For example, in a process that depends on weather, it is not the same a case in winter that a case in summer. The *season* variable can be considered the context variable, while ambient temperature can be considered a behavioral variable.

Table 1 shows the common techniques utilized in anomaly detection, and some examples of applications that can be found in the literature.

Table 1. Techniques used for anomaly detection and applications where they have been applied according to [3].

Techniques	Applications
Classification based	Cyber-intrusion detection
Clustering based	Fraud detection
Nearest neighbor based	Medical anomaly detection
Statistical	Industrial damage detection
Information theoretic image processing	Text anomaly detection
Spectral	Sensor networks

This paper presents a contribution for the anomaly detection community with the use of Bayesian networks. Also, an application in renewable energy is presented in this paper.

3   Sensor Validation Theory

The sensor validation theory was initially designed to find errors in the readings of sensors in industrial processes [7]. The basic idea is to calculate the probability of the values that a sensor provides, given the current values of the most related variables or sensors. Comparing the estimated value with the current variable reading, it is possible to detect failures between sensors readings.

The sensor validation theory follows the two phase approach that industry uses for diagnosis, namely fault detection and isolation (FDI) [5]. The first phase detects that there is a failure between the variables, and the second phase isolates and discovers the faulty variables.

The first requirement of the sensor validation theory is the construction of a model that represents the probabilistic relations between variables in the application. This can be done with the learning algorithms for Bayesian networks available in the community. Using historical data corresponding to a normal behavior, the model is constructed. Figure 1 shows the network learned for the wind turbine. Once the model is defined, the sensor validation algorithm indicates that for all variables (or nodes in the model), instantiate all other nodes and propagate to calculate a posterior probabilistic distribution that indicates the probability of the real value. If there is a coincidence, then no failure is detected. Otherwise, an apparent failure is detected.

The sensor validation algorithm can be expressed as follows:

Algorithm 1. Detection algorithm

Require: A node n.
Ensure: Either correct or faulty.
 1: assign a value (instantiate) to all nodes except n
 2: propagate probabilities and obtain a posterior probability distribution of node n
 3: read real value of variable represented by n
 4: **if** $P(real\_value) \geq p\_value$ **then**
 5:     return(correct)
 6: **else**
 7:     return(faulty)
 8: **end if**

Where p_value is a threshold that can be adjusted to calibrate preferences in the failure detection. If the application requires the detection of all faults, even with the risk of false alarms, assign a p_value high. On the other hand, for the ability to catch the important failures even with the risk of unrecognized failures, assign p_value lower.

The detection algorithm can only specify a set of variables that contains an apparent failure. The failures are considered apparent since correct variables can be validated with faulty ones producing incorrect failure detections. In order to isolate the real faulty variable, a second stage is required. This is called the isolation algorithm. It is based in a property of Bayesian networks called the Markov blanquet (MB). The MB of a node in a BN is the set of nodes that when instantiated, isolates that node from changes in the nodes outside the MB. The MB of a node is formed by the set of parents, children and spouses of a node in a network [9]. Thus, when the set of apparent nodes coincides with the MB of a node, then this node has a real fault. Utilizing this property, if a fault exists in one of the variables, it will be exposed in all the sensors on its MB. On the contrary, if a fault exists outside a sensors' MB, it will not affect the estimation of that sensor. It can be said then, that the MB of a sensor acts as its protection against others faults, and also protects others from its inside failures. Thus, the

MB is utilized to create a *fault isolation* module that distinguishes the *real faults* from the apparent faults. This isolation stage utilizes a second Bayesian network to identify the real faulty variable.

Figure 2 shows the isolation network corresponding to the network of Fig. 1. The upper layer of nodes represents the vector of real failure. The lower layer represents the apparent failure. All nodes are binary representing {Fault, OK} values. The arcs in the network correspond to the MB of each node. Real faults cause apparent failures in all the variables MB, and the existence of an apparent fault indicate the existence of a real fault in one node in its MB. The apparent faulty nodes are instantiated with the detection cycle, and the propagation calculates the probability of real faults in the variables.

The isolation algorithm can be expressed as follows:

Algorithm 2. Isolation algorithm using the isolation network.

Require: A sensor n and the state of sensor n.
  1: assign a value (instantiate) to the apparent fault node corresponding to n
  2: propagate probabilities and obtain a posterior probability of all nodes *Real fault*
  3: update vector $P_f(sensors)$

The sensor validation theory was successfully utilized in the validation of temperature sensors in a gas turbine of a power plant [6]. However, a fair question is still unanswered: what happen if a failure is detected but the sensor is working properly? The system identifies an improper behavior even if the sensors are working properly. The next section discusses the proposed model to detect deviations in a process behavior based on the sensors behaviors.

4   Proposed Model for Anomaly Detection in the Behavior

This paper utilizes the Sensor Validation Theory and Anomaly Detection mechanisms to diagnose wind turbines. The diagnosis is based on detecting anomalies in the behavior of turbines. Three steps are required in this diagnosis process:

1. Create a behavioral model, using Bayesian networks considering all the variables that may influence in the behavior.
2. Complete the isolation model and run the sensor validation algorithm to generate a pattern of faulty variables once that a fault is presented in the turbine.
3. Recognize a pattern of faulty variables that represents a failure in the turbine. This is learned with historical data and logbook of the wind turbine.

To create and use a model for anomaly detection in behavior, the next methodology is defined:

1. Select the participating variables among the complete historical SCADA data set. Notice that if many variables are used, more complex models can result and higher computational effort would be necessary for the diagnosis. Expert in the domain may advice in an adequate variables data set that represents the behavior.
2. Clean and discretize the data set. Most of the information is obtained through sensors that are prone to noise and failures. Discretization is required for using Bayesian networks. The number of intervals in the discretization should be chosen to balance computer power.
3. Identify a subset of variables that conform contexts in the process, and define the number of combinations that the context variables form. In this paper, the context variables are wind speed and power generation. The wind turbine behave different with high winds and hence high power, with respect low winds. Four contexts were defined in this work.
4. Separate a training data set for every context from the complete data set.
5. For every one of the contexts, utilize a learning algorithm and construct a Bayesian network that represents the probabilistic relations between the variables. In this paper, the *Greedy and search* algorithm [4] of the *Hugin* [1, 2] package was used. Figure 1 shows the model obtained for medium speed winds. This is called the detection network.
6. For all the detection networks, identify the Markov blankets of all nodes and construct the isolation network as shown in Fig. 2 [7]. The isolation network produces a vector with the probability of fault in all variables considered in the model.

One important issue in this methodology is the amount of data recollected to learn the behavior model. The SCADA data set must be enough that most of the combinations of proper behavior are included in the learning data set. There could be a wide variety of normal behaviors between the variables. This is also the reason to separate the history of the system in contexts. Thus, enough data should be considered for every context in the diagnosis. If less data is available, some instances of normal operation of the turbine may be interpreted as abnormal. In this paper, 3 years of data are included in the learning process of the model. However, large periods of inactivity of the turbine produce useless data.

When the models have been defined for a specific application, the system is ready for diagnosis. The following procedure is followed:

1. For validation of the system, establish a data set for testing with the historical data of some known failure in the application.
2. From the testing data set, or from the on–line SCADA, read the current value of all variables.
3. Apply the sensor validation algorithm and identify the real faulty variable(s).
4. Register the pattern of real faulty variables for the diagnosed failure in the process.

The diagnosis is executed on-line, i.e., the response time should be enough for detecting insipient and unexpected failures in the wind turbine. The execution time depends mainly on the interconnection of the behavior model. If some nodes contain many parents, the CPT tables grow exponential and the Markov blankets become also larger. If this happen, both inferences, on the detection and isolation networks, take longer time to execute. The recommendation is to use a learning algorithm that produces less interconnected possible model, as the Greedy algorithm used in this project. This algorithm allows limiting the maximum number of parents for nodes in the model. The time spent in one diagnosis cycle was 27 ms for the experiments of this paper. Nevertheless, the data set collected corresponds to periods of 5 min. If the diagnosis cycle corresponds to 5 min, and if the execution time of the system is below a second, then the response time is appropriate even with more complex behavior models.

The next section exemplifies the proposal in this paper for the diagnosis of a wind turbine.

5   Experiments and Results

The anomaly behavior detection was applied to a case study: the diagnosis of wind turbines.

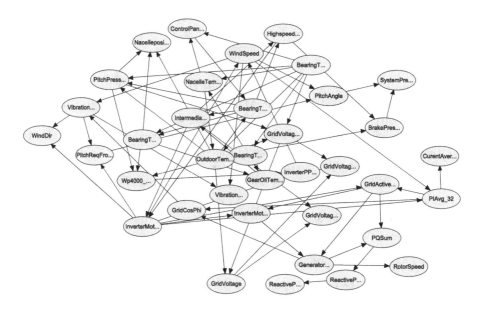

Fig. 1. Behavior model learned from historical data. It is also the detection network.

Wind turbines are devices that capture the kinetic energy of the wind and convert it first to mechanic energy, when the blades rotate, and then convert it in electric power. This power generation represents the higher percentage of renewable energy generation in the world. In Mexico, 1 % of the total generation was produced by wind farms in 2013 and it is predicted that the produced power coming from clean energies, mainly wind energy, will reach 35 % by 2024.

The Electrical Research Institute (IIE in Spanish) possesses an experimental field with one wind turbine with the capacity to generate 300 kW. The wind turbine is controlled through a SCADA (supervisory control and data acquisition) system. The SCADA program has the function to store historical data of all variables values every 5 min. The total number of variables stored is 76. From those 76 variables, only 34 variables can be used to represent the turbine behavior.

In order to create the probabilistic behavior model, a specific context is chosen. The context variables are wind speed and power generation. For this experiment, data from normal behavior were filtered for the context of wind speed from 3 to 12 m/s, and a generation from 10 to 200 kW. Thus, historical data consisting of 3,300 registers from March 2013 to July of 2014 were selected to train the model. The network shown in Fig. 1 was learned using the *Greedy and search* algorithm [4] of the *Hugin* [1, 2] software package. Using this network, the isolation network of Fig. 2 was obtained.

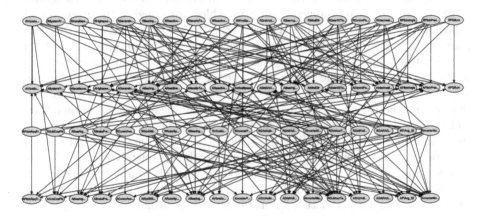

Fig. 2. Isolation network resulted from the model of Fig. 1.

To validate the system, historical data from August 2014 were selected for testing. Figure 3 shows the behavior of the wind speed and the power generation values. As can be noticed in this figure, the turbine was fired and protected to stop generation after a failure was detected by the operator. The turbine was generating around 60 kW when it drops to zero generation. The wind turbine operator informs that the failure detected at that time was the break of a screw causing the yaw mechanism unbalanced.

Figure 4 shows the results of the anomaly detection in the behavior of the turbine. The vertical axis represents the probability of failure of the variables. The horizontal axis represents the numbered instances of data selected for testing. According to Fig. 3, the turbine was behaving properly until something happen

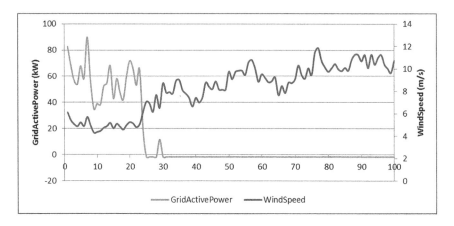

Fig. 3. Historic data from December 2014. Notice the difference in the behavior of the power generation variable and the wind speed.

and drop the generation to zero. This is shown around instance 25 in the graph of Fig. 3. At this same moment, some variables are found with an abnormal behavior. They are for example {BearingTempB, NacelleTemperature, Nacelle-Position, Vibration2WP4084_1}. According to the wind park operator and his logbook, the failure reported was a decalibration of the yaw break caused due to the loss of a screw. The yaw is the mechanism that faces the turbine to the direction of the wind in order to catch most of the wind energy. With a failure in this system, the nacelle could face a different angle, some extra vibration occurs and the bearings may increment the temperature. Therefore, this case study shows that most of the variables with failure detected as shown in Fig. 4 refer to the yaw system. Other faulty variables may not be completely related like the nacelle temperature. However, the behavior corresponds to a failure in the yaw position. Several tests are needed to complete the relations between changes in the normal behavior and actual failures in the wind turbine.

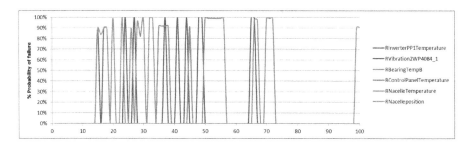

Fig. 4. Results of the anomaly detection. The vertical axis represents the probability of failure of the variables. The results are at the same time than the graph in Fig. 3

6 Conclusions and Future Work

This paper has described the use of sensor validation theory and anomaly detection techniques combined to build an online diagnosis system for wind turbines. The contribution of this project includes the generation of behavioral models based on Bayesian networks and an original form to detect anomalies in the behavior. The case study is the diagnosis of wind turbines where a model was constructed using SCADA historical data and filtering the different contexts proposed. Experiments show that it is possible to identify incipient deviations of normal behavior and the identification of the wind turbine failure.

Even with the promising results obtained in the experiments, several questions remain and require future work. Some of these are the following:

- Can a difference be identified between a failure in a sensor or a failure in an equipment?
- Which is the best way to identify the contexts in an application?
- Do different contexts require different models?
- Is it possible to guarantee a pattern of identified faulty variables for each fault in equipment?
- Is it possible to separate an application in sub–modules and apply this technique for each module?
- Is it worth to use dynamic Bayesian networks to consider time in the failure detection?

Acknowledgements. This work is a preliminary result of the P12 project of the Mexican Center of Innovation in Energy (CEMIE-Eólico), partially sponsored by Fund (FSE) CONACYT-SENER Energy Sustainability, and at the IIE, under the project 14629. Authors also thank the anonymous referees for their insightful comments.

References

1. Hugin expert, hugin expert A/S. Aalborg, Denmark (2000)
2. Andersen, S.K., Olesen, K.G., Jensen, F.V., Jensen, F.: Hugin: a shell for building bayesian belief universes for expert systems. In: Proceedings of the Eleventh Joint Conference on Artificial Intelligence, IJCAI, pp. 1080–1085, Detroit, Michigan, USA, 20–25 August 1989
3. Chandola, V., Banerjee, A., Kumar, V.: Anomaly detection: a survey. Technical report TR 07-107, University of Minnesota, USA (2007)
4. Chickering, D.M.: Optimal structure identification with greedy search. J. Mach. Learn. Res. **3**, 507–554 (2002)
5. Frank, P.M.: Fault diagnosis in dynamic systems using analytical and knowledge based redundancy- a survey and some new results. Automatica **26**, 459–470 (1990)
6. Ibargüengoytia, P.H., Sucar, L.E., Vadera, S.: Real time intelligent sensor validation. IEEE Trans. Power Syst. **16**(4), 770–775 (2001)
7. Ibargüengoytia, P.H., Vadera, S., Sucar, L.E.: A probabilistic model for information and sensor validation. Comput. J. **49**(1), 113–126 (2006)
8. Márquez, F.P.G., Tobias, A.M., Pérez, J.M.P., Papaelias, M.: Condition monitoring of wind turbines: techniques and methods. Renew. Energy **46**, 169–178 (2012)

9. Pearl, J.: Probabilistic Reasoning in Intelligent Systems: Networks of Plausible Inference. Morgan Kaufmann, San Francisco (1988)
10. Schlechtingen, M., Santos, I.F., Achiche, S.: Wind turbine condition monitoring based on scada data using normal behavior models. Part 1: system description. Appl. Soft Comput. **13**, 259–270 (2013)
11. Zhou, A., Yu, D., Zhang, W.: A research on intelligent fault diagnosis of wind turbines based on ontology and FMECA. Adv. Eng. Inform. **32**, 255–270 (2014)

Explanatory Relations Revisited: Links with Credibility-Limited Revision

María Victoria León and Ramón Pino Pérez[(⊠)]

Departamento de Matemáticas, Facultad de ciencias,
Universidad de Los Andes, Mérida, Venezuela
{mleon,pino}@ula.ve

Abstract. We study binary relations $\triangleright$ over propositional formulas built over a finite set of variables. The meaning of $\alpha \triangleright \gamma$ is that γ is a preferred explanation of the observation α. These relations are called Explanatory or abductive relations. We find two important families of abductive relations characterized by his axiomatic behavior: the ordered explanatory relations and the weakly reflexive explanatory relations. We show that both families have tight links with the framework of Credibility limited revision. These relationships allow to establish semantical representations for each family. An important corollary of our representations results is that our axiomatizations allow us to overcome the background theory present in most axiomatizations of abduction.

1 Introduction

We study binary relations $\triangleright$ over propositional formulas built over a finite set of variables which try to capture the abductive reasoning. The expression $\alpha \triangleright \gamma$ will mean that the observation α is well explained by the explanation γ.

In the pioneer logical work of Levesque [13] there is a background theory Σ such that when we have $\alpha \triangleright \gamma$ then the explanation γ has to entail together with Σ the observation α. Actually, the good explanations, in that framework, are in some sense some preferred formulas of the set

$$\{\gamma : \gamma \text{ consistent with } \Sigma \text{ and } \Sigma \cup \{\gamma\} \vdash \alpha\}.$$

This is the view adopted also in [15].

In this work we don't make use of an explicit background theory Σ. We take an axiomatic presentation of our explanatory relations. Following the works [15,17] we make a systematic study of two families of explanatory relations. In particular, we show that our families are very deeply related with the credibility limited revision operators introduced by Hansson et al. [11] These operators are a generalization of the AGM revision operators introduced by Alchourron et al. in [2].

Different families of explanatory relations and their properties have been studied in [3,9,10,15,16]. Our main goal is to continue to have a better understanding of these abstract explanatory relations (behavior, axioms, constructions) and their links with belief revision. There are works relating the

© Springer International Publishing AG 2016
M. Montes-y-Gómez et al. (Eds.): IBERAMIA 2016, LNAI 10022, pp. 25–36, 2016.
DOI: 10.1007/978-3-319-47955-2_3

explanatory reasoning with belief revision [5,7,8,14,15,17], but to our knowledge this is the first time that some explicit relationships between abduction and credibility limited revision operators are established. However, we have to point out that in the work of Fermé and Rodríguez [8], where they establish deep relationships between belief revision and the defeasible model (DFT) of Alchourrón [1], some typical postulates of credibility limited revision appear.

In the logic of abduction, some postulates -for instance (Right strengthening) (see Sect. 4) or the (Reflexivity)- are not well viewed. Using (Right strengthening) we could have that if Good_coffee▷colombian_coffee then Good_coffee▷ colombian_coffee_with_pepper which is not very convincing. In this paper we will see a family of explanatory relations which doesn't satisfy (Right strengthening).

Concerning the reflexivity of the explanatory relations, one can consider the relations satisfying this property like proto-explanatory relations. And then using them for constructing explanatory relations where (Reflexivity) doesn't hold. A way to do that is considering the proto-explanatory relation and then to withdraw the diagonal (*i.e.* the pairs $\alpha \rhd \alpha$).

A natural feature to consider is the fact of having *impossible observations*, that is, observations which don't have any explanation. Imagine an observation like A_pink_elephant_driving_a_Fiat_500. One is not leaning to belief that such observation has an explanation. Considering impossible observations has interesting consequences like we will see in Sect. 4.

This work is organized as follows. Section 2 is devoted to fix some notation and some results about credibility limited revision operators. Section 3 is devoted to give a very abstract view of explanatory relations. Section 4 is devoted to give our two families of explanatory relations the links with credibility limited revision operators and the main representation theorems. We finish with Sect. 5 in which we make some final remarks and point out some future work.

2   Preliminaries

Let $\mathcal{L}$ be the set of propositional formulas built over a finite set $\mathcal{P}$ of atomic propositions. A consistent formula is a formula which does not entail the formula $\alpha \wedge \neg\alpha$ (a contradiction). $\mathcal{L}^*$ will denote the set of consistent formulas. An interpretation ω is a total function from $\mathcal{P}$ to $\{0,1\}$. The set of all interpretations is denoted $\mathcal{W}$. An interpretation ω is a model of a formula $\varphi \in \mathcal{L}$ if and only if it makes it true in the usual truth functional way. If φ is a formula, we denote by $[\![\varphi]\!]$ the set of models of φ, *i.e.* $[\![\varphi]\!] = \{\omega \in \mathcal{W} : \omega \models \varphi\}$. As usual, $\top$ and $\bot$ denote two fixed propositional formulas such that $[\![\top]\!] = \mathcal{W}$ and $[\![\bot]\!] = \emptyset$. If I is a nonempty set of interpretations, we denote by α_I a formula such that $[\![\alpha_I]\!] = I$. We write α_ω instead of $\alpha_{\{\omega\}}$. When A is a finite set of formulas we denote by $\bigvee A$ the disjunction of all the formulas of A. We denote by $\vdash$ the classical entailment relation, that is $\alpha \vdash \beta$ when α entails β (this happens exactly when $[\![\alpha]\!] \subseteq [\![\beta]\!]$). The logical equivalence will be denoted by $\equiv$, that is $\alpha \equiv \beta$ when $\alpha \vdash \beta$ and $\beta \vdash \alpha$.

Let A be a set. A binary relation $\preceq$ over A is a total preorder if it is total (therefore reflexive) and transitive. Let $\preceq$ be a total preorder over A. We define the strict relation $\prec$ and the indifference $\simeq$ associated to $\preceq$ as follows: $a \prec b$ if and only if $a \preceq b$ and $b \npreceq a$; $a \simeq b$ if and only if $a \preceq b$ and $b \preceq a$.

Let $\preceq$ be a total preorder over A. Let C be a subset of A. We say that c is a minimal element of C with respect to $\preceq$ if $c \in C$ and for all $x \in C$, $x \nprec c$. The set of minimal elements of C will be denoted $min(C, \preceq)$. The minimal elements of the whole set A in which the total preorder is defined will be denoted $min(\preceq)$.

Let $\rhd$ be a binary relation over $\mathcal{L}$. This relation is supposed to represent a well behaved explanatory relation. We use infix notation $\alpha \rhd \gamma$ to express that the pair (α, γ) is in the relation $\rhd$. The set $\{\gamma : \alpha \rhd \gamma\}$ is called the set of explanations of α, and it is denoted $Expl(\alpha)$. The set Expl is the set of all explanations, that is $\mathsf{Expl} = \{\gamma : \exists \alpha \text{ such that } \alpha \rhd \gamma\}$. An explanation having one unique model is called a complete explanation.

2.1   Credibility-Limited Revision

We give a recall of propositional credibility-limited revision. The first version of credibility-limited revision was proposed by Hansson et al. in [11]. This is a generalization of revision operators proposed by Alchourrón et al. [2]. The version we proposed here, is a more compact version which has been presented by Booth et al. in [4] in the Katsuno-Mendelzon [12] style.

Definition 1. *An operator* $\circ : \mathcal{L}^* \times \mathcal{L} \longrightarrow \mathcal{L}$ *is called a credibility-limited revision operator (CL revision operator for short) if the following postulates[1] hold for any consistent formula* φ *and every formulas* α *and* β:

$$\varphi \circ \alpha \equiv \varphi \quad \text{or} \quad \varphi \circ \alpha \vdash \alpha \qquad \text{(Relative success)}$$

$$\text{If } \alpha \wedge \varphi \nvdash \bot \text{ then } \varphi \circ \alpha \equiv \varphi \wedge \alpha \qquad \text{(Vacuity)}$$

$$\varphi \circ \alpha \nvdash \bot \qquad \text{(Strong coherence)}$$

$$\text{If } \varphi \equiv \psi \text{ and } \alpha \equiv \beta \text{ then } \varphi \circ \alpha \equiv \psi \circ \beta \qquad \text{(Syntax independence)}$$

$$\text{If } \varphi \circ \alpha \vdash \alpha \text{ and } \alpha \vdash \beta \text{ then } \varphi \circ \beta \vdash \beta \qquad \text{(Success monotony)}$$

$$\varphi \circ (\alpha \vee \beta) \equiv \begin{cases} \varphi \circ \alpha \text{ or} \\ \varphi \circ \beta \text{ or} \\ (\varphi \circ \alpha) \vee (\varphi \circ \beta) \end{cases} \qquad \text{(Trichotomy)}$$

Definition 2 (Assignment). *A CL-faithful assignment is a function mapping each consistent formula* φ *into a pair* $(C_\varphi, \leq_\varphi)$ *where* $[\![\varphi]\!] \subseteq C_\varphi \subseteq \mathcal{W}$, $\leq_\varphi$ *is a total preorder over* C_φ, *and the following conditions hold for all* $\omega, \omega' \in C_\varphi$:

1. *If* $\omega \models \varphi$, *then* $\omega \leq_\varphi \omega'$
2. *If* $\omega \models \varphi$ *and* $\omega' \nvDash \varphi$, *then* $\omega <_\varphi \omega'$
3. *If* $\varphi \equiv \varphi'$, *then* $(C_\varphi, \leq_\varphi) = (C_{\varphi'}, \leq_{\varphi'})$

The interpretations in C_φ will be called the *credible worlds* relative to φ. Next we state the Representation Theorem in [4].

---

[1] As usual we use the infix notation $\alpha \circ \beta$ instead of $\circ(\alpha, \beta)$.

Theorem 1 (Booth et al.). *A mapping* $\circ : \mathcal{L}^* \times \mathcal{L} \longrightarrow \mathcal{L}$ *is a CL revision operator iff there exists a CL-faithful assignment* $\varphi \mapsto (C_\varphi, \leq_\varphi)$ *such that*

$$[\![\varphi \circ \alpha]\!] = \begin{cases} \min([\![\alpha]\!], \leq_\varphi) & \text{if } [\![\alpha]\!] \cap C_\varphi \neq \emptyset \\ [\![\varphi]\!] & \text{otherwise} \end{cases}$$

3   An Abstract Approach for Abduction

The main idea is to think that the meaning of $\alpha \rhd \gamma$ is actually $\pi_1(\gamma) \vdash \pi_2(\alpha)$, where the functions $\pi_i : \mathcal{L} \longrightarrow \mathcal{L}$ are some sort of "core" functions. That is, they are functions which give the more relevant part of their input. In particular, they have to satisfy $\pi_i(\beta) \vdash \beta$. Note that, when π_1 and π_2 are the identity function the relation $\rhd$ is simply reverse deduction which is not, in general, a very interesting explanatory relation [6].

Actually, we will give two families of explanatory relations such that $\rhd$ satisfies a set of postulates iff π_1 and π_2 are precisely determined and moreover we have the following equivalence:

$$\alpha \rhd \gamma \Leftrightarrow \pi_1(\gamma) \vdash \pi_2(\alpha)$$

In order to see how this relations are working, let us give the following illustrative example.

Example 1. *Consider a propositional language with four propositional variables: c, s, p, g meaning Colombian coffee, coffee with sugar, coffee with pepper and good coffee respectively. We are reasoning about good coffee. Thus, the worlds in which the coffee is not good are impossible worlds. In particular, the worlds in which there are pepper in the coffee are incompatible with good coffee. Thus, the only credible worlds are $0101, 0001, 1101, 1001$ (the bits follow the order c, s, p, g). Suppose now that the preferences of these worlds are*

$$\begin{array}{c} 0101 \\ 1101 \qquad 0001 \\ 1001 \end{array}$$

that is, the worlds in lowest levels are preferred ($1001 \prec 1101 \sim 0001 \prec 0101$). Suppose that π_1 is the identity and π_2 is "taking the minimal models". Then we would have

– $g \rhd c \wedge \neg s \wedge \neg p$
– $g \wedge s \rhd c \wedge s \wedge \neg p$
– $g \wedge s \wedge p$ *has no explanations!*

4   Credibility Limited Explanatory Relations

In this section we define two classes of explanatory relations which will characterized in terms of credibility limited revision operators. Then as a corollary of our characterization we will obtain representation theorems for these two classes of explanatory relations.

Weakly Reflexive Explanatory Relations

We begin with the definition of our first class of explanatory relations.

Definition 3. *Let* $\triangleright$ *be a binary relation over* $\mathcal{L}$. *The relation* $\triangleright$ *is called a weakly reflexive explanatory relation if the following postulates hold:*

$Expl(\top) \neq \emptyset$	**(Strong non triviality)**
$\alpha \triangleright \gamma \Rightarrow \gamma \nvdash \bot$	**(Coherence)**
$\alpha \triangleright \gamma, \delta \vdash \gamma, \delta \nvdash \bot \Rightarrow \alpha \triangleright \delta$	**(Right strengthening)**
$\alpha \wedge \beta \triangleright \delta, \exists \gamma (\alpha \triangleright \gamma \text{ and } \gamma \vdash \beta) \Rightarrow \alpha \triangleright \delta$	**(Weak cut)**
$\alpha \triangleright \gamma \Rightarrow \gamma \vdash \alpha$	**(Infra-classicality)**
$\alpha \triangleright \gamma, \alpha \triangleright \delta \Rightarrow \alpha \triangleright \gamma \vee \delta$	**(Right or)**
$\alpha \triangleright \gamma, \gamma \vdash \beta \Rightarrow \alpha \wedge \beta \triangleright \gamma$	**(Cautious monotony)**
$\alpha \equiv \alpha', \gamma \equiv \gamma' \Rightarrow (\alpha \triangleright \gamma \Leftrightarrow \alpha' \triangleright \gamma')$	**(Congruence)**
$Expl(\alpha) \neq \emptyset, \alpha \vdash \beta \Rightarrow Expl(\beta) \neq \emptyset$	**(Explanatory monotony)**

The postulates above can be seen as a minimal set of inference rules whose meaning we explain as follows. *Strong non triviality* says (together with Congruence) that every tautology has at least one explanation. *Coherence* says that the explanations are consistent. *Right strengthening* says that consistent formulas stronger than one explanation are also explanations. *Weak cut* is a cut rule for the observations under some conditions. *Infra-classicality* is the way to impose that explanations have to be formulas classically entailing the observation. *Right or* says that the disjunction of explanations of one observation is still one explanation of this observation. *Cautious monotony* establish the conditions under which we can strength one original observation and maintain the same explanation of this original observation. *Congruence* says that the explanatory relation is independent of the syntax representation. *Explanatory monotony* says that weaker formulas than formulas having explanations, have also explanations.

The following Theorem can be interpreted as a justification for the choice of the set of postulates appearing in Definition 3.

Theorem 2. $\triangleright$ *is a weakly reflexive explanatory relation iff there exists a consistent formula* φ *and a credibility-limited revision operator* $\circ$ *such that*

$$\alpha \triangleright \gamma \Leftrightarrow (\gamma \vdash \varphi \circ \alpha), (\varphi \circ \alpha \vdash \alpha) \text{ and } \gamma \nvdash \bot \qquad (1)$$

Definition 4. *We define* φ, *associated to the relation* $\triangleright$ *as follows:*

$$\varphi \equiv \bigvee \{\alpha_\omega : \top \triangleright \alpha_\omega\}$$

Observation 1. *Note that if $\rhd$ satisfies (Strong non triviality), (Coherence) and (Right strengthening) the formula φ of the Definition 4 is consistent.*

Proposition 1. *Let $\rhd$ be a weakly reflexive explanatory relation. If $\alpha \rhd \gamma$, then $\gamma \rhd \gamma$.*

Proof. If $\alpha \rhd \gamma$ then, by (Coherence) and (Infra-classicality), we have $\gamma \nvdash \bot$ and $\gamma \vdash \alpha$. By (Cautious mootony) we have $\alpha \wedge \gamma \rhd \gamma$. Since $\gamma \vdash \alpha$ we have $\alpha \wedge \gamma \equiv \gamma$. Then, by (Congruence), $\gamma \rhd \gamma$. ∎

Proof of Theorem 2: First we prove the *only if part*. Assume that $\rhd$ is a weakly reflexive explanatory relation. Define an operator $\circ$ in the following way:

$$\varphi \circ \alpha \equiv \begin{cases} \varphi & \text{si } Expl(\alpha) = \emptyset \\ \bigvee\{\alpha_\omega : \alpha \rhd \alpha_\omega\} & \text{si } Expl(\alpha) \neq \emptyset \end{cases}$$

(i) First, we prove the following equivalence:

$$\alpha \rhd \gamma \Leftrightarrow (\gamma \vdash \varphi \circ \alpha), \ (\varphi \circ \alpha \vdash \alpha) \ y \ \gamma \nvdash \bot$$

($\Leftarrow$) Suppose that $\gamma \vdash \varphi \circ \alpha$ and $\varphi \circ \alpha \vdash \alpha$. If $Expl(\alpha)$ is nonempty, then, it is easy to see that there are elements of the form α_{ω_i} in $Expl(\alpha)$ and for all such elements, $\gamma \vdash \bigvee \alpha_{\omega_i}$. Then by (Right or), $\alpha \rhd \bigvee \alpha_{\omega_i}$. From this, by (Right strengthening), $\alpha \rhd \gamma$. If α has not explanations, then there are elements α_{ω_i} of $Expl(\top)$ such that $\gamma \vdash \bigvee \alpha_{\omega_i}$ (the disjunction of all such α_{ω_i}). Then, by (Right or) and (Right strengthening), we have $\top \rhd \gamma$. Since $\gamma \vdash \alpha$ and $\top \wedge \alpha \nvdash \bot$, by (Cautious monotony), we have $\top \wedge \alpha \rhd \gamma$ and therefore, by (Congruence), $\alpha \rhd \gamma$.

($\Rightarrow$) If $\alpha \rhd \gamma$, then, by (Coherence), γ is consistent. Thus,

$$[\![\gamma]\!] = \{\omega_1, \dots, \omega_n\} \quad \text{where} \quad \gamma \equiv \bigvee_{i=1}^n \alpha_{\omega_i}$$

Then $\alpha_{\omega_i} \vdash \gamma$ for all i. By (Right strengthening) we have $\alpha \rhd \alpha_{\omega_i}$. Then $\alpha_{\omega_i} \vdash \varphi \circ \alpha$ for all i. Thus, $\bigvee_{i=1}^n \alpha_{\omega_i} \vdash \varphi \circ \alpha$, and therefore $\gamma \vdash \varphi \circ \alpha$.

We have also that $\alpha \rhd \alpha_\omega$ for any α_ω satisfying $\alpha_\omega \vdash \varphi \circ \alpha$. Then, by (Infra-classicality) $\alpha_\omega \vdash \alpha$, and then $\bigvee \alpha_\omega \vdash \alpha$, that is $\varphi \circ \alpha \vdash \alpha$.

(ii) Now we prove that the operator $\circ$ satisfies all the postulates defining a credibility limited revision operator relative to φ:

(Relative success) $\varphi \circ \alpha \vdash \alpha$ or $\varphi \circ \alpha \equiv \varphi$.

If $Expl(\alpha) = \emptyset$, then $\varphi \circ \alpha \equiv \varphi$. If $Expl(\alpha) \neq \emptyset$, there exists γ such that $\alpha \rhd \gamma$ and +for (i), we have $\varphi \circ \alpha \vdash \alpha$.

(Vacuity) $\varphi \wedge \alpha \nvdash \bot \Rightarrow \varphi \circ \alpha \equiv \varphi \wedge \alpha$.

If $\alpha_\omega \vdash \varphi \wedge \alpha$, then $\top \rhd \alpha_\omega$ and $\alpha_\omega \vdash \alpha$. Since $\alpha \nvdash \bot$ then $\top \wedge \alpha \nvdash \bot$, and, by (Cautious monotony) and (Congruence), we have $\alpha \rhd \alpha_\omega$. Thus, $\alpha_\omega \vdash \varphi \circ \alpha$.

Now suppose that $\alpha_\omega \vdash \varphi \circ \alpha$. Since $Expl(\alpha) \neq \emptyset$, by definition of $\circ$, we have $\alpha \rhd \alpha_\omega$, so $\top \wedge \alpha \rhd \alpha_\omega$. Since $\varphi \wedge \alpha \nvdash \bot$, there exists $\alpha_{\omega'}$ such that

$\top \vartriangleright \alpha_{\omega'}$ and $\alpha_{\omega'} \vdash \alpha$. From this, by (Weak cut), we have $\top \vartriangleright \alpha_\omega$, then $\alpha_\omega \vdash \varphi$. From the fact that $\alpha \vartriangleright \alpha_\omega$, by (Infraclassicality), $\alpha_\omega \vdash \alpha$. Therefore $\alpha_\omega \vdash \varphi \wedge \alpha$.

(Strong coherence) $\varphi \circ \alpha \nvdash \bot$.

If $Expl(\alpha) = \emptyset$ then $\varphi \circ \alpha \equiv \varphi$. Since $\varphi \nvdash \bot$ (see Observation 1), we have $\varphi \circ \alpha \nvdash \bot$. If $Expl(\alpha) \neq \emptyset$, there exists γ tal que $\alpha \vartriangleright \gamma$. By (Coherence) $\gamma \nvdash \bot$, thus, there exists α_ω such that $\alpha_\omega \vdash \gamma$. Then, by (Right strengthening) $\alpha \vartriangleright \alpha_\omega$; in particular, $\alpha_\omega \vdash \varphi \circ \alpha$. Therefore, $\varphi \circ \alpha$ is consistent.

(Syntax independence) $\alpha \equiv \alpha' \Rightarrow \varphi \circ \alpha \equiv \varphi \circ \alpha'$.

If α has not explanations, then α' has not explanations. Then $\varphi \circ \alpha \equiv \varphi$ and $\varphi \circ \alpha' \equiv \varphi$, therefore $\varphi \circ \alpha \equiv \varphi \circ \alpha'$.

If α has explanations, there exists α_ω such that $\alpha_\omega \vdash \varphi \circ \alpha$ and $\alpha \vartriangleright \alpha_\omega$. Then, by (Congruence), $\alpha' \vartriangleright \alpha_\omega$. Therefore, $\alpha_\omega \vdash \varphi \circ \alpha'$. Thus, we have proven $\varphi \circ \alpha \vdash \varphi \circ \alpha'$. The converse entailment can be proven in an analogous way.

(Success monotony) $\varphi \circ \alpha \vdash \alpha$ and $\alpha \vdash \beta \Rightarrow \varphi \circ \beta \vdash \beta$.

Assume $\varphi \circ \alpha \vdash \alpha$ and $\alpha \vdash \beta$. Suppose that α has no explanations. Then $\varphi \circ \alpha \equiv \varphi$. From this and the assumptions $\varphi \vdash \alpha$. Then, if $\alpha_\omega \vdash \varphi$ we have $\top \vartriangleright \alpha_\omega$ and $\alpha_\omega \vdash \alpha$, thus $\top \wedge \alpha \nvdash \bot$. Then, by (Cautious monotony) $\top \wedge \alpha \vartriangleright \alpha_\omega$, thus, by (Congruence) $\alpha \vartriangleright \alpha_\omega$, contradiction. Therefore, α has explanations. Then, by (Explanatory monotony), β has explanations. If $\alpha_\omega \vdash \varphi \circ \beta$, by definition of $\circ$, $\beta \vartriangleright \alpha_\omega$. Then, by (Infraclassicality), $\alpha_\omega \vdash \beta$. Therefore, $\varphi \circ \beta \vdash \beta$.

(Trichotomy)

$$\varphi \circ (\alpha \vee \beta) \equiv \begin{cases} \varphi \circ \alpha & \text{or} \\ \varphi \circ \beta & \text{or} \\ (\varphi \circ \alpha) \vee (\varphi \circ \beta) \end{cases}$$

If $Expl(\alpha \vee \beta) = \emptyset$, by (Explanatory monotony,) we have $Expl(\alpha) = \emptyset$ and $Expl(\beta) = \emptyset$. Thus, $\varphi \circ (\alpha \vee \beta) \equiv \varphi$, $\varphi \circ \alpha \equiv \varphi$ and $\varphi \circ \beta \equiv \varphi$. In particular, $\varphi \circ (\alpha \vee \beta) \equiv \varphi \circ \alpha$.

Now suppose that $Expl(\alpha \vee \beta) \neq \emptyset$. Thus, there exists α_ω such that $\alpha \vee \beta \vartriangleright \alpha_\omega$ and, by (Infraclassicity), $\alpha_\omega \vdash \alpha \vee \beta$. Note that, by (Strong coherence), $\varphi \circ (\alpha \vee \beta) \nvdash \bot$.

Case 1: $Expl(\alpha) \neq \emptyset$ and $Expl(\beta) = \emptyset$. In this case we prove that $\varphi \circ (\alpha \vee \beta) \equiv \varphi \circ \alpha$.

Let $\alpha_{\omega'}$ be such that $\alpha_{\omega'} \vdash \varphi \circ (\alpha \vee \beta)$, then, by definition of $\circ$, $\alpha \vee \beta \vartriangleright \alpha_{\omega'}$. Then, by (Infraclassicity), $\alpha_{\omega'} \vdash \alpha \vee \beta$. If $\alpha_{\omega'} \vdash \beta$, since $(\alpha \vee \beta) \wedge \beta \nvdash \bot$, we have, by (Cautious monotony), $(\alpha \vee \beta) \wedge \beta \vartriangleright \alpha_{\omega'}$. From this and (Congruence), $\beta \vartriangleright \alpha_{\omega'}$, a contradiction. Therefore, $\alpha_{\omega'} \nvdash \beta$, thus, $\alpha_{\omega'} \vdash \alpha$. From this, (Cautious monotony) and (Congruence), we have $\alpha \vartriangleright \alpha_{\omega'}$. Therefore $\alpha_{\omega'} \vdash \varphi \circ \alpha$.

Now suppose that $\alpha_{\omega'} \vdash \varphi \circ \alpha$, then, by definition of $\circ$, $\alpha \vartriangleright \alpha_{\omega'}$. If $\alpha_\omega \vdash \beta$, then, since $\alpha \vee \beta \vartriangleright \alpha_\omega$, by (Cautious monotony) and (Congruence) we would have $\beta \vartriangleright \alpha_\omega$, contradiction. Therefore, $\alpha_\omega \nvdash \beta$ and, since $\alpha_\omega \vdash \alpha \vee \beta$, necessarily $\alpha_\omega \vdash \alpha$. Since $\alpha \vartriangleright \alpha_{\omega'}$, by (Congruence), $(\alpha \vee \beta) \wedge \alpha \vartriangleright \alpha_{\omega'}$.

Moreover, since $\alpha \vee \beta \rhd \alpha_\omega$ and $\alpha_\omega \vdash \alpha$, by (Weak cut), we have $\alpha \vee \beta \rhd \alpha_{\omega'}$, that is $\alpha_{\omega'} \vdash \varphi \circ (\alpha \vee \beta)$.

Case 2: $Expl(\alpha) = \emptyset$ and $Expl(\beta) \neq \emptyset$. In this case we prove that $\varphi \circ (\alpha \vee \beta) \equiv \varphi \circ \beta$.
The proof is analogous to the previous case.

Case 3: $Expl(\alpha) \neq \emptyset$ and $Expl(\beta) \neq \emptyset$. Let $\alpha_{\omega'}$ and $\alpha_{\omega''}$ be such that $\alpha \rhd \alpha_{\omega'}$ and $\beta \rhd \alpha_{\omega''}$.

(a) Suppose $\alpha \vee \beta \rhd \alpha_{\omega'}$ and $\alpha \vee \beta \not\rhd \alpha_{\omega''}$. In this case we prove that $\varphi \circ (\alpha \vee \beta) \equiv \varphi \circ \alpha$.
Suppose $\alpha_\omega \vdash \varphi \circ (\alpha \vee \beta)$, then $\alpha \vee \beta \rhd \alpha_\omega$. From this, by (Infraclassicity) $\alpha_\omega \vdash \alpha \vee \beta$.
Suppose $\alpha_\omega \vdash \beta$. Since $(\alpha \vee \beta) \wedge \beta \equiv \beta$ and $(\alpha \vee \beta) \wedge \beta \rhd \alpha_{\omega''}$, by (Weak cut), we have $\alpha \vee \beta \rhd \alpha_{\omega''}$, contradiction. Thus, $\alpha_\omega \vdash \alpha$. Since $(\alpha \vee \beta) \wedge \alpha \not\vdash \bot$, by (Cautious monotony), we have $(\alpha \vee \beta) \wedge \alpha \rhd \alpha_\omega$. From this and (Congruence), we obtain $\alpha \rhd \alpha_\omega$. Therefore, $\alpha_\omega \vdash \varphi \circ \alpha$. Thus, we have proven $\varphi \circ (\alpha \vee \beta) \vdash \varphi \circ \alpha$.
Now suppose that $\alpha_\omega \vdash \varphi \circ \alpha$. Then $\alpha \rhd \alpha_\omega$. From the assumptions and (Infraclassicity), we have $\alpha_{\omega'} \vdash \alpha$. Since $(\alpha \vee \beta) \wedge \alpha \equiv \alpha$, by (Congruence), $(\alpha \vee \beta) \wedge \alpha \rhd \alpha_\omega$. Since $\alpha \vee \beta \rhd \alpha_{\omega'}$ and $\alpha_{\omega'} \vdash \alpha$, by (Weak cut) $\alpha \vee \beta \rhd \alpha_\omega$. Therefore, $\alpha_\omega \vdash \varphi \circ (\alpha \vee \beta)$. Thus, we have proven $\varphi \circ \alpha \vdash \varphi \circ (\alpha \vee \beta)$. This finishes the proof that $\varphi \circ (\alpha \vee \beta) \equiv \varphi \circ \alpha$.

(b) $\alpha \vee \beta \not\rhd \alpha_{\omega'}$ and $\alpha \vee \beta \rhd \alpha_{\omega''}$. In this case we prove that $\varphi \circ (\alpha \vee \beta) \equiv \varphi \circ \beta$.
The proof is similar to the one of case (a).

(c) $\alpha \vee \beta \rhd \alpha_{\omega'}$ and $\alpha \vee \beta \rhd \alpha_{\omega''}$. In this case we prove that $\varphi \circ (\alpha \vee \beta) \equiv (\varphi \circ \alpha) \vee (\varphi \circ \beta)$.
Suppose $\alpha_\omega \vdash \varphi \circ (\alpha \vee \beta)$. Then $\alpha \vee \beta \rhd \alpha_\omega$ and, by (Infraclassicity) $\alpha_\omega \vdash \alpha \vee \beta$. Suppose that $\alpha_\omega \vdash \alpha$. Since $(\alpha \vee \beta) \wedge \alpha \not\vdash \bot$, by (Cautious monotony) $(\alpha \vee \beta) \wedge \alpha \rhd \alpha_\omega$. From this, by (Congruence), we have $\alpha \rhd \alpha_\omega$. Thus, $\alpha_\omega \vdash (\varphi \circ \alpha)$ and therefore $\alpha_\omega \vdash (\varphi \circ \alpha) \vee (\varphi \circ \beta)$. Suppose $\alpha_\omega \not\vdash \alpha$. Then $\alpha_\omega \vdash \beta$, with a analogous reasoning we get $\alpha_\omega \vdash \varphi \circ \beta$. Therefore $\alpha_\omega \vdash (\varphi \circ \alpha) \vee (\varphi \circ \beta)$. This proves that $\varphi \circ (\alpha \vee \beta) \vdash (\varphi \circ \alpha) \vee (\varphi \circ \beta)$.
Suppose $\alpha_\omega \vdash (\varphi \circ \alpha) \vee (\varphi \circ \beta)$. If $\alpha_\omega \vdash \varphi \circ \alpha$ then $\alpha \rhd \alpha_\omega$. From the assumption and (Infraclassicity), we have $\alpha_{\omega'} \vdash \alpha$. From this and (Weak cut), $\alpha \vee \beta \rhd \alpha_\omega$. Therefore, $\alpha_\omega \vdash \varphi \circ (\alpha \vee \beta)$. If $\alpha_\omega \vdash \varphi \circ \beta$, the proof is made in a similar way. This proves that $(\varphi \circ \alpha) \vee (\varphi \circ \beta) \vdash \varphi \circ (\alpha \vee \beta)$.

The proof of the *if* part of the Theorem is easy using Theorem 1 and it is left as an exercise for the reader.   ∎

From the previous Theorem and Theorem 1 we obtain the following result:

Corollary 1. $\rhd$ *is a weak reflexive explanatory relation iff there exists a non-empty set $C \subseteq \mathcal{W}$ and a total preorder $\preceq$ over C such that $\alpha \rhd \gamma$ iff $[\![\alpha]\!] \cap C \neq \emptyset$ and $[\![\gamma]\!] \subseteq \min(\alpha \cap C, \preceq)$.*

Note that the Example 1 is an illustration of this kind of explanatory relation at work. Actually, in that example the set C is $\{1001, 1101, 0001, 0001\}$ and the total preorder over C is given by $1001 \prec 1101 \sim 0001 \prec 0001$. The only observations α having explanations are the formulas α satisfying $[\![\alpha]\!] \cap C \neq \emptyset$. Here π_1 is the identity and π_2 is a function such that $[\![\pi_2(\alpha)]\!] = \min([\![\alpha]\!] \cap C, \preceq)$ for any α having explanations.

4.1   Ordered Explanatory Relations

Now we define our second class of explanatory relations.

Definition 5. *Let $\rhd$ be a binary relation over $\mathcal{L}$. The relation $\rhd$ is called an ordered explanatory relation if the following postulates hold:*

$\rhd \neq \emptyset$	(Non triviality)
$Expl(\alpha) \neq \emptyset \Rightarrow \alpha \rhd \alpha$	(Limited reflexivity)
$\alpha \rhd \gamma \Rightarrow \alpha \wedge \gamma \nvdash \bot$	(Weak infra-classicality)
$\alpha \rhd \gamma, \delta \vdash \gamma, \delta \nvdash \bot \Rightarrow \alpha \rhd \delta$ or $\alpha \wedge \neg \delta \rhd \gamma$	(Weak
$\alpha \rhd \gamma, \gamma \rhd \delta \Rightarrow \alpha \rhd \delta$	(Transitivity)
$\alpha \rhd \gamma, \beta \rhd \gamma \Rightarrow \alpha \wedge \beta \rhd \gamma$	(Left and)
$\alpha \rhd \gamma, \alpha \rhd \delta \Rightarrow \alpha \rhd \gamma \vee \delta$	(Right or)
$\alpha \rhd \gamma, \gamma \vdash \beta \Rightarrow \alpha \wedge \beta \rhd \gamma$	(Cautious monotony)
$\alpha \equiv \alpha', \gamma \equiv \gamma' \Rightarrow (\alpha \rhd \gamma \Leftrightarrow \alpha' \rhd \gamma')$	(Congruence)
$Expl(\alpha) \neq \emptyset, \alpha \vdash \beta \Rightarrow Expl(\beta) \neq \emptyset$	(Explanatory monotony)

As in Definition 3, the postulates above can be seen as a minimal set of inference rules whose meaning we can explain as follows. Actually, the meaning of *Right or, Cautious monotony, Congruence* and *Explanatory monotony* have been explained after Definition 3. *Non triviality* says that there exists something that can be explained. *Limited reflexivity* says that one explanation is explained by itself. *Weak infra-classicality* says that the explanations are at least consistent with the observation. *Weak right strengthening* says that if a formula stronger than one explanation of a given observation is not an explanation of the given observation, the reason is that the original explanation is also an explanation of the conjunction of the given observation and the negation of the stronger formula. The meaning of *Transitivity* is explained by itself. *Left and* says that one explanation of two observations is also an explanation of the conjunction of the two observations.

The following Theorem explain the rationale behind the choice of postulates in Definition 5.

Theorem 3. *$\rhd$ is an ordered explanatory relation iff there exists a consistent formula φ and a credibility-limited revision operator $\circ$ such that*

$$\alpha \rhd \gamma \Leftrightarrow (\varphi \circ \gamma \vdash \varphi \circ \alpha), (\varphi \circ \alpha \vdash \alpha) \text{ and } (\varphi \circ \gamma \vdash \gamma) \qquad (2)$$

The proof of this theorem uses the following two results whose proofs are not included here because of space limitations:

Lemma 1. *If $\triangleright$ is an ordered explanatory relation then $Expl(\alpha) \neq \emptyset \Leftrightarrow \alpha \in$* Expl

Proposition 2. *If $\triangleright$ is an ordered explanatory relation then the following postulates hold:*

(P1) $\alpha \in$ Expl $\Rightarrow \alpha \triangleright \alpha$.
(P2) $\alpha \triangleright \alpha_\omega \Rightarrow \alpha_\omega \vdash \alpha$.
(P3) $\alpha \triangleright \alpha$ *and* $\beta \vdash \alpha \Rightarrow \alpha \triangleright \beta$ *or* $\alpha \triangleright \alpha \wedge \neg\beta$.
(P4) $Expl(\alpha) \neq \emptyset \Rightarrow \alpha$ *has complete explanations.*
(P5) $\alpha \triangleright \gamma$ $\beta \vdash \alpha$ *and* $(\alpha \not\triangleright \beta) \Rightarrow \alpha \wedge \neg\beta \triangleright \gamma$.
(P6) $\top$ *has explanations.*
(P7) $Expl(\alpha) \neq \emptyset \Rightarrow \alpha \vee \beta \triangleright \alpha \vee \beta$ *for any formula* β.
(P8) $\alpha \triangleright \gamma$ *and* $\beta \triangleright \delta \Rightarrow \alpha \vee \beta \triangleright \gamma$ *or* $\alpha \vee \beta \triangleright \delta$.
(P9) $\alpha \triangleright \gamma$, $\gamma \vdash \beta$ *and* $\alpha \wedge \beta \triangleright \delta \Rightarrow \alpha \triangleright \delta$.
(P10) $Expl(\alpha \vee \beta) \cap Expl(\alpha) = \emptyset \Rightarrow Expl(\alpha \vee \beta) = Expl(\beta)$.

Now we can sketch the proof of Theorem 3.

Key elements of the proof of Theorem 3: The *if* part is easy using the Theorem 1. For the *only if* part we informally proceed as follows:

1. Define $\varphi = \bigvee\!\!\!\bigvee\{\alpha_\omega : \top \triangleright \alpha_\omega\}$
2. Define
$$\varphi \circ \alpha \equiv \begin{cases} \varphi & \text{if} \quad Expl(\alpha) = \emptyset \\ \bigvee\!\!\!\bigvee\{\alpha_\omega : \alpha \triangleright \alpha_\omega\} & \text{if} \quad Expl(\alpha) \neq \emptyset \end{cases}$$
3. Establish the representation equivalence (2).
4. Then prove that $\circ$ satisfies the CL revision operator postulates.

Corollary 2. *$\triangleright$ is an ordered explanatory relation iff there exists a nonempty set $C \subseteq \mathcal{W}$ and a total preorder $\preceq$ over C such that $\alpha \triangleright \gamma$ iff $[\![\alpha]\!] \cap C \neq \emptyset$, $[\![\gamma]\!] \cap C \neq \emptyset$ and $\min(\gamma \cap C, \preceq) \subseteq \min(\alpha \cap C, \preceq)$.*

Example 2. *We take the same data that in Example 1, that is, four propositional variables: c, s, p, g meaning Colombian coffee, coffee with sugar, coffee with pepper and good coffee respectively. The set C is $\{1001, 1101, 0001, 0001\}$ and the total preorder over C is given by $1001 \prec 1101 \sim 0001 \prec 0001$. The only observations α having explanations are the formulas α satisfying $[\![\alpha]\!] \cap C \neq \emptyset$. The only possible explanations γ are also formulas γ satisfying $[\![\gamma]\!] \cap C \neq \emptyset$. Here π_1 and π_2 have the same behavior, that is, $[\![\pi_i(\beta)]\!] = \min([\![\beta]\!] \cap C, \preceq)$ for any β having explanations. Thus, taking $\triangleright$ the ordered explanatory relation given by $(C, \preceq)$, using Corollary 2, we have* good_coffee $\triangleright$ (Colombian_coffee_without_pepper)*; that is, the coffee could have or not sugar but if it is a Colombian one and it has not pepper this is still an explanation of the coffee goodness. However, this explanation of* good_coffee *is not an explanation in the framework of the weakly reflexive explanatory relation associated to same pair $(C, \preceq)$.*

It is interesting to note that the set C and the total preorder $\preceq$ given by Corollaries 1 and 2 can be completely different in case we have two explanatory relations $\rhd_1$ and $\rhd_2$, the first one a weakly reflexive explanatory relation and the second one an ordered explanatory relation. But if we start from a subset C and a total preorder $\preceq$ Corollaries 1 and 2 define explanatory relations $\rhd_1$ and $\rhd_2$, the first one weakly reflexive and the second one ordered, such that $\rhd_1 \subseteq \rhd_2$, that is for all observation α if $\alpha \rhd_1 \gamma$ then $\alpha \rhd_2 \gamma$. Thus, for a given pair $(C, \preceq)$ the weakly reflexive explanatory relation associated is more precise than the ordered explanatory relation associated. Moreover, the weakly reflexive explanatory relation associated is an explanatory relation in a classical sense, that is the explanation entails the observation. That is not, in general, the case with the ordered explanatory relations. The ordered explanatory relations are more flexible than the weakly reflexive relations. They accept more explanations. Some explanations can be conceived as imprecise explanations. A complete knowledge of the one ordered explanatory relation can lead to a more precise explanatory relation: a weakly reflexive explanatory relation.

5   Final Remarks

We have obtained two families of explanatory relations, namely the weakly reflexive explanatory relations and the ordered explanatory relations. We have established tight links with credibility limited revision operators, given by Theorems 2 and 3. As corollaries we obtain semantical representations.

In both families of explanatory relations studied, the background theory Σ is, actually, implicit. It is, in fact, the theory of C, where the set C is the set given by Corollaries 1 and 2. The weakly reflexive explanatory relations are in fact an alternative view of the E-rational relations defined in [15].

The fact that the theory Σ is implicit allows us to give a simpler and purer logical presentation of the explanatory relations than the logical presentations where the postulates have to mention the background theory.

Note that, there are, in general, formulas without explanations. In particular the formulas whose set of models has no intersection with C.

By the semantical representation, it is easy to see that the family of ordered explanatory relations doesn't satisfy (Right strengthening).

Due to the fact that CL revision operators are generalizations of AGM revision operators, our Theorems 2 and 3 are, actually, generalizations of Walliser et al. results in [17].

There are many future research lines in perspective. In particular, to study more schemas (π_1, π_2) in order to define more families of explanatory relations. To introduce dynamics in the explanatory relations. To study the minimality of the set of postulates defining the families of explanatory relations considered in this work.

Acknowledgements. Thanks to the anonymous referees for their helpful remarks. The second author was partially supported by the research project CDCHT-ULA N° C-1451-07-05- A.

References

1. Alchourrón, C.E.: Detachment and defeasibility in deontic logic. Stud. Logica. **51**, 5–18 (1996)
2. Alchourrón, C.E., Gärdenfors, P., Makinson, D.: On the logic of theory change: partial meet contraction and revision functions. J. Symbolic Logic **50**, 510–530 (1985)
3. Bloch, I., Pino-Pérez, R., Uzcátegui, C.: Explanatory relations based on mathematical morphology. In: Benferhat, S., Besnard, P. (eds.) ECSQARU 2001. LNCS (LNAI), vol. 2143, pp. 736–747. Springer, Heidelberg (2001)
4. Booth, R., Fermé, E., Konieczny, S., Pino Pérez, R.: Credibility limited revision operators in propositional logic. In: Proceedings of the Thirteenth International Conference on Principles of Knowledge Representation And Reasoning, pp. 116–125 (2012)
5. Boutilier, C., Becher, V.: Abduction as belief revision. Artif. Intell. **77**, 43–94 (1995)
6. Mayer, M.C., Pirri, F.: Abduction is not deduction-in-reverse. J. IGPL **4**(1), 1–14 (1996)
7. Falappa, M.A., Kern-Isberner, G., Simari, G.R.: Explanations, belief revision and defeasible reasoning. Artif. Intell. **143**, 1–28 (2002)
8. Fermé, E., Rodríguez, R.: DFT and belief revision. Análisis Filosófico **XXVII**(2), 373–393 (2006)
9. Flach, P.A.: Rationality postulates for induction. In: Shoam, Y. (ed.) Proceedings of the Sixth Conference of Theoretical Aspects of Rationality and Knowledge (TARK-96), pp. 267–281 (1996)
10. Flach, P.A.: Logical characteristics of inductive learning. In: Gabbay, D.M., Kruse, R. (eds.) Abductive Reasoning and Learning, pp. 155–196 (2000)
11. Hansson, S.O., Fermé, E.L., Cantwell, J., Falappa, M.A.: Credibility limited revision. J. Symbolic Logic **66**, 1581–1596 (2001)
12. Katsuno, H., Mendelzon, A.: Propositional knowledge base revision and minimal change. Artif. Intell. **52**, 263–294 (1991)
13. Levesque, H.J.: A knowledge level account of abduction. In: Proceedings of the Eleventh International Joint Conference on Artificial Intelligence, Detroit, pp. 1061–1067 (1989)
14. Pagnucco, M.: The Role of Abductive Reasoning Within the Process of Belief Revision. Ph.D. thesis, Department of Computer Science, University of Sydney, February 1996
15. Pino Pérez, R., Uzcátegui, C.: Jumping to explanations vs jumping to conclusions. Artif. Intell. **111**(2), 131–169 (1999)
16. Pino Pérez, R., Uzcátegui, C.: Preferences and explanations. Artif. Intell. **149**(1), 1–30 (2003)
17. Walliser, B., Zwirn, D., Zwirn, H.: Abductive logic in a belief revision framework. J. Logic Lang. Inf. **14**, 87–117 (2005)

Semantic Enrichment of Web Service Operations

Maricela Bravo[✉], José A. Reyes-Ortiz, Roberto Alcántara-Ramírez,
and Leonardo Sánchez

Systems Department, Autonomous Metropolitan University,
02200 Azcapotzalco, DF, Mexico
{mcbc,jaro,raar,ldsm}@correo.azc.uam.mx

Abstract. In this paper we describe the process by which semantic relatedness assertions are discovered and defined between Web service operations. The general approach relies on a global ontology model that describes Web services. Obtaining semantic similarities between operations is performed by calculating eight semantic relatedness measures between all operations pairs. The entire process consists of: Web service parsing, Web service data extraction, automatic Web service ontology population, similarity measures calculation, similarity discovery; and finally, object property assertion between web service operations.

Keywords: Semantic web services · Automatic discovery · Ontology axioms

1 Introduction

There are many Web services available and executable on the Web, most of these services are published in WSDL, a language for the description of services which offers syntactic information of the functionality of the service. In order to reuse and take advantage of these programming resources, there is the need to implement computational infrastructures that allow software developers and integrators to foster automatic search, selection and invocation of public services. Towards this need, many researchers have promoted the use of semantic technologies to enhance and discover the service functionality. However, until now semantic solutions have not been widely adopted, this is mainly due to the difficulty of using semantic tools.

In this paper we describe a semantic enrichment of Web service operations. The general approach relies on a global ontology model to describe Web services. Obtaining semantic similarities between operations is performed by calculating eight semantic relatedness measures between all operations pairs. The entire process is shown in Fig. 1, consists of: Web service parsing, Web service data extraction, automatic Web service ontology population, similarity measures calculation, similarity discovery; and finally, object property assertion between web service operations. The resulting ontology model has been designed to allow the interoperability with multiple versions of service description languages and semantic description models.

© Springer International Publishing AG 2016
M. Montes-y-Gómez et al. (Eds.): IBERAMIA 2016, LNAI 10022, pp. 37–48, 2016.
DOI: 10.1007/978-3-319-47955-2_4

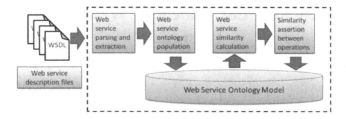

Fig. 1. Semantic enrichment of Web services operations.

Experimentation with SDWS[1] tool shows that the translation of syntactic description to semantic descriptions is feasible, automated and the resulting ontologies can be used to obtain useful information for software developers and integrators.

2 Web Service Parsing and Extraction

Over the last decade, numerous initiatives have been proposed to describe and represent Web services. These initiatives fall into two categories: syntactic and semantic description models. Syntactic models refer to the core set of languages used to describe service functionality such as XML and WSDL. Semantic models are those based on ontology languages to annotate and describe service functionality and facilitate automatic service discovery, invocation and composition.

A Service Description Language (SDL) can be defined as a mean by which a service vendor provides the technical description (document or contract) of the service programmatic interface. Such technical description must include the network address where the executable file of the service resides and accepts requests, the name of the operations (functions or methods) available to be remotely invoked, parameter names and data types. The World Wide Web Consortium[2] (W3C) recommended the Web Service Description Language (WSDL), which is currently a well-established W3C standard. WSDL defines an XML grammar for describing networked services as collections of communication endpoints capable of exchanging messages[3].

Semantic description models rely on the design and incorporation of ontologies to annotate or enhance service descriptions. In this case, the use of ontologies for the semantic enhancement of service descriptions have gained much attention among research and software communities. For the purpose of this work, the following semantic technologies and languages are considered: SAWSDL, OWL-S, and WSDL.

In order to automatically retrieve and extract data from Web service description files, the SDL heterogeneity has to be addressed. This requirement is derived from the existence of various service description models and languages: WSDL, OWL-S, WSML, and SAWSDL. This problem was solved by identifying the common elements

---

[1] http://aisii.azc.uam.mx:8080/sdws/.

[2] The international standard organization for the World Wide Web.

[3] http://www.w3.org.

that all service description models provide, which are: a communication interface that the client uses to create a proxy object to invoke the service remotely. This communication interface must describe information about the functions that the service offers (operation in WSDL, profile in OWL-S or capability in WSML) as well as the descriptions (name and data types) of input and output parameters. In order to make the parsing module easily expansible, an Interface - Implements architecture was used. With this architecture, a general WSDL Interface was defined and particular WSDL languages versions were implemented, where peculiarities of each SDL were incorporated.

3   Web Service Ontology Model

The general ontology model shown in Fig. 2 was designed for the representation of Web services. In particular, this general model allocates the representations of the following service description implementations: WSDL 1.1, WSDL 2.0 and OWL-S. The following classes are the main constitutive concepts of this general ontology model:

Service Provider. The service provider is the concept used to represent the contact data and network address of the organization or institution responsible for the deployment and maintenance of a Web service or a set of Web services.

Service Client. The service client concept is used to represent a software entity description with its physical address (a local or remote application, programmed agent or autonomous intelligent agent, or a simple service tester) that has been invoking operations or methods of a specific Web service.

General Service. A general service concept represents an endpoint reference or address of a given service name.

General Operation. A general operation defines the methods or operations that are provided by a given general service instance. In this general model any operation requires a set of input parameter instances to be executed and produces an output parameter after execution.

General Parameter. The parameter concept representation at a high level of abstraction is relevant for supporting type compatibilities. The general parameter is defined as a super-classification of the input parameter and output parameter classes.

Input Parameter. The input parameter concept represents pairs of parameter name and parameter type that are required for the execution of a given service operation.

Output Parameter. The output parameter concept represents the resulting data or object after the execution of the operation.

General Precondition. The general precondition concept is used to specify (whenever this information is available) a required state or condition of the invoking program or client before the execution of the service operation.

General Effect. The general effect concept is used to specify the resulting state or situation of the invoking program after the execution of the operation.

A similar model was reported in [22]; authors described a service reference model (SRM) to represent the following concepts: Service, Service Input, Service Output, Service Context, Service Logic, Service Provider, Service Client and Service Feedback.

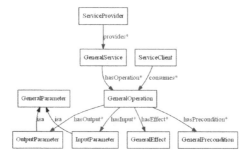

Fig. 2. General ontology model for the representation of different Web service languages and models.

One disadvantage of the SRM model is that it is not possible to define different operations or methods associated with the same service, in real Web services described in WSDL any service defines more than one operation or method. Another difference is the lack of a precondition and effect explicit declaration. SRM defines a service logic concept, but this element does not map to the precondition and effect concepts used in semantic models.

4 Web Service Similarity Calculation

Every Web service description consists of one or more service operations which are the programming application interfaces through which any client invokes particular service functionality. Therefore, in order to get a semantic relatedness between Web services, it is necessary to calculate similarity between service operations. In particular, in this work, semantic relatedness is measured by retrieving and text-processing service operation names (see Fig. 3).

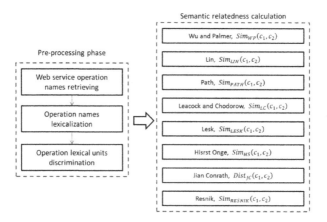

Fig. 3. Web service semantic relatedness calculation.

Web service operation names are short texts that combine one to seven words. Operation names are written in several formats following no general convention or nomenclature. So, in order to obtain the semantic relatedness between operation names a pre-processing step is required.

Pre-processing involves obtaining lexical units that are part of the operation name. First, text normalization is performed in order to transform operations names into a single canonical form. For example, considering the operation name getFlightPrice, its lexicalization results in [get][Flight][Price]. Finally, pre-processing phase includes discrimination of words that do not offer important information, for instance: http, for, return, result, soap. These words are filtered out prior the calculation of semantic relatedness between operations.

According with Budanitsky and Hirst [1], semantic relatedness is a more general concept than similarity; similar concepts are assumed to be related by their likeness, but dissimilar concepts may also be semantically related by other relationships such as meronymy, antonymy or any kind of relationship.

The following measures of relatedness were used to calculate a global semantic similarity relationship between Web service operations. These semantic relatedness measures utilize WordNet database and exploit additional non-hierarchical relations.

Wu and Palmer [2] introduced a relatedness measure that finds the path length to the root node from the least common subsumer (LCS) of two concepts, which is the most specific common concept $lso(c_1, c_2)$ they share as an ancestor. This value is scaled by the sum of the path lengths from the individual concepts to the root (see Formula 1).

$$Sim_{WP}(c_1, c_2) = \frac{2 * depth(lso(c_1, c_2))}{len(c_1, c_2) + 2 * depth(lso(c_1, c_2))} \qquad (1)$$

Lin [3] presents a measure that calculates semantic relatedness between two concepts. Lin stated that "the similarity between A and B is measured by the ratio between the amount of information needed to state the commonality of A and B and the information needed to fully describe what A and B are." This measure uses the amount of information needed to state the commonality between the two concepts and the information needed to describe these terms (see Formula 2).

$$Sim_{LIN}(c_1, c_2) = \frac{2 * IC(lso(c_1, c_2))}{IC(c_1) + IC(c_2)} \qquad (2)$$

Path [4] semantic relatedness is a node-counting scheme (path). The relatedness score is inversely proportional to the number of nodes along the shortest path between the synsets. The shortest possible path occurs when the two synsets are the same, in which case the length is 1. Thus, the maximum relatedness value is 1.

This measure only takes the longitude between concepts c1 and c2 into consideration. It assumes that the similarity between concepts c1 and c2 depends on how close the two concepts are in the taxonomy. A longer path length indicates less relatedness; therefore the semantic relatedness is calculated as follows (Formula 3):

$$Sim_{PATH}(c_1, c_2) = \frac{1}{Path\ Length(c_1, c_2)} \tag{3}$$

Leacock and Chodorow semantic relatedness [4] is a measure that finds the shortest path length between two concepts, and scales that value by the maximum path length in the is-A hierarchy in which they occur. It considers that the conceptual distance between two nodes is proportional to the number of edges separating the two nodes in the hierarchy (see Formula 4).

$$Sim_{LC}(c_1, c_2) = -log\frac{len(c_1, c_2)}{2D} \tag{4}$$

Lesk [5, 6] proposed that the relatedness of two words is proportional to the extent of overlaps of their dictionary definitions. Lesk algorithm disambiguates words in short phrases. The definition of each sense of a word in a phrase is compared to the definitions of every other word in the phrase. A word is assigned the sense whose definition shares the largest number of words in common with the definitions of the other words.

Hirst and St-Onge [7] is a semantic relatedness measure that states that two lexicalized concepts are semantically close if their WordNet synsets are connected by a path that is not too long and it is not changing its direction frequently. This relationship is measured by Formula 5.

$$Sim_{HS}(c_1, c_2) = C - path\ length - k * d \tag{5}$$

Jian and Conrath [8] is an IC distance measure that uses the conditional probability of encountering an instance of a subclass synset given an instance of a superclass synset. Thus the information content of the two nodes, as well as that of their most specific subsume are considered. Jian Conrath Formula 6 measures the semantic distance, the inverse of similarity.

$$Dist_{JC}(c_1, c_2) = 2\log(p(lso(c_1, c_2))) - (\log(p(c_1)) + \log(p(c_2))) \tag{6}$$

Resnik [9] presents a semantic relatedness approach that uses the information of concepts, computed from their frequency of occurrence in a large corpus. Considers that the similarity between a pair of concepts may be judged by "the extent to which they share information", Resnik (see Formula 7) calculates the semantic relatedness between two lexicalized concepts.

$$Sim_{RESNIK}(c_1, c_2) = -\log p(lso(c_1, c_2)) \tag{7}$$

4.1 Similarity Decision

Considering a collection of n service operations, the number of comparison pairs (**cp**) between them is n^2. For each **cp** between service operations each semantic relatedness

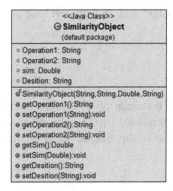

Fig. 4. Similarity object structure.

is calculated. As a result a similarity object array is created, which has the structure depicted in Fig. 4.

$$< operation_1, operation_2, sim, decision >$$

The results of semantic similarities calculations are calculated and recorded into this structure. In order to define a final decision for each **cp** into the array of similarity results, the mean and standard deviation are calculated. A lower and upper thresholds are calculated as follows:

$$lowerThresHold = mean - stdDev$$

$$upperThresHold = mean + stdDev$$

Decision of final similarity function returns "*isDifferent*" if the similarity measure is less than the *lowerThresHold*, "*isSimilar*" if the similarity measure is greater than *upperThresHold*, and "*unDefined*" if the similarity measure is greater than the *lowerThresHold* and less than the *upperThresHold*.

As a result of the decision procedure, a set of eight similarity object arrays are defined, each with their respective decision. Finally, for each similarity object array,

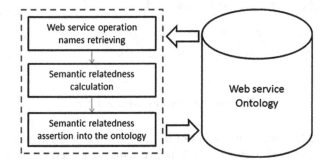

Fig. 5. Semantic enrichment of Web service descriptions.

those that resulted in *"isSimilar"* and *"isDifferent"* are asserted between individuals into the ontology (Fig. 5).

5 Evaluation

For evaluation we calculated the mean of all similarity results, selecting those values that were over an arbitrary threshold of 0.5. This filter returned a collection of 45 comparison pairs. Figure 6 shows this filtered set of similarity calculations.

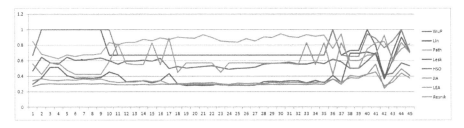

Fig. 6. Graph showing similarity results of a filtered set of comparison pairs

In order to get a reference value for evaluation, we implemented a human evaluation of the similarity between the 45 comparison pairs. Then we calculated the average distance between these eight similarity results against the human evaluations. Results of these distances are shown in Fig. 7, where the Leacock and Chodorow [4] semantic relatedness shows the higher distance average, and the Lin semantic relatedness [3] is closer to human evaluations.

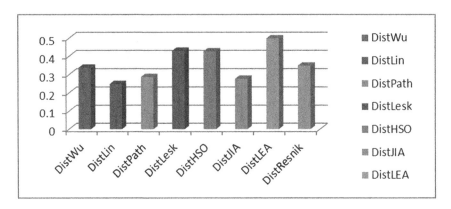

Fig. 7. Semantic similarity average distance

However the final decision consists of a voting schema, where a majority value is established and the final decision of a definitive semantic relatedness is defined considering all results.

6 Related Work

Semantic annotation or enrichment of Web services has been studied for a long time, for instance in [10], authors enhanced ebXML registries with OWL ontologies and used ebXML query functions to discover services. In [11] Paolucci et al. proposed a matchmaking engine for an UDDI registry to match services described with DAML-S. In [12], Kourtesis and Paraskakis proposed the FUSION Semantic Registry which offers publication and discovery facilities based on SAWSDL and OWL-DL for an UDDI-based registry. Another similar approach was proposed by Pilioura and Tsalgatidou [13]. In this work authors describe the PYRAMID-S framework that uses the hybrid peer-to-peer topology to organize different service registries. In [14] Lee and Kim address the problem of annotating Web services from the Deep Web. Deep Web refers to Web pages that are not accessible to search engines. In particular, authors consider Web forms interface pages as Deep Web services that reflect the real content types of the Deep Web. Their proposed solution consists of the automatic generation of a domain ontology for semantic annotation of Web services. Such domain ontology is built based on Web page attributes (any items of descriptive information about the site). Their research main goal is improving the automatic search and discovery of public Web services. However, their service description sources are different as they are using Web form interface pages instead of a formal service description language.

In [15] Sabou and Pan presented a study of the major problems with Web service repositories (some of them are no longer available, however the result of the study is still relevant). They concluded that Web service repositories use simple techniques of accessing the content of Web services, browsing across services listings relies on few and low quality metadata, and metadata is not fully exploited for presentation. Authors also proposed various semantic-based solutions to enhance semantically service repositories. Retaking the early ideas of these authors, the solution that is reported in this article is to lay the foundations for the automatic construction of public Web services repositories based on ontologies.

In [16] Linna proposed a model for semantic Web service discovery based on proxy technology. An important element of this model is a broker that provides semantic interoperability between semantic Web service provider and proxys by translating WSDL to DF description for semantic Web services and DF description to WSDL for FIPA multi proxys. Linna model has uses a translation approach which is similar to the mapping strategy used in the general ontology model. Mapping and translation between concepts are two of the most promising uses of ontologies, as they allow semantic integration and interoperability between systems.

In [17] Farrag et al. presented a mapping algorithm from traditional WSDL definition to OWL-S profiles. This algorithm starts obtaining WSDL Types (primitive XSD Type and complex XSD Type), then extracts the service name and text description, and finally, it extracts service inputs and outputs. An additional manual process is required to enhance the OWL-S profiles to define preconditions and effects of the provided service. The main idea of this algorithm is very similar to the approach presented in this paper, which is to provide a general repository of Web services and support for automated service tasks. However, the mapping algorithm was not designed and implemented to support mappings between different service models and languages.

In [18] Pedrinaci et al. presented iServe, a Semantic Web Service search engine, which collects semantic descriptions of Web services through a crawler. This search engine is capable of allocating and discovery of HREST, SAWSDL and OWL-S services. The main difference of iServe search engine is that it does not translates different models to a general service model, nor it is capable of dealing with traditional syntactic models.

In [19] Yousefipour et al. propose an ontology-based framework for the discovery of semantic Web services (SWS) using a QoS approach. Their framework describes an ontology manager component which handles the provider and the requester domain or general ontologies. This component merges these ontologies with general ontologies and creates a new generalized ontology, which is used for ranking the resulting list of SWS. Even though authors address automatic discovery of SWS by means of ontologies, their ontologies are domain-oriented. In contrast, in this paper ontologies are used for modeling service programmatic interfaces aiming at supporting automatic search and discovery of public available Web services.

In [20] Yu et al. presented a semantically enhanced service repository for user-centric service discovery and management. Their approach consists of a service repository that facilitates users search, access and management of service descriptions based on facets. In particular, the faceted repository requires that the description of services includes information about each facet related to the service providing the facet type, and facet reference based on a specification of facets. The idea of including faceted search mechanisms for Web services is promising and valuable. However, their main drawback is that the requirement for adding facet information imposes a new difficulty for the automatic acquisition and representation of pre-existing Web services. Nor they consider the incorporation of different service descriptions implementations.

In [21] Xu et al. presented their SMiner, a multimodal-query Web services search system. Their search mechanisms uses a structural and semantic similarity measure that ranks Web services according to their similarities. The multimodal query mechanism allows the user to refine the query according to his needs and obtaining better results. The main drawback of this approach is the lack of ontologies as a representation mechanism of processed and mined services.

Loutas et al. [22] described a semantic search engine that facilitates the collection of the service descriptions, the extraction of the semantic information and the translation of this information to a Reference Service Model (SRM). This search engine is closely related to the general model reported in this paper. However, as mentioned before, this reference model does not allow the granular treatment of specific operations related to a given service (as the WSDL specification does) and the definition of precondition and effect relations. From a technological point of view the service model is based on RDF ontological definition not on OWL ontology language, together with a different query language, which is SPARQL vs SQWRL.

Related work present different approaches for enhancing semantically Web services or Web service registries. Main attempts rely on the incorporation of ontologies as a mechanism to achieve semantic interoperability. However, none of reported works have presented mechanisms to enhance services repositories represented by a general ontology model capable of adhering to any sort of service description model either syntactic or semantic. The main goal of the general ontology design is to enable a large set of service description languages and models to coexist and interoperate over a simple and general service registry through which different services are discovered, correlated, etc.

7 Conclusions

We have described a semantic relatedness discovery process which calculates eight semantic relatedness measures between all operations pairs, then calculates the arithmetic mean and standard deviation; sets the upper and lower thresholds; and identifies operation pairs that are over and under respective thresholds. For the final decision of the new semantic relationships that are to be asserted back into the ontology, a voting scheme is used. The use of a voting scheme ensures that the establishment of new semantic relations in the ontology is sufficiently reliable since it is based on a majority vote.

Even though few similarity relations are found and asserted, they represent an important step towards the automatic discovery of information that was previously unknown and that can be very useful during automatic search, selection and invocation of Web services based on the operation names.

These similarity measurements will be extended with more elements of the Web service descriptions, such as input and output parameter names and parameter types, and texts of the document tags.

References

1. Budanitsky, A., Hirst, G.: Semantic distance in WordNet: an experimental, application-oriented evaluation of five measures. In: Workshop on WordNet and Other Lexical Resources, vol. 2, p. 2, June 2001
2. Wu, Z., Palmer, M.: Verbs semantics and lexical selection. In: Proceedings of the 32nd Annual Meeting on Association for Computational Linguistics, pp. 133–138. Association for Computational Linguistics (1994)
3. Lin, D.: An information-theoretic definition of similarity. In: Proceedings of the 15th International Conference on Machine Learning, pp. 296– 304. Morgan Kaufmann, San Francisco (1998)
4. Leacock, C., Chodorow, M.: Combining local context and WordNet similarity for word sense identification. WordNet: Electron. Lex. Database **49**(2), 265–283 (1998)
5. Banerjee, S., Pedersen, T.: An adapted Lesk algorithm for word sense disambiguation using WordNet. In: Gelbukh, A. (ed.) CICLing 2002. LNCS, vol. 2276, pp. 136–145. Springer, Heidelberg (2002)
6. Lesk, M.: Automatic sense disambiguation using machine readable dictionaries: how to tell a pine cone from a ice cream cone. In: Proceedings of SIGDOC 1986 (1986)
7. Hirst, G., St-Onge, D.: Lexical chains as representations of context for the detection and correction of malapropisms. In: Fellbaum, C. (ed.) pp. 305–332 (1998)

8. Jiang, J.J., Conrath, D.W.: Semantic similarity based on corpus statistics and lexical taxonomy. In: Proceedings of International Conference on Research in Computational Linguistics, Taiwan (1997)

9. Resnik, P.: Using information content to evaluate semantic similarity. In: Proceedings of the 14th International Joint Conference on Artificial Intelligence, Montreal, pp. 448–453 (1995)

10. Dogac, A., Kabak, Y., Laleci, G.: Enriching ebXML registries with OWL ontologies for efficient service discovery. In: Proceedings of the 14th International Workshop on Research Issues on Data Engineering (RIDE 2004), pp. 69–76 (2004)

11. Paolucci, M., Kawamura, T., Payne, T.R., Sycara, K.: Semantic matching of web services capabilities. In: Horrocks, I., Hendler, J. (eds.) ISWC 2002. LNCS, vol. 2342, pp. 333–347. Springer, Heidelberrg (2002). doi:10.1007/3-540-48005-6_26

12. Kourtesis, D., Paraskakis, I.: Combining SAWSDL, OWL-DL and UDDI for semantically enhanced web service Discovery. In: Simperl, E., Cimiano, P., Polleres, A., Corcho, O., Presutti, V. (eds.) ESWC 2012. LNCS, vol. 7295, pp. 614–628. Springer, Heidelberg (2008). doi:10.1007/978-3-540-68234-9_45

13. Pilioura, T., Tsalgatidou, A.: Unified publication and discovery of semantic web services. ACM Trans. Web **3**(3), 11–44 (2009)

14. Lee, Y.-J., Kim, C.-S.: A learning ontology method for RESTful semantic web services. In: Proceedings of the IEEE International Conference on Web Services (2011)

15. Sabou, M., Pan, J.: Towards semantically enhanced Web service repositories. J. Web Semant.: Sci. Serv. Agents World Wide Web **5**, 142–150 (2007)

16. Linna, L.: The design of semantic web services discovery model based on multi proxy (2009)

17. Farrag, T., Saleh, A., Ali, H.: Mapping from traditional WS definitions into SWS Profile. In: 2nd International Conference on Computer Technology and Development (2010)

18. Pedrinaci, C., Liu, D., Maleshkova, M., Lambert, D., Kopecky, J., Domingue, J.: iServe: a linked services publishing platform. In: Paper Presented at the Workshop: Ontology Repositories and Editors for the Semantic Web at 7th Extended Semantic Web Conference, Heraklion, Crete (2010)

19. Yousefipour, A., Mohsenzadeh, M., Ghari Neiat, A., Sadegzadeh, M.: An ontology-based approach for ranking suggested semantic web services (2010)

20. Yu, J., Sheng, Q., Han, J., Wu, Y., Liu, C.: A semantically enhanced service repository for user-centric service discovery and management. Data and Knowl. Eng. **72**, 202–218 (2011)

21. Xu, B., Luo, S., Sun, K.: Towards multimodal query in web service search (2012)

22. Loutas, N., Peristeras, V., Zeginis, D., Tarabanis, K.: The Semantic Service Search Engine (S3E). J. Intell. Inf. Syst. **38**, 645–668 (2012)

Agent Technology and Multi-agent Systems

The ICARO Goal Driven Agent Pattern

Francisco Garijo and Juan Pavón[(⊠)]

Department of Ingeniería del Software e Inteligencia Artificial,
Universidad Complutense Madrid, 28040 Madrid, Spain
fgarijo@ucm.es, jpavon@fdi.ucm.es

Abstract. ICARO is an open source platform for the implementation of multi-agent systems (MAS), which provides architectural patterns for several types of agent models, following well established software engineering principles. This paper describes a pattern of cognitive agent, whose main characteristic is to be goal-driven, and its logic described as a rule based system. This has been used in different real projects and as a tool in a master course on the development of intelligent agent applications. Some of these are used to illustrate its use and explain some of the conclusions derived from these experiences, mostly from a software engineer point of view.

Keywords: Multi-agent systems (MAS) · Cognitive agent · Goal-driven agent · Agent pattern · Production systems · ICARO

1 Introduction

ICARO [1] is a framework for the development of multi-agent systems (MAS), whose purpose is to introduce well established principles from software engineering (such as the use of architectural patterns, componentization at macro and micro levels with well-defined interfaces, dynamic control of configuration, error management, support for agile development, etc.) ICARO has evolved for more than ten years with the experience on different domains such as telecommunication systems, natural language processing, coordination of teams of robots, and ambient assisted living [2–5]. The software is open and available at GitHub (https://github.com/fgarijo/ICARO), and some manuals exist for an extensive documentation of the framework. This paper bases on this rich set of experiences, which can lead to explain the decisions taken during its design, so some learned lessons can be useful in other frameworks as well. In fact, the conclusions present an example of the application of ICARO patterns in another agent-based tool, MASSIS [7], in order to demonstrate that these patterns can be also applied by others.

Note that ICARO is not dealing with the use of specific agent protocols (such as FIPA, which is the main focus of other frameworks, such as the well-known JADE) or agent specification languages. In ICARO, agent communication can be performed using the most convenient protocols or middleware (FIPA, CORBA, RMI, or others), and several programming languages can be used (for instance, the Goal Driven Agent pattern is currently implemented with Java and Drools [7], but others have been used previously, in concrete Jess [8] and ILOG JRules [9]). The main concerns of ICARO

© Springer International Publishing AG 2016
M. Montes-y-Gómez et al. (Eds.): IBERAMIA 2016, LNAI 10022, pp. 51–62, 2016.
DOI: 10.1007/978-3-319-47955-2_5

are the management of the MAS organization and the provision of several patterns for the design of agents and "resource components", which form the MAS environment. Currently there are two agent patterns in ICARO: *reactive agent* (finite state machine based) and *cognitive agent* (or Goal Driven Agent). This last is the purpose of this paper.

The Goal Driven Agent pattern provides computational elements (factories, domain classes, infrastructure classes, libraries, threads, rules) that represent and implement the following concepts:

- *Goals*, or *Objectives*, a symbolic representation of the goals that the agent attempts to realize.
- *Tasks*, which are needed to achieve the goals. Tasks are procedures or processes that are selected and executed in order to obtain information for achieving agent's goals.
- *Internal memory of the agent*, it is set up of information elements that are stored and managed to achieve the goals.
- *Decisions* that determine the behavior of the agent. Decisions are expressed as situation-action rules.
- *Incoming Information Processing*, information flow received by means of messages or by external events, and their inclusion in the internal memory of the agents.

Some of these concepts are inspired on human behavior, but it is important to note that we are dealing with an information processing machine, whose behavior is far from psychological or biological models. Initially, the pattern was named "cognitive" agent, and was claimed to follow the BDI (Beliefs, Desires, Intentions) model. However, the experience with the utilization of this pattern has made us think about the drawbacks derived from the mistaken introduction of psychological and anthropomorphic concepts in engineering, and the consequences of creating exaggerated expectations to developers, which led to deep disappointment. As a result we finally decided to use more modest but realistic terms and rename the pattern as "goal driven" agent pattern, but describe its design and implementation model with a sound computational model according to current software engineering practice.

There are relevant differences with respect to BDI model such as the following:

- Instead of beliefs (which may have some religious connotations for developers), we use the term *Agent Information Model*. This model consists of a set of classes with their relationships. It can be also considered as an Ontology but, as most of the applications have not used all the elements that characterize an ontology, we consider more appropriate to call it simply an *information model*.
- There are not desires, because it is not clear what that means from a computational point of view. Instead, there is a set of *goals* that the agent can achieve, and a life-cycle for goals.
- There is not an explicit support for intentions. Implicit support for specifying and implementing agent's intentions is provided by means of decision rules that define when to create new goals and stipulate criteria for *modifying the current intention* of the agent by changing the focus from the current goal into a different goal.

Many other proposals for agent architectures focus on specific domains (see for instance [10–12]). A key concern about most of existing MAS systems, with the significant exception of JADEX [13], is the lack of componentization and then the difficulty for integration, interaction with heterogeneous middleware, and deployment. ICARO share with JADEX the effort for providing engineering tools for integrating MAS concepts in the development process for complex distributed applications, however while JADEX put the emphasis on facilitating specific constructs for implementing BDI concepts, in ICARO the focus is on providing architectural patterns, behavior models, and clear organization principles for developing the MAS and its environment with standard software engineering IDEs and open source middleware.

Regarding our teaching experiences, it is essential to give students tools where they can see clearly what an agent is, which are the interactions with a real and heterogeneous environment, and the challenges for implementing most of the requirements that are supposed to meet an intelligent agent such as deep reasoning, negotiation, learning, self-organization and so on. ICARO only provides a modest collection of components to build the MAS but it avoids the students to reinvent the wheel, by facilitating ready to use code to model agent's architecture and basic behavior, This facilitate the development and integration of more sophisticated components, and make them aware of the complexity of implementing intelligent agent's features.

The rest of the paper is organized as follows. Section 2 presents main concepts behind the ICARO framework. Section 3 describes the Goal Driven Agent pattern. Section 4 illustrates an example of its application for a leader agent in a team of robots. The results of the experience working with ICARO Goal Driven Agent pattern and its adoption in other agent-based framework are discussed in the last section.

2 Overview of the ICARO Framework

ICARO is a software framework that conceives applications as agent's organizations populate by two types of basic components: agents and resources. The framework provides agent patterns, currently two: Reactive Agent and Goal Driven Agent. The behavior of the first is based on a state machine, whilst the second works as a rule production system.

All components in ICARO, either agents or resources, interact through two interfaces:

- *Management interface*, with operations for start, stop, monitoring and control the component.
- *Use interface*, a communication interface that allows the reception of events and messages.

This structure of interfaces is similar to other component models. Interfaces of agents and resources are registered in an interface repository for the organization. Agents and resources can obtain references to the interfaces of others by querying that repository. This is the typical mechanism in distributed systems.

The implementation of the *management interface* is similar for all components, so the one that is provided with the platform should suffice. The purpose of the management interface is to monitor and manage the correct working and configuration of all the components. One of the goals of ICARO is to support the operation of the configuration, taking special care of the issues of initialization and finalization of the system. These operations are normally prone to errors in distributed systems, so ICARO tries to facilitate the developer, whose main task should be to define the configuration of the organization in a XML description file (there are plans to provide a graphical user interface to facilitate this task).

Application management is performed by ICARO reactive agents. When the system starts, an *Initiator* agent takes care of the creation of an *OrganizationManager* agent with the file describing the organization. The *OrganizationManager* will request to the *AgentManager* and the *ResourceManager* agents at each node the creation of the application agents and resources that are specified in the organization description. They will check that, besides being created, agents have access to the required resources. All this process is supported by a *Trace* resource, which shows the logs of agents in the system, so the developer can follow the flow of actions in the system. The *OrganizationManager, AgentManager* and *ResourceManager* agents, as well as the *Trace* resource, are provided by ICARO and normally they can be used as they are.

The *use interface* has to be defined for each component, as it is specific of the application. It is the point where messages and external events are given to the component.

Resources (e.g., communication software, data bases, visualization, language processor, etc.) usually rely on some existing software that has to be integrated in the ICARO platform and this is done by the provision of a *use* interface. The adapter pattern can be applied in many of these cases to get the requests received at the use interface and call the specific methods on the underlying software. ICARO does not put any other specific constraint on the implementation of resources, only the availability of the management and use interfaces. In this way, it is relatively easy to integrate other software packages and tools in the application, so they can be used by the agents.

Communication between agents and resources can be:

- Synchronous, when the agent make requests to the resource with the operations that are defined in the use interface and receive a return result.
- Asynchronous, when the agent makes a request to the resource and receives the response as an event through the agent's use interface.

Agents are control components: they receive information, process it, and make decisions on actions on other agents or resources in order to get new information. They receive messages and events through the use interface. These are taken by a *Perception* component, which is in charge of extracting relevant information from messages and events. This information is then provided to the *Control* element. The next section explains how these work with the Goal Driven Agent.

In summary, developing an agent-based application using the ICARO framework consists of identifying the agents and resources that are required for an application, specifying their configuration as an organization, specifying the use interface for agents and resources, implementing adapters for resources to reuse existing tools or libraries, and implementing the agents following one of the patterns: reactive or goal driven.

3   The Goal Driven Agent Pattern

Similarly to other ICARO components, a Goal Driven Agent has, from an external point of view, a management and a use interfaces. Internal components are in Fig. 1:

- *Agent Management.* It implements the operations of the management interface. This component takes care during agent initialization that all components are well configured and running, and that the agent has access to the required resources.
- *Agent Perception.* It implements the operations of the use (communication) interface. It is in charge of storing and processing the events and messages that the agent receives. This consists basically of filtering, extracting information from the contents of messages, and sending it to the Goals processor.
- *Goals Processor.* It takes the information coming from the perception in order to resolve the goals. It consists of the following two components:
  - A *Rule Engine*, which implements the basic processing of goals, by generating goals and controlling their achievement with the execution of tasks and management of information.
  - A *Tasks Manager*, which executes those tasks that the rule engine requests.

This component relies on a set of infrastructure classes and tasks (*basicEntities*) that implement the life-cycle of the processor. They are also the basis for the definition of goals, tasks and domain classes.

A Goal Driven Agent processes external events and messages, extracts information from them, which is asserted as facts in the rule engine and can be passed as argument

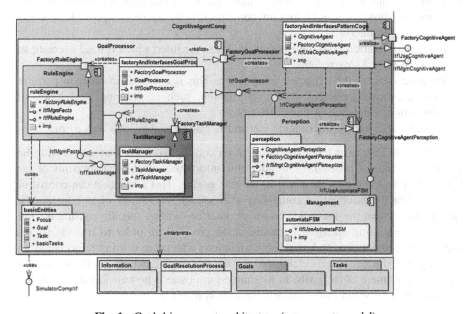

Fig. 1. Goal driven agent architecture (components model)

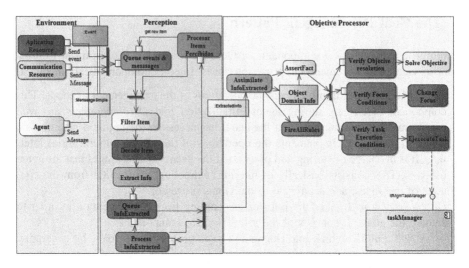

Fig. 2. Goals processing

to the tasks. Figure 2 shows the information flow and activities that are performed between the Perception and the Goals Processing components.

The perception performs the following cycle:

- Queue events or messages that are received through the use interface of the agent.
- Process next element in queue.
- Filter elements in order to eliminate those messages that have not the appropriate format or whose receiver id does not correspond to the agent. It is possible to define other specific filters.
- Extract information from the elements that pass the filter, which is used to create an object of class *ExtractedInfo*.
- Pre-processing of the extracted information. This is optional, in order to perform some additional treatment of the information, such as ordering, correlation of information, validation or transformations.
- Finally, the extracted information element is passed to the Goals Processor.

The Goals Processor takes the information coming from the Perception with the activity *Assimilate ExtractedInfo* to adapt the information to the requirements of the Goals Processor, and puts the result in the rule engine. This adaptation can consist of verifying that the information complains with specific conditions or transforming it taking into account the semantic context or other application specific constraints. For instance, when using an agent communication protocol, in order to translate information between domains and establish correlations with other existing information. In this way the information gets elaborated for the rule engine, where it will be used for the accomplishment of the goals. In its simple form there is no transformation and the information is provided to the rule engine as it is received.

The rule engine is an abstract machine with the following features:

- It provides interfaces to store and manage information in a working memory, which can be considered the agent memory (or global state).
- It interprets the definition of the pending goals of the agent and the processes to achieve the goal.

Each time the engine receives new information, it performs a new cycle to check the impact of the received information in the process to achieve the goals, which consists of the following steps:

- Interpret the definition of the pending goals of the agent and the processes to achieve the goal.
- Storing the information in the working memory of the rule engine.
- Verification of conditions, to check whether the new information satisfies the requirements for achievement of a goal or the conditions that allow the execution of actions that make progress the process for achievement of some goal.
- Execution of actions, which can be:
  - Generation of new goals.
  - Selection of a goal to achieve (*focus*). The focus operation selects agoal to resolve in a context.
  - Execution of tasks, which can be autonomous threads. Tasks produce results that are received by the agent perception (external tasks) or directly by the Goals Processor (internal tasks).
  - Change the state of a goal. Goals have a life-cycle that evolves following a process to resolve the goal. Initially goals are in the state *Pending*. They get to the state *Solved* when there are evidences, but it normally passes through intermediate states such as *Solving*, or may get to *Failure*.

Currently the engine is implemented as a production rule system with Drools [7], so the process for generation and achievement of goals is expressed by rules of the form *<condition>* → *<action>*. Previous versions of the pattern were implemented with other rule engine systems such as JESS and ILOG JRules. Current choice for Drools is because it is open access, well documented and supported by an active community.

4   Example of Development of a Goal Driven Agent

The case study to illustrate the design and implementation of Goal Driven Agents is taken from the ROSACE project [4]. A team of robots are situated in an environment to perform tasks such as localization of victims after an accident, and have to rescue them. A control center coordinates the assignment of available resources, including the robots that are deployed in the environment. Robots are able to self-organize, form teams to perform the tasks that are requested by the control center, and distribute them among them.

ICARO allows the simulation of the behavior of the robots in a realistic way by considering they are physically distributed entities, with their own autonomous behavior, and asynchronous communication through messages. In order to work together, agents should be aware of the concept of *Team*, and use it to communicate and cooperate, to define the cooperation and the form in which they achieve their common goals. In this point, the organization model is relevant, as it provides the concept of role to support the definition of different agent behaviors in a context: an agent role in a team defines the goals that the agent can generate and achieve, as well as communication with other agents, depending on their role.

For the illustration in this case, consider a team of robots with an initial organization configuration. On creation, each robot (agent) knows the identifier of the team it belongs to, and its role, although it does not know which are the other members of the team. One of the first goals to start working together is to know its teammates.

The steps for designing and implementing a Goal Driven Agent are the following:

1. *Generate goals.* In this case the first goal to define is *DefineMyTeam*.

Each goal requires the definition of its life-cycle. This implies:

- The conditions for its generation.
- The conditions for its satisfaction (when goal state becomes *Solved*).
- The conditions for its failure (when goal state becomes *Failed*).
- The actions that can contribute to achieve the goal.

In this case, the goal will be considered as *Solved* when the agent gets all the information concerning the members of the team. A deadline for achieving the goal is also considered.

Note that this requires the definition of domain information classes, in this case, *TeamInfo*, to model the concept of team, and *TeamRole*, to model the role of the agent in the team.

When agents start working, they initialize their working memory by executing a task *InitWorkingMemoryInfo*, where a new object *TeamInfo* is created, with attributes the team identifier, and other attributes *initialContactWithTeam = false* and *teamMates* list empty.

Initially, the goal*DefineMyTeam*starts in state *Pending*.

2. *Select a goal to achieve.* The agent may have different goals in the working memory, so it has to *focus* on one of them.
3. *Determine the actions and the process to achieve the goal.* Agents can perform their own actions, ask other agents to do something, or use some resources.

In this example, agents can get information on whether another agent is on the same team or not by asking it. However, when using asynchronous messages the answer may not come back or arrives late (see as an example what happens with emails). This may be due to the fact that the message does not arrive to destination or it is ignored, or the receiver is busy or does not want to answer. A way to avoid the sender to stay blocked is to use timers. All these situations make something that in principle is simple, much more complicated.

Another way to do it is to assume that other agents are also trying to find their teammates. So all agents can broadcast a message indicating who they are, their team identifier and their role. This is the solution that is taken.

4. *Check with new information whether the goal has been achieved.* In case this is not yet achieved, the agent should check whether other actions can be performed or if the goal has failed (by checking the failure conditions).

During the design, an activity diagram for each goal can be drawn in order to organize the different conditions and actions that should be considered for this cycle. Figure 3 shows an example for this case. The tasks that are required are well identified:

- *InitWorkingMemoryInfo*: creates the initial objects (*:TeamInfo, :MyGoals, :Focus*).
- *GenerateAndFocusGoal*: creates an object of class *DefineMyTeam* and focus on it.
- *ContactTeammates*: obtains the references to the agents in the organization through the interface repository and sends to all a message indicating its team and its role. This task starts a timer for the period of time to obtain the information from the teammates.
- *ProcessInfoAgentRole*: processes information received from other agents indicating their team and role. This process consists on updating the attributes in the *:TeamInfo* object, such as the *numberOfTeamMembers*, or the list of identifiers of teammates (*teamRobotsIds*).

Once these elements are defined, ICARO provides a package structure to implement the code for each agent. Note that the ICARO framework already provides the infrastructure for the Management, the Perception and the Process Infrastructure. The developer only needs to create one package for each type of agent and declare the agents in the Organization Description, as it was explained in Sect. 2.

So, to implement a Goal Driven Agent the developer only has to create a package with four sub-packages:

- *information*: It contains the classes of objects that the agent is using, which have not been defined in a more general information model package. This are typical Java classes, such as *TeamInfo* and *TeamRole* in the example.
- *goals*: It contains the classes that define the goals. All of these inherit from a class Goal, which provides the methods for managing the goal life-cycle. In the example above, the class *DefineMyTeam* will be defined here.
- *tasks*: It contains the classes that define the tasks, which inherit from a class *Task* (a task which will run on its own thread) or *SynchronousTask* (a task running on the same thread as the caller). The code of the task is written in the method *execute (Object... params),* which is provided by the super class.
- *goalResolutionProcess*: It contains the files with the rules that are executed by the engine to process the goals. In this version, these are Drools files.

For the interested reader, the source code for this example is available at the GitHub repository (https://github.com/fgarijo/rosaceSIM), with instructions to download and execute the example.

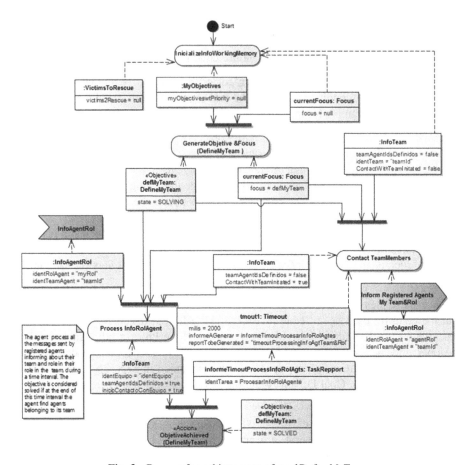

Fig. 3. Process for achievement of goal*DefineMyTeam*

5 Conclusions

A key weakness of the MAS as a research area is the unbalance between theoretical/formal proposals and their validation through reproducible experimentation. In order to minimize this gap, a MAS should be developed according to sound software engineering practices. This is shown here for the Goal Driven Agent, which is a software architectural pattern. It is based on a software component approach to facilitate the integration of a variety of tools (asynchronous messaging, external events, production system, etc.), as well as the organization of the design and the code. The support of a software infrastructure, such as ICARO, also simplifies the work of the developer by providing a real computational representation of the design concepts. For instance, with the Goal Driven Agent the developer only has to focus on concrete issues: identifying the goals of the agent, the life-cycle of the goals, the tasks that can contribute to achieve the goals, the definition of domain information classes, and the process for achievement of the goals. Only these elements have to be packaged to deliver a type of Goal Driven Agent, because the underlying infrastructure takes care of how to interpret and execute the defined process and the low level processing of events and messages. In this way the developer is more focused on the purpose of the agent.

The experience using this pattern has shown two issues that the developer has to take into account:

- From a conceptual point of view, to determine whether a Goal Driven Agent or other kind of agent is needed. Also, how many goals to consider, their break down in a set of sub-goals, the complexity of the tasks, etc.
- From an implementation point of view, the use of a rule engine as a computational mechanism to implement the control of the agent requires learning a formalism that is not always common to programmers. Also, the agents receive events and messages, obtain information and generate new information with tasks. The asynchronous information flow can be a source of problems.

To check the applicability of the Goal Driven Agent pattern in other frameworks, this has been implemented in another agent-based framework, MASSIS [6]. This is an agent-based system for crowd simulation in indoor scenarios. The Goal Driven Agent pattern has been introduced for the high-level control part of the agents in MASSIS, and this was made in less than one week. This has been easy because the components have well defined interfaces, and the agent itself has also sound components, which have been easily installed and integrated in the other platform.

Our teaching experience in a master course about developing of intelligent systems shows that it is essential to provide the students with a framework that allows them to focus on how to develop the intelligence of their system. This would need to specify, design and implement the behavior of the agents and the resources in the environment. In past experiences without ICARO the students passed most of their time fighting against infrastructure and integration issues and they never arrive to face the most relevant issues of implementing intelligent features. With ICARO students may focus on the design and implementation of those "intelligent capabilities" and develop them as a software team where components might be easily developed, tested and integrated.

More information, the source code, and extensive documentation, can be found at the GitHub repository of the ICARO project at https://github.com/fgarijo/ICARO.

Acknowledgements. This work has been partially supported by the project "Collaborative development of AAL solutions (ColoSAAL)", with grant TIN2014-57028-R by the Spanish Ministry for Economy and Competitiveness.

References

1. Garijo, F.J., Bravo, S., Gonzalez, J., Bobadilla, E.: BOGAR_LN: an agent based component framework for developing multi-modal services using natural language. In: Conejo, R., Urretavizcaya, M., Pérez-de-la-Cruz, J.-L. (eds.) CAEPIA 2003. Lecture Notes in Artificial Intelligence (LNAI), vol. 3040, pp. 207–220. Springer, Heidelberg (2004). doi:10.1007/978-3-540-25945-9_21
2. Gascueña, J., Garijo, F.J., Fernández-Caballero, A., Gleizes, M.P., Machonin, A.: Deliberative control components for eldercare robot team cooperation. J. Intell. Fuzzy Syst. **28**(1), 17–28 (2015)
3. Gascueña, J.M., Navarro, E., Fernández-Sotos, P., Fernández-Caballero, A., Pavón, J.: IDK and ICARO to develop multi-agent systems in support of ambient intelligence. J. Intell. Fuzzy Syst. **28**(1), 3–15 (2015)
4. Lacouture, J., Gascueña, J.M., Gleizes, M.P., Glize, P., Garijo, F.J., Fernández-Caballero, A.: ROSACE: agent-based systems for dynamic task allocation in crisis management. In: Demazeau, Y., Müller, J.P., Corchado Rodríguez, J.M., Bajo Pérez, J. (eds.) PAAMS 2012: Advances in Intelligent and Soft Computing, vol. 155, pp. 255–259. Springer, Heidelberg (2012)
5. Georgé, J.P., Gleizes, M.P., Garijo, F.J., Noël, V., Arcangeli, J.P.: Selfadaptive coordination for robot teams accomplishing critical activities. In: Demazeau, Y., Dignum, F., Corchado, J.M., BajoPérez, J. (eds.) PAAMS 2010: Advances in Intelligent and Soft Computing, vol. 70, pp. 145–150. Springer, Heidelberg (2010)
6. Pax, R., Pavón, J.: Agent-based simulation of crowds in indoor scenarios. In: Novais, P., Camacho, D., Analide, C., El Fallah Seghrouchni, A., Badica, C. (eds.) IDC'2015. SCI, vol. 616, pp. 121–130. Springer, Heidelberg (2016). doi:10.1007/978-3-319-25017-5_12
7. Drools. http://www.drools.org/. Accessed 3 May 2016
8. Jess. http://www.jessrules.com. Accessed 3 May 2016
9. ILOG JRules. http://www-01.ibm.com/software/integration/business-rule-management/jrules-family/. Accessed 3 May 2016
10. Corchado, J.M., Laza, R.: Constructing deliberative agents with case-based reasoning technology. Int. J. Intell. Syst. **18**(12), 1227–1241 (2003)
11. Camacho, D., Aler, R., Borrajo, D., Molina, J.M.: A multi-agent architecture for intelligent gathering systems. AI Commun. **18**(1), 15–32 (2005)
12. Sánchez, N., Molina, J.M.: A multi-agent approach for provisioning of E-Services in U-Commerce environments. Internet Res. **20**(3), 276–295 (2010)
13. Braubach, L., Pokahr, A.: The jadex project: simulation. Multiagent Syst. Appl. **45**, 107–128 (2013). doi:10.1007/978-3-642-33323-1_5

A Study for Self-adapting Urban Traffic Control

P.S. Rodríguez-Hernández, J.C. Burguillo$^{(\boxtimes)}$, Enrique Costa-Montenegro,
and Ana Peleteiro

Department of Telematics Engineering, University of Vigo, 36310 Vigo, Spain
{pedro,jrial,kike,apeleteiro}@det.uvigo.es

Abstract. Nowadays, managing traffic in cities is a complex problem involving considerable physical and economical resources. However, traffic can be simulated by multi-agent systems (MAS), since cars and traffic lights can be modeled as agents that interact to obtain an overall goal: to reduce the average waiting times for the traffic users. In this paper, we present a self-organizing solution to efficiently manage urban traffic. We compare our proposal with other classical and alternative self-organizing approaches, observing that ours provides better results. Then, we present the main contributions of the paper that analyze the effects of different traffic conditions over our cheap and easy-to-implement method for self-organizing urban traffic management. We consider several scenarios where we explore the effects of dynamic traffic density, a reduction in the percentage of sensors needed to support the traffic management system, and the possibility of using communication among cross-points.

Keywords: Urban traffic control · Multi-agent system · Self-organization

1 Introduction

Self-organizing systems are not controlled in a top-down manner by some central unit, such as an intelligent operator. Instead, they are based on internal, bottom-up processes, and they solve problems by means of local interactions among a large number of components (agents). Among the advantages of self-organization we may cite robustness, scalability, flexibility, adaptivity, simplicity of the components and reduced cost of deployment. Therefore the applications of self-organization techniques are especially relevant in the case of decentralized control problems, such as urban traffic control (UTC).

UTC tries to decrease congestion by using traffic signals. In this way, traffic control algorithms take into account measured and predicted traffic data as input variables. Traditional approaches for traffic management attempt to optimize the solution for static situations, normally considering time basic configurations of traffic density. For instance, centralized traffic control methods are based on adaptation of some pre-calculated signalization schedules. However, these configurations can abruptly change and traditional methods are not

© Springer International Publishing AG 2016
M. Montes-y-Gómez et al. (Eds.): IBERAMIA 2016, LNAI 10022, pp. 63–74, 2016.
DOI: 10.1007/978-3-319-47955-2_6

able to automatically adapt to those changes [1]. In fact, many classical methods are usually based in optimization approaches, considered for a set of typical traffic densities and topologies, since the space of solutions to model every possible situation is very large. Classical works in UTC consider models taken from physics, using kinetics [2], fluids [3], interacting "physical particles" [4], and even considering cellular automata frameworks [5].

Advanced Traffic Management Systems (ATMS) use learning methods to adapt the phases of traffic lights, normally using a central computer [6] or a hierarchical computer level [7]. However, ATMS have several drawbacks, as their high cost, their complexity and the use of proprietary methodologies that difficult the migration from a particular methodology to another. Moreover, those systems usually require maintenance done by specialists and provide a different particularized solution for each city.

There have been several AI models, for instance, in [8] the authors propose a model based on logistic regression and neural networks to better control the traffic signals. From the multi-agent perspective, Guerrero-Ibáñez et al. [9] uses a dynamic urban traffic-lights coordination that relies on locally available traffic data, and the traffic condition of the neighboring intersections, to propose an intelligent algorithm with adaptive control and learning capabilities.

Other approaches try to achieve a more realistic simulation. [10] proposes the Real Trace Vehicular Mobility Model (RTVMM) to reproduce the urban traffic scenario using a more realistic speed control method. [11] proposes an intelligent traffic light control scheme in terms of dynamic and real-time traffic information, weights of all the serial lanes are dynamically computed, so as to automatically manage the current status and duration of traffic lights. Vehicle communication is a new trend, but still not generally available.

The limited performance of centralized and adaptive methods in the traffic management has motivated the interest in self-organizing traffic control [12]. In comparison with adaptive traffic control algorithms, the self-organizing strategy is more flexible with respect to local demands and more robust to variations in the traffic flow. A self-organizing traffic control system can respond to actual real-time traffic conditions without using any pre-determined signalization schedules that are based on average traffic characteristics. Within this *selforg* framework, the proposal in [13] treats traffic as an adaptation problem, where every traffic light adapts its green cycle to its current local conditions. This way, macro-level patterns emerge from the self-organized behavior of these elements. A more realistic simulation, under the same principles, has been carried away in [14] where the authors found that, on average, for different traffic densities, travel waiting times are reduced by 50 % compared to the current green wave method. These two works have been the basis for our approach, which, in our opinion, provides interesting results with a simple algorithm and at a very low cost.

The main contribution of this paper is a study of a history-based self-organizing solution (HB-SOTL), based in à preliminary work described in [15], and the analysis of the influence that different parameters have over it. Our results validate such model in different scenarios, i.e., considering dynamic traffic densities (to simulate rush hours), analyzing the impact of reducing the number of sensors in the city (reducing also the cost) and the possibility of using communication among several managing nodes.

The rest of the paper is organized as follows: Sect. 2 presents our city model, and Sect. 3 describes a previous proposal, which is the basis of our approach. In Sect. 4 we present our model, and in Sect. 5 a detailed comparison of the results obtained using different alternatives. Then, in Sect. 6, we consider the effect of dynamic traffic densities over our model. Section 7 studies the effect of reducing the number of cross points that have measurement devices. In Sect. 8 we explore the use of communication among traffic lights. Finally, Sect. 9 concludes the paper, and draws some future work.

2   The Netlogo City Model

A Self-Organizing Traffic Lights (SOTL) model was presented in [13] running under the NetLogo Multi-agent platform [16]. In this model, cars flow in a straight line, eastbound or southbound by default. Each crossroad has traffic lights that only allow traffic flow in one of the arteries that intersect it with a green light. Yellow or red lights stop the traffic. The light sequence for a given artery is green-yellow-red-green. Cars have a size of a patch and simply try to drive at a maximum speed of a "patch" per time step, but they stop when a car or a red or yellow light is in front of them. Time is discrete, but space is continuous. Roads have a single lane and direction, and there are only two directions by default, south and east, although we can obtain four by adding north and west. A car only changes road or direction depending on a probability (*prob-turn*), which means that the cars move at random. Also, in the previous model, a torus world was used by default, i.e., when a given car arrives at the end of the scenario, the same car will appear at the opposite side.

Given the previous traffic model of a Manhattan-like city, described in [13], we have made some improvements to represent a more realistic scenario (see Fig. 1). We remove the torus world and impose four directions (north, east, south and west). Also, a by-pass road was created to improve traffic, which is the outermost in the scenario. We have also changed the car creation and elimination scheme. Now, for every car, we define a source (a random road patch) and a destination (another random road patch), such that every car is created at a source, and it moves (following the shortest path) to its destination, where it is eliminated. The source and destination may be outside the world. Besides, cars can park at a certain place, and then drive to another destination from such parking place. We have added the possibility of bidirectional roads and roads with two lanes in the same direction, configured by the Netlogo interface.

At setup, a user-defined number of cars is created at once. Then, our simulator tries to keep the number of cars stable, creating new cars on demand in order to compensate for the cars exiting the world.

In order to avoid deadlocks at the intersections, a deadlock algorithm has been implemented. If a given car at an intersection is stuck, it tries to change

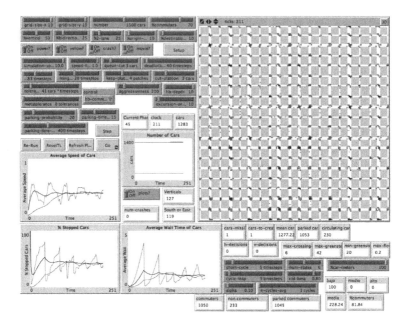

Fig. 1. Snapshot of the simulator interface (Color figure online)

direction in order to keep moving, and to leave the deadlock. This movement could affect other cars, helping to finish a potential deadlock.

Due to all these improvements, specially the possibility of an origin and a destination and bidirectional roads, a complex algorithm to guide the cars is needed. Whenever a car is in a patch that belongs to an intersection (it belongs simultaneously to a horizontal and a vertical road), it runs the guiding algorithm in order to know if a change of direction is necessary, before moving on. If not, the car will keep the same direction, at least until the next intersection.

With all these changes we obtain a more complex Manhattan-like scenario (see Fig. 1), where we can notice the different widths of the streets, depending if they are bidirectional and single or dual-lane streets. We can also see the distribution of the traffic lights (represented in cross-roads by green and red dots), and the by-pass road surrounding the city. This extended model has been used in the scenarios simulated by us, and described in the next sections.

3   Self-organizing Traffic Lights (SOTL)

The proposal in [13] treats traffic as an adaptation problem, as traffic conditions constantly change, and every traffic light adapts to its current local conditions. This way, macro-level patterns emerge from the self-organized behavior of these elements. In such model, the self-organization of traffic lights to improve traffic flow is based on a simple measure of local conditions: in every intersection, the number of cars approaching the red light (k_i) are counted.

From this measure, two adaptive self-organized methods are proposed:

- *SOTL-request*: when k_i reaches a threshold, lights change.
- *SOTL-phase*: to prevent fast switching of lights, an additional condition is set: a minimum time must have elapsed since last change.

These two methods are compared with two popular non-adaptive methods:

- *marching*: all green lights are either "vertical" or "horizontal".
- *optim* (green wave): this method is an improvement of the *marching* one. It sets phases to traffic lights, so "green corridors" are created, i.e., a car driving at maximum allowed speed will find the next traffic lights in green.

Among the results presented in [13], *SOTL-request* provides the best performance for low traffic densities, whereas it is very inefficient for high traffic ones, due to constant switching. *SOTL-phase* performance is nearly as good as *SOTL-request* for low densities, and very similar to *marching* for high ones.

4   History-Based SOTL (HB-SOTL)

The results attained by the self-organized methods described in the previous section, and proposed in [13], are encouraging. However, their implementation is not easy, as precisely counting the cars approaching the red light could be very expensive or even unfeasible depending on the weather or daylight conditions.

From such approach, we have proposed another self-organized method [15], in which the duration of the next green cycle of every light is proportional to the traffic density in the intersection, i.e., the quotient of the number of crossing cars in the previous green cycle and the duration of that cycle. Due to memory concerns, we denote our proposal as History-based Self-Organized Traffic Lights (HB-SOTL)[1]. To measure the number of crossing cars is much easier to implement with current low-cost sensor technologies.

- The rubber band is installed perpendicular to the traffic flow direction, and it sends a burst of air pressure when a vehicle's tires passes over.
- The inductive loop detector is used often for permanent installations. It consists of a coil with a few windings of copper wire, which are installed under the road surface.

5   History-Based SOTL Results

We implemented our method using NetLogo in [15], and we compared it with the four methods described in [13]. We tried to keep the same simulation parameters, but with some improvements to have a more flexible and adaptive model:

---

[1] The source code is available at URLs: http://hb-sotl.sourceforge.net and http://ccl.northwestern.edu/netlogo/models/community/HB-SOTL_8_10.

– An improved city model, described in Sect. 2, was used, where 25 % of the
  streets are bidirectional, 25 % of the arteries (both belonging to unidirectional
  and bidirectional streets) have two lanes of traffic running in the same direction, 10 % of the cars have their origin out of the city, and 10 % of the cars
  have their destination out of the city.
– Instead of a grid of 10 × 10 arteries used in [13], a greater model was used: a
  grid of 19 × 19, which corresponds to an area of $4\,km^2$. This is a big area for
  simulation, compared with most of the published literature in UTC.
– The same number of vertical and horizontal streets were set, among the horizontal streets, the same number of east and westbound; and, among the vertical streets, the same number of north and southbound.

The results show that the performance of HB-SOTL is in all cases similar
or better than the other self-organized or classical approaches. In particular,
compared to SOTL-phase:

– As shown in Fig. 2a, the average speed of cars is higher (40 % of maximum
  speed vs. 33.6 % in an 80 cars scenario) for low traffic densities, and only
  slightly higher (10 % vs. 9.4 % in a 1280 cars scenario) for higher ones.
– The average number of stopped cars is significantly lower (37.5 % of the cars
  vs. 49.2 % in an 80 cars scenario) for low densities, and only slightly lower
  (79.5 % vs. 80.6 % in a 1280 cars scenario) for higher densities (Fig. 2b)..
– One of the most relevant result is the average waiting time of cars, which with
  HB-SOTL is much better (2.9 time steps vs. 6.9 in an 80 cars scenario) for
  low densities and only slightly better for higher densities (97.2 vs. 113.3 in a
  1280 cars scenario), as seen in Fig. 3a.

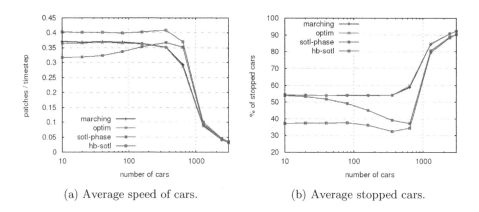

(a) Average speed of cars.   (b) Average stopped cars.

Fig. 2. Statistics for speed and stopped cars.

Besides, in all cases, the variance we have obtained is similar to the other
approaches, as shown in Fig. 3b for the speed.

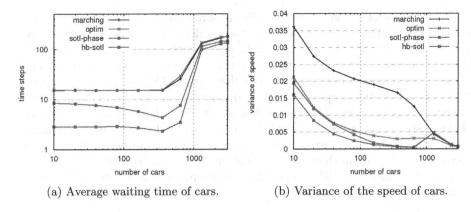

(a) Average waiting time of cars.   (b) Variance of the speed of cars.

Fig. 3. Statistics for waiting times and speed variance.

Finally, we also considered the effect of using a larger memory buffer (longer history) to count the average number of cars over the last $n \in [1, 10]$ cycles. We have found no observable effect. This is very relevant since it means that using a very simple device, able to count the number of cars over the last cycle, is enough to obtain good results.

6   Adaptability to Changing Traffic Conditions

In this section, using as base our method and the results from [15], we consider the capabilities of the different traffic management strategies to adapt to changing traffic conditions. Traffic density is not usually uniform along the day, and there are times when the number of cars in the city has a strong variation considering normal traffic and rush hours. The conditions we have selected for testing the different strategies under these variable conditions are extreme. However, we believe that they allow to discriminate more efficiently among the different algorithms. We consider that there is a rush hour every four hours, and that it last for one hour, i.e., there is reduced traffic (80 cars) during the 80 % of time and high traffic (3000 cars) during the remaining 20 % of time.

Table 1. Results under dynamic conditions (normal traffic in 80 % of time and rush hour during 20 % of time)

Strategy	Avg. speed	Avg. % of stopped cars	Avg. waiting time
Marching	0.26	62.62	10.58
Optim	0.26	62.84	11.10
SOTL-phase	0.27	54.38	6.50
HB-SOTL	0.33	43.76	5.92

Table 1 shows the results for the different strategies, and the different values under consideration. We can see that the results obtained by HB-SOTL are better for all the elements considered. Concerning the average speed of cars in this dynamic scenario, we see that the results of marching, optim and SOTL-phase strategies are pretty similar, and that HB-SOTL has an average speed 20 % higher than its competitors. As for the average percentage of stopped cars in the city, we see how SOTL-phase has a value 15 % lower than the first two classical strategies, while HB-SOTL has a value 20 % lower than SOTL-phase and a 30 % lower than those classical strategies. Finally, concerning the average waiting time for cars in the traffic lights of the city, again HB-SOTL obtains the best result, which is a 10 % better than SOTL-phase and around 45 % better than the two classical strategies.

7   Reducing the Sensor Density

When adapting the systems of UTC in a city, it is usually difficult to migrate from a previous solution to a new one, and to implement a new approach in a very short period of time. This is specially hard when the size of the city is not small. In order to facilitate this process, in this section we explore the effect of reducing the car counting sensor density, considering its effects on the average speed of cars and waiting time.

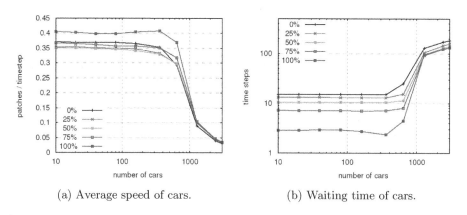

(a) Average speed of cars.   (b) Waiting time of cars.

Fig. 4. Statistics with reduced number of controllers.

In the simulations presented next, only a selectable percentage of intersections have our traffic controllers. On the one hand, the intersections with controllers select the duration of the next vertical or horizontal green cycle directly proportional to traffic measured in the last n vertical or horizontal cycles, where $n \in [1, 10]$. On the other hand, we decided that an intersection with no controller will use the classical and non-adaptive method *marching*.

We have used three different grid sizes: 5×5, 10×10 and 19×19. Each size has been tested with different percentages of controllers in intersections: 0%, 25%, 50%, 75% and 100% (0% corresponding to *marching* and 100% to *HB-SOTL*). The results measured are similar, so we present only the results and graphics for the bigger case, i.e., a grid of 19×19 blocks corresponding to an area of $4\,\mathrm{km}^2$.

Figure 4a shows the average speed of cars with different percentage of reduction in the number of controllers. We can see that the average speed is incremented if we have more controllers. However, there is no big difference, as most of the curves are around 0.4 patches/timestep. Figure 4b shows how the average waiting time of the cars is strongly influenced by the number of controllers, passing from 2 timesteps in a city with 100% of controllers to 20 timesteps in case we have no controllers at all. In the last two figures we also see that when the number of controllers goes beyond the 50% the effects have a clear impact, and as we approach the 100% the effect is more remarkable.

8 Self-organizing UTC with Communication

In this section, we consider the possibility of having communication among the controllers in one intersection with other controllers in its nearby intersections, in order to predict and react to the changes happening in the incoming traffic. This is not an unrealistic assumption as this information could be sent in the power line of the traffic lights using PLC. Therefore, we want to study if we can improve the model described in previous section, introducing communication among traffic lights to take decisions on the duration of the next green cycle.

Each intersection controller has horizontal and vertical neighbors, in the closest intersections, that send and receive traffic data. We assume that a particular controller knows if the neighbor intersections have measurement controller devices. We consider four different cases to calculate the next green cycle duration:

- If the traffic light and any of its neighbors have controllers, the traffic light will use the next formula to calculate the next cycle: $\beta \times own\_measurement + (1 - \beta) \times neighbors\_measurement$, being β a selectable value. After testing several values, we decided to use $\beta = 0.8$ as a reasonable one.
- If the traffic light has a controller but none of its neighbors has one, in the next cycle it will use *own_measurement*.
- If the traffic light has no controller, but any of its neighbors has one, it will use *neighbors_measurement*.
- If neither the traffic light nor the neighbors have controllers, it will use the non-adaptive method *marching*.

Firstly, we want to study if with communication the results are better, in the case that we have controllers in every possible intersection, and having all of them connected. Figures 5a and b show the effect of adding communication to the *HB-SOTL* method. Surprisingly, the influence is only around a 10 % in the case that there is no dense traffic, and has no influence at all if we have a higher traffic density.

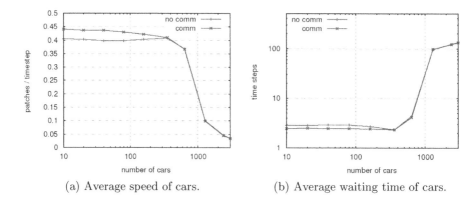

(a) Average speed of cars.   (b) Average waiting time of cars.

Fig. 5. HB-SOTL, communication vs. no communication.

Secondly, we consider the communication extension and test its influence in previous scenario. We do this to check if having a reduction in the number of controllers in the city can be compensated by introducing communication among the nodes. As in previous section, we have used three different grid sizes: 5×5, 10×10 and 19×19. Each size has been tested with different percentages of controllers in intersections: 0 %, 25 %, 50 %, 75 % and 100 % (0 % corresponding to *marching* and 100 % to *HB-SOTL*). As the results were similar for all scenarios, we only present the results for the bigger one of 19×19 corresponding again to an area of $4 \, \text{km}^2$, a big area for simulation, considering the UTC literature.

Figures 6a and b present the results obtained for the average speed of cars and the average waiting time of cars. We can observe that the results are similar to the ones presented in the previous section. Therefore, we can conclude that communication does not play a significant role in the behavior of our method, as a difference with other experiences or methods used in practice. For instance, [17] uses a similar model and relies on communication to provide good results. Nevertheless, first they count the number of car approaching (like in SOTL), and second they do not quantitatively described the influence of communication.

9   Conclusions

The main contribution of the paper is a performance analysis of a multi-agent history-based self-organizing solution (HB-SOTL) for managing urban traffic, which is cheap and easy to implement with current technology. We compare our proposal with some other classical ones, and with other self-organized approaches in different scenarios; considering dynamic traffic densities, reducing the number of sensors in the city, and including communication among the traffic lights.

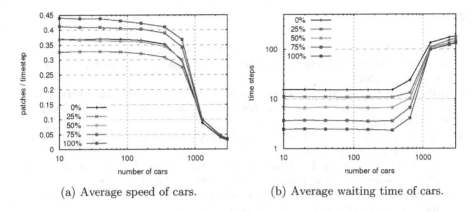

(a) Average speed of cars.   (b) Average waiting time of cars.

Fig. 6. Statistics with communication.

We have seen that our approach outperforms other mechanisms in adaptability to changing traffic conditions, from normal traffic density to very high traffic densities (see Sect. 6). In fact, when we dynamically change the traffic flow to simulate the effect of rush hours, we have seen that the results of the HB-SOTL proposal are better, in practically all the conditions measured, compared with all the other classical and self-organizing methodologies presented in the paper.

Concerning the infrastructure and the cost needed to enhance the UTC in a city, our model requires minimal infrastructure for providing very good results (see Sect. 5). We also have seen how it is possible to setup traffic lights with sensors only in certain parts of the city. In these conditions, our method easily adapts to such situation without extra requirements. Considering a reduction in the number of sensors in the city, we have devised that as soon as the percentage keeps over the 50 % of the city intersections, then the effects also keep clearly visible; having a stronger impact as the percentage gets closer to 100 %.

As a final contribution, we have also considered the effects of using communication among traffic lights to infer future traffic conditions. We have seen that adding communication into HB-SOTL does not have a strong impact. This is good mainly for two reasons: firstly, because it keeps the investments needed to support the HB-SOTL system very low. Secondly, because it means that the indirect communication done by means of the car platoons (stigmergy), sent and received from one to another traffic light, solves the communication needs; managing the system in a self-organized style.

As future work, we will consider scenarios where cars receive real time information from the UTC center (congestion, accidents, etc.) in order to negotiate with other cars real time solutions for achieving a more efficient driving in cities.

Acknowledgment. The authors are supported by the Galician Regional Government under project CN 2012/260 (Consolidation of Research Units: AtlantTIC).

References

1. Hamilton, A., Waterson, B., Cherrett, T., Robinson, A., Snell, I.: The evolution of urban traffic control: changing policy and technology. Transp. Plan. Technol. **36**(1), 24–43 (2013)
2. Helbing, D., Treiber, M.: Gas-kinetic-based traffic model explaining observed hysteretic phase transition. Phys. Rev. Lett. **81**(14), 3042–3045 (1998). doi:10.1103/PhysRevLett.81.3042
3. Klar, A., Wegener, R.: Enskog-like kinetic models for vehicular traffic. J. Statist. Phys. **87**(1–2), 91–114 (1997)
4. Chowdhury, D., Santen, L., Schadschneider, A.: Statistical physics of vehicular traffic and some related systems. Phys. Rep. **329**(4), 199–329 (2000)
5. Rosenblueth, D.A., Gershenson, C.: A model of city traffic based on elementary cellular automata. Complex Syst. **19**(4), 305 (2011)
6. Federal Highway Administration, Traffic Control Systems Handbook. U.S. Department of Transportation (1998)
7. TYCO Integrated Systems, The Sydney co-ordinated adaptive traffic system (SCATS) (2008). http://www.traffic-tech.com/pdf/scatsbrochure.pdf
8. Box, S., Waterson, B.: An automated signalized junction controller that learns strategies from a human expert. Eng. Appl. Artif. Intell. **25**(1), 107–118 (2012)
9. Guerrero-Ibañez, A., Contreras-Castillo, J., Buenrostro, R., Marti, A., Muñoz, A.: A policy-based multi-agent management approach for intelligent traffic-light control. In: Intelligent Vehicles Symposium (IV), 2010 IEEE, pp. 694–699 (2010). doi:10.1109/IVS.2010.5548133
10. Wei-dong, Y., Ji-zhao, L., Hong-song, Z.: A mobility model based on traffic flow theory for vehicular delay tolerant network. In: 2010 IEEE Fifth International Conference on Bio-Inspired Computing: Theories and Applications (BIC-TA), pp. 1071–1075 (2010). doi:10.1109/BICTA.2010.5645106
11. Hu, C., Wang, Y.: A novel intelligent traffic light control scheme. In: 2010 9th International Conference on Grid and Cooperative Computing (GCC), pp. 372–376 (2010). doi:10.1109/GCC.2010.78
12. Ferreira, M., Fernandes, R., Conceicao, H., Viriyasitavat, W., Tonguz, O.K.: Self-organized traffic control. In: Proceedings of the Seventh ACM International Workshop on VehiculAr InterNETworking, ACM, pp. 85–90 (2010)
13. Gershenson, C.: Self-organizing traffic lights. Complex Syst. **16**, 29 (2005)
14. Cools, S., Gershenson, C., D'Hooghe, B.: Self-organizing traffic lights: a realistic simulation. In: Prokopenko, M. (ed.) Self-organization: Applied Multi-agent Systems. Chap. 3, pp. 41–49. Springer, London (2007)
15. Burguillo-Rial, J.C., Rodríguez-Hernández, P.S., Costa-Montenegro, E., Gil-Castiñeira, F.: History based self-organizing traffic lights. Comput. Inf. (CAI) **28**, 1001–1012 (2009)
16. Wilensky, U., NetLogo: center for connected learning and computer-based modeling, Northwestern University. Evanston, IL (1999). http://ccl.northwestern.edu/netlogo/
17. Smith, S.F., Barlow, G.J., Xie, X.-F., Rubinstein, Z.B.: Smart urban signal networks: initial application of the surtrac adaptive traffic signal control system. In: ICAPS, Citeseer (2013)

Planning and Scheduling

A Constraint-Based Approach for the Conciliation of Clinical Guidelines

Luca Piovesan$^{(\boxtimes)}$ and Paolo Terenziani

DISIT, Institute of Computer Science, Università del Piemonte Orientale,
Alessandria, Italy
luca.piovesan@uniupo.it, paolo.terenziani@unipmn.it

Abstract. The medical domain often arises new challenges to Artificial Intelligence. An emerging challenge is the support for the treatment of patients affected by multiple pathologies (comorbid patients). In the medical context, clinical practice guidelines (CPGs) are usually adopted to provide physicians with *evidence-based* recommendations, considering only single pathologies. To support physicians in the treatment of comorbid patients, suitable methodologies must be devised to "merge" CPGs. Techniques like *replanning* or *scheduling*, traditionally adopted in AI to "merge" plans, must be extended and adapted to fit the requirements of the medical domain. In this paper, we propose a novel methodology, that we term "conciliation", to merge multiple CPGs, supporting the treatments of comorbid patients.

Keywords: Computer interpretable clinical guidelines · Comorbidities · Combining medical guidelines · Constraint satisfaction problems

1 Introduction

The medical domain is a challenging application field for Artificial Intelligence, since heterogeneous forms of knowledge (e.g., ontological, experiential, evidence based) and of reasoning techniques (e.g., diagnosys, planning, case-based, scheduling) are involved. One of the most interesting areas is the one of clinical practice guidelines (CPGs). CPGs are defined as *"systematically developed statements to assist practitioner and patient decisions about appropriate health care in specific clinical circumstances"* [5]. They are one of the major decision tools introduced in the medical field to promote *evidence-based medicine*.

In the last two decades, the research in Artificial Intelligence in Medicine has shown that Computer Science can improve the impact of CPGs through the introduction of computer-based approaches that acquire, represent, execute, and reason with Computer Interpretable Guidelines (CIGs), and that provide physicians with different forms of decision support facilities (see, e.g., [13,18]).

In the area of CIGs, a challenging and "hot" topic has recently emerged, i.e. the use of CIGs in the treatment of *comorbid* patients (i.e., patients affected by two or more diseases, see [6]). CPGs (and CIGs) are prevalently designed

© Springer International Publishing AG 2016
M. Montes-y-Gómez et al. (Eds.): IBERAMIA 2016, LNAI 10022, pp. 77–88, 2016.
DOI: 10.1007/978-3-319-47955-2_7

for the treatment of specific diseases and they loose their effectiveness when coping with *comorbidities*. Although some CPGs covering the most frequent co-occurring diseases might be devised, the approach of considering all the possible combinations of diseases is *"not only difficult, but also impractical"*, so that *"there is a need for formal methods that would allow combining several disease-specific clinical practice guidelines in order to customize them to a patient"* [10].

The main problem in the treatment of comorbid patients is that actions in different CIGs may interact and be inconsistent with each other. For instance, when coping with patients affected by venous thrombosis, an anticoagulant (e.g., Warfarin) is recommended to avoid the development of clots. However, if a patient suffering from thrombosis has to undergo a surgery (for another disease), the anticoagulant therapy must be stopped because of the high risk of bleeding.

Roughly speaking, CIGs can be conceived as (conditional) plans. As a consequence, in the area of Artificial Intelligence in Medicine, several approaches have resorted to AI approaches to merge plans, spacing from *replanning* to *scheduling* techniques. For instance, the approach in [21] uses constraint logic programming to identify and address adverse interactions. A constraint logic programming (CLP) model is derived from the combination of logical models that represent each CIG and a mitigation algorithm is applied to detect and mitigate interactions. On the other hand, Sánchez-Garzón et al. [17] propose an agent-based approach to guideline merging. Each CIG is modeled by an agent with hierarchical planning capabilities. The result is obtained through the coordination of all the agents and respects the recommendations of each CIG. Riaño et al. represent CIGs as sets of clinical actions that are modelled into an ontology [4]. Treatments are first unified in a unique treatment and then a set of "combination rules" is applied to detect and avoid possible interactions. Jafarpour and Abidi [9] use semantic-web rules and an ontology for the merging criteria. Given such knowledge, an execution engine merges CIGs according to merge criteria. Unfortunately, such approaches, though theoretically valuable, do not properly take into account two main constraints imposed by the medical domain (and such a limitation limits the practical applicability of such approaches):

– the "merge" of CIGs must be as close as possible to the original CIGs. This is due to the fact that CIGs contains only *evidence-based* recommendations, and the goal is to provide an output as evidence-based as possible; (ii)
– the "merge" of two CIGs cannot be achieved though the application of arbitrary modifications: only a limited set of management (of interactions) options are indeed adopted in the medical practice.

To comply the above requirements, the use of "traditional" AI techniques is not sufficient. We thus propose a new methodology, that we term "conciliation", which is based on the *Constraint Satisfaction Problem* (CSP) framework [11].

This paper is organized as follows. In Sect. 2, we briefly describe the GLARE approach [19]. Section 3 is the core of this paper, describing our methodology to conciliate two or more CIGs and its integration with GLARE. Finally, in Sect. 4 we propose conclusions.

2   GLARE and Comorbidities: Overview of the Approach

The approach described in this paper is part of a project devoted to the extension of a well known framework to manage CIGs (GLARE) to deal with patients affected by comorbidities. The GLARE (Guideline Acquisition, Representation and Execution) framework [19] is the result of a long-term project started in 1997 to provide a user-friendly and domain-independent tool to support CIG acquisition, representation and execution, and to provide physicians with a wide range of facilities, helping them in decision making or CIG analysis tasks. In GLARE, a CIG is represented by a conditional and hierarchical graph (we will show two examples of CIGs in Fig. 2a), whose nodes represent *actions* (to execute on patients) or *decisions* (which represent branches in the CIGs) and whose arcs model the *control flow relations* and the *temporal constraints* between them. GLARE distinguishes between *atomic actions* (simple steps in a CIG) and *composite actions* (plans), which are defined in terms of their components.

GLARE architecture is composed by two main modules: an acquisition and an execution module. The acquisition module provides a user-friendly graphical interface to support expert-physicians in the acquisition of CIGs, and provides as output a formal representation of CIGs. The execution module takes in input an acquired CIG, and the patient's data, and support user-physicians in the application of the CIG to the patient. A *log* of the executed actions is provided as output. In the last ten years GLARE has been extended with several modules, based on advanced AI techniques, to provide further support to physicians: automatic resource-based contextualization [20], temporal reasoning [1], decision-making support [12], and model-based verification [3].

Starting from 2013, we began to extend GLARE to provide support to the management of comorbidities. Our final goal is to provide physicians with a set of methodologies that, taken as input the original CIGs developed for single diseases, helps physicians in the treatment of comorbid patients. Given a specific patient, and two or more CIGs under execution on the patient (and the log of such executions), the tool supporting comorbidities manages the interactions between the CIGs, to suggest to physicians how to go on with the patient's treatment. GLARE provides interactive support to the three different phases in the treatment of comorbidities: (1) the detection of interactions (occurring between CIGs actions or actions and patient's status), (2) the management of the detected interactions, and (3) the final merging of the CIGs into an unique treatment for a specific patient. Phase 1 (interaction detection) requires a comparison of actions belonging to different CIGs, and a deep knowledge about them, to detect their interactions. GLARE provides a tool that, on the basis of a medical ontology, detects a wide range of interactions and supports physicians in the detection of the remaining ones [15]. Once detected, interactions have to be evaluated, and a management (if needed) has to be applied (phase 2 above). Managements are small local changes in the original CIGs, which make the original therapeutic plans executable avoiding undesirable interactions and promoting desirable ones. Analyzing the medical literature, we have identified the managing operations adopted by physicians in the medical practice [16]: *Safe Alternative* (to avoid an interaction by proposing alternative paths in the CIG, or adding new actions), *Temporal Avoidance* (to avoid an interaction by imposing additional temporal constraints), *Effect Monitoring* (which maintains the interaction, but introduces a monitoring action, to monitor the possible effects of the interaction itself), *Dosage Adjustment* (to mitigate an interaction by changing drug dosage), *Interaction Mitigation* (to mitigate an interaction through the introduction of a new action), *Interaction Alignment* (to force the occurrence of an interaction, by imposing that specific paths are executed at specific times in the two CIGs), and *Intention Alignment* (to manage actions aiming to the same intention, by avoiding redundancy). GLARE provides physicians with the possibility of choosing the most appropriate management for every interaction, and a semi-automatic tool for the application of such managements to the original CIGs. In particular, it provides three types of operations to implement the above managements (in some cases, a combination of some of them is required): (1) forcing the execution (or the non-execution) of actions in the CIGs (2) adding/removing actions to/from CIGs, and (3) adding new temporal constraints on CIGs.

Once the interactions have been identified (phase 1), and the management options to treat them have been selected (phase 2), the merge of the original CIGs and the new interaction managements must result in an executable CIG. Notice that (1) the management options lead to changes to the original CIGs that are "locally" consistent, but possibly not consistent with each other, and that (2) the conjunctive application of multiple management options can lead to the introduction of additional constraints. The goal of phase 3 (that we term "conciliation") is to provide as output a "merged" (and consistent) CIG resulting from the conjunctive application of all the selected management options. While we have already faced phases 1 and 2, this paper addresses the conciliation phase.

3   CIG Conciliation

Our conciliation module (see Fig. 1) is integrated in the pre-existent architecture of GLARE. Once detected the interactions (through the *Interaction Detection* module, not shown), the physician decides how to manage them (*Interaction Management* module). After that, the system performs an automatic step of *Conciliation*, to merge the CIGs and to check the consistency of their union.

We base our approach to CIG conciliation on the CSP framework. The first step of the procedure consists in the translation of the problem into a CSP (TO_CSP function). Such a task is divided into two sub-tasks. First, the original CIGs and their execution log are translated into a CSP (CLT function, see CIGs and Log Transformation in Subsect. 3.1), obtaining a CSP called CSP_{CL}. CSP_{CL} represents the CIGs and the log *without* the changes imposed by the managements chosen by the physician during the Interaction Management phase. After that, CSP_{CL} must be modified to consider the managements chosen by the user. Such a task is performed by the AOM function (Application of Managements in Subsect. 3.1) and, technically speaking, it is achieved by removing some constraints and adding other ones to CSP_{CL}. We call the resulting CSP

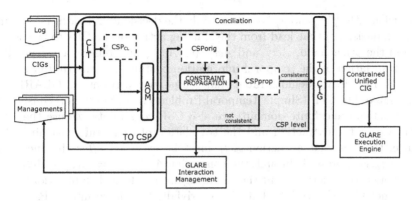

Fig. 1. Integration of our approach in GLARE and its architecture. Notice that Interaction Management and Execution Engine are GLARE modules.

CSPorig. After that, we verify the consistency of CSPorig and propagate it, to obtain a new CSP (*CSPorig*) that represents an executable CIG in which all the interactions are managed (see Subsect. 3.2). If CSPprop is consistent, our system transforms it in a CIG executable in GLARE (see Subsect. 3.3). Otherwise, an execution in which the interactions are managed as chosen is not possible. Therefore, there is the need of changing the chosen managements (going back to the interaction management phase) and repeating the conciliation phase, until the result is satisfying.

3.1 The TO CSP Function

Our transform module translates the CIGs, the execution log and the management options into a CSP. The resulting CSP is built as follows. First, a CSPorig representing the original CIGs and the log is built (see "CIGs and Log Transformation"). Then, CSPorig is modified by adding the constraints deriving from the management options (see "Application of Managements"). The resulting CSP represents the union of original and management constraints.

CIGs and Log Transformation (CLT Module in Fig. 1). First, the module CLT associates three variables of CSPorig to each action Act_i belonging to the original CIGs: a Boolean variable $Act_i^{EXEC} \in \{true, false\}$, and two positive integer variables $Act_i^S, Act_i^E \in \mathbb{N}$. Intuitively speaking, Act_i^{EXEC} is true when Act_i will be executed, while Act_i^S and Act_i^E represent the starting and ending points of its execution time. For punctual actions, we use an unique variable $Act_i^P \in \mathbb{N}$. In addition, each decision action Dec_j is associated with an enumerative variable $Dec_j^{OUT} \in D_j$. Each element of the set D_j represents one of the possible alternative paths starting from Dec_j. An assignment of a particular value for Dec_j^{OUT} means that such a path will be executed after Dec_j. The execution of an action (i.e., the value of the variable Act_i^{EXEC}) depends on the values of the decisions preceding it. For this reason, the function CLT adds to CSPorig the constraint $Act_i^{EXEC} \equiv (true \wedge Dec_1^{OUT} = d_1^i \wedge \cdots \wedge Dec_n^{OUT} = d_n^i)$ for each action, where Dec_j^{OUT} and d_j^i are all and only the decisions and paths (stemming from such decisions) that lead from the starting action of the CIG to the action Act_i. On the other hand, Act_i^S and Act_i^E (or Act_i^P) are constrained depending on the translation of the temporal information contained in the *control* arcs of the CIGs. In [14], we demonstrated that temporal constraints in GLARE can be translated into STP (Simple Temporal Problem) ones. Since STP constraints (over an integer and finite domain) are also CSP constraints, we use the same techniques for the extraction and the translation of temporal constraints. For instance, the fact that an action Act_b must be executed *after* the end of an action Act_a is expressed through the constraint $Act_b^{EXEC} \Rightarrow Act_a^E - Act_b^S < 0$. Please notice that here and in the following we use logical implications "$\Rightarrow$": they are not CSP constraints, but can be trivially translated into CSP.

Finally, if an action has been already executed, the log contains information about its execution. In such a case, the log (temporal) constraints and the fact that such action has been executed are added to the CSP. For instance, given the log entry "Act_b has been executed 5 time unit after the end of Act_a", the constraints $Act_b^S - Act_a^E \equiv 5$ and $Act_b^{EXEC} \equiv true$ are added to CSPorig.

Application of Managements (AOM in Fig. 1**).** In Sect. 2 we have elicited the management options provided by GLARE, and pointed out three basic types of operations to implement them. The translation of the management options depends on the types of operations adopted to implement them. Hereinafter, for our examples, we consider two generic interacting actions A and B in the paths identified by the decisions outputs $\{DecA_1^{OUT} = dA_1 \wedge \cdots \wedge DecA_n^{OUT} = dA_n\}$ and $\{DecB_1^{OUT} = dB_1 \wedge \cdots \wedge DecB_n^{OUT} = dB_m\}$.

(1) Forcing the execution (or the non-execution) of specific paths of actions. Such an operation imposes limitations on the execution values (Act_i^{EXEC}) of the decisions preceding the interacting actions. For instance, the Safe Alternative option between actions A and B, imposes that the two actions cannot be executed together. This management is translated through Eq. 1.

$$\text{add\_to\_CSPorig } \langle A^{EXEC} = false \vee B^{EXEC} = false \rangle \tag{1}$$

(2) Adding/removing actions. Such an operation adds new actions to the CIGs, so that new variables are added to CSPorig. In Eq. 2, we show an example (deriving from the application of the Safe Alternative management option): since A interacts with B, if B has to be executed, C must be executed instead of A.

$$\text{remove\_from\_CSPorig } \langle A^{EXEC} \equiv DecA_1^{OUT} = dA_1 \wedge \cdots \wedge DecA_n^{OUT} = dA_n \rangle$$
$$\text{add\_to\_CSPorig } \langle C^{EXEC} \equiv B^{EXEC} = true \wedge DecA_1^{OUT} = dA_1 \wedge \cdots \wedge DecA_n^{OUT} = dA_n \rangle$$
$$\text{add\_to\_CSPorig } \langle A^{EXEC} \equiv B^{EXEC} = false \wedge DecA_1^{OUT} = dA_1 \wedge \cdots \wedge DecA_n^{OUT} = dA_n \rangle$$
$$\text{for each temporal constraint involving } A^S | A^E | A^P$$
$$\text{add\_to\_CSPorig the same temporal constraint for } C^S | C^E | C^P \tag{2}$$

(3) Adding new temporal constraints. Temporal constraints may be added to avoid (or enforce) a given interaction (e.g., Temporal Avoidance can be used to avoid the interaction between A and B). Since actions in CIGs are usually temporally related with each other, the addition/update of a temporal constraint can modify other constraints. Thus, temporal constraint propagation is needed, giving as output the set of strictest constraints between each pair of actions in the CIGs (also called "minimal network"; see, e.g., [14]). The temporal constraints in the minimal network (MN) must be asserted into CSPorig, stating that such constraints must hold just in case both A and B are executed (i.e., if $A^{EXEC} = true \land B^{EXEC} = true$). In Eq. 3, we show how the above example of Temporal Avoidance is translated.

$$\text{for each temporal constraint constr} \in MN$$
$$\text{add\_to\_CSPorig} \langle A^{EXEC} = true \land B^{EXEC} = true \Rightarrow \text{constr} \rangle \qquad (3)$$

To deal with management options involving two or more implementation operations, the above treatments must be combined.

3.2 CSP Propagation

Once the CIGs, log and managements have been mapped onto CSPorig, some forms of reasoning must be applied to verify its consistency and/or to get its solutions. Basically, if CSPorig is consistent (i.e., at least one solution exists), each one of its solutions corresponds to a possible execution of the merged CIGs. Thus one could solve the CSP and return an unique solution to users. However, in the medical domain, this approach would be too restrictive, since it would impose to physicians a specific execution of the CIGs (while many other executions would be usually possible). On the contrary, there is the need to obtain a new CSP (CSPprop in Fig. 1) representing (implicitly, without eliciting them) all and only the possible solutions (i.e., all the possible executions of the CIGs in which the interactions are managed). Technically speaking, this amounts to aiming at obtaining a *backtrack-free* CSP [8]. Indeed, in a CSP backtrack-free, if a value x_i (belonging to the domain of a variable X) is allowed by the constraints, it belongs to at least a solution. This ensures that, if the physician chooses to valorize X with x_i (e.g., taking a decision), there exists at least an execution consistent with such a valorization. Several approaches exist to obtain backtrack-free CSPs. Among them, we based our implementation (which is out of the scope of this paper) on [8]. Given CSPorig, we choose an ordering according to which the variables can be valorized (orderings can be chosen by following the order of the actions in the CIGs, without significantly loosing generality). Given an ordering, the *width* k (of the ordering) is defined as the maximum number of constraints involving a variable and the preceding ones in the ordering. Freuder demonstrated that a CSP is backtrack-free if it is *strongly (k+1)-consistent* (i.e., n-consistent for $1 \leq n \leq k + 1$, where n-consistency [7] is a generalization of arc-consistency). To obtain a CSPprop that is strongly (k+1)-consistent, one must apply k+1 steps of constraint propagation, considering, for each step n, the subsets of n constraints, with $1 \leq n \leq k + 1$. Notice that each step of propagation may add constraints involving $n - 1$ variables to CSPprop.

3.3   The TO CIG Function

The function TO_CIG (see Fig. 1) interprets the propagated CSP CSPprop, updating the original CIGs and the managements and adding new ones. The result is a new CIG, union of the original ones and of the updated/new managements, that we call *Constrained Unified CIG*. Algorithm 1 describes TO_CIG.

Algorithm 1. TO CIG

Input : CSPorig, CSPprop, CIGs, Managements
Output: a Constrained Unified CIG *result*

result $\leftarrow$ createConstrainedUnifiedCIG(CIGs, Managements);
if $\neg$ *isConsistent(CSPprop)* **then**
  return *getContradiction(CSPprop)*;
else
  foreach $c_i^{PROP} \in CSPprop$ **do**
    if $\exists c_i^{ORIG} \in CSPorig \wedge c_i^{PROP} \neq c_i^{ORIG}$ **then**
      update(c_i^{ORIG}, c_i^{PROP}, result);
    else
      createNewManagement(c_i^{PROP}, result);
    end
  end
  return *result*;
end

After the propagation (see Subsect. 3.2), the resulting CSPprop can be consistent or not. In case it is not consistent, the propagation procedure reaches a point in which the domain of a variable (say x) is restricted to be empty[1]. Such a variable can concern the execution of an action (type $EXEC$), the output of a decision (type OUT), or the time(s) of execution (types S, E or P). In these cases, there is no execution of the CIGs consistent with all the constraints. In most of the cases, the inconsistency can be avoided by changing the management of some interactions. To provide support to solve the conflicts, we provide users with an error output similar to

```
Propagation failed because the domain of X, {starting point |
ending point | output | executability} of Act_X became empty.
```

In case CSPprop is consistent, the system interprets it in order to update the original CIGs and managements, or add new constraints to the CIGs. Notice that, to obtain CSPprop, the propagation procedure (see Subsect. 3.2) adds new constraints or modifies existing ones. For this reason, in our implementation, the constraints are annotated with a unique identifier (e.g., the "i" in c_i^{ORIG}),

---

[1] Actually, there are some domain-specific cases in which we consider a CSP "non-consistent" also in case all the domains are not empty. See, for instance, the example of Subsect. 3.4.

so that given a constraints c_i^{PROP} belonging to CSPprop, one can check whether it belongs to CSPorig (and it has been eventually modified by the propagation procedure) or it has been added by the propagation. For the constraints c_i^{PROP} belonging to CSPprop such that a c_i^{ORIG} exists in CSPorig, TO_CIG checks whether c_i^{PROP} has been modified by the propagation ($c_i^{PROP} \neq c_i^{ORIG}$). If so, we use the identifier to identify the original element that has been translated into c_i^{ORIG}, and we update it (procedure *update*).

Example. For instance, a constraint in CSPprop can restrict the domain of a Dec_i^{OUT} variable to assume only the values $\{d_i^1, d_i^3\}$. In case i identifies a constraint in CSPorig, our algorithm checks whether the original constraint represents the same restriction. If so, TO_CIG does not perform any operation. Otherwise, it compares the two constraints. For instance, the original constraint can derive from the translation of a CIG, in which the action Dec_i has three paths starting from it: $\{d_i^1, d_i^2, d_i^3\}$. In such a case, the procedure *update* has to modify the original CIG by removing the path corresponding to value d_i^2.

If a c_i^{ORIG} does not exist, the constraint c_i^{PROP} has been added by the propagation. Thus, a new management is added (procedure *createNewManagement*), representing the new constraint. *createNewManagement* is modular: different "modules" can be added to recognize specific patterns (see, e.g., the disjunctions added in the example of Subsect. 3.4).

3.4   Case Study

To demonstrate our approach, we combined a (simplified) CIG for Peptic Ulcer (PU, shown, at an high level of abstraction, in the rectangle on the top of Fig. 2a) with a (simplified) CIG for deep Venous Thrombosis (VT, rectangle on the bottom) taken from the British National Institute for Health and Care Excellence (www.nice.org.uk). To distinguish between the CIGs before and after the conciliation, in Fig. 2 we represented actions and paths belonging to the original CIGs and removed by the conciliation with the gray color, and elements added/updated by the conciliation process with red labels. Using the interaction detection tool provided by GLARE, we detected two interactions between the action Amoxicillin therapy, part of each sub-path of the composite action H. pylori eradication (belonging to PU), and the actions Warfarin therapy and Fondaparinux therapy belonging to VT (the CIGs at such a level of detail are shown in Fig. 2b). As a first experiment, we managed both the interactions with the application of Safe Alternative management options proposing alternative paths (as shown in Fig. 2b). Then, we added to the execution log the fact that the action Anticoagulant decision of VT has been executed. Finally, we tried to combine the two CIGs. The conciliation process failed, returning the output Propagation failed because the domain of $HPREv^{OUT}$, output of H. pylori evaluation (diagnostic decision) has been restricted. This error output is returned by our procedure when a variable not under the control of the physician is restricted. H. pylori evaluation is a diagnostic decision, and its output depends only on

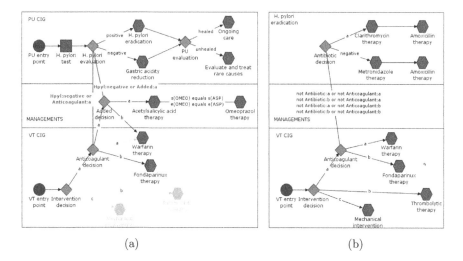

(a)   (b)

Fig. 2. Example of application of our approach. Figure 2a represents a constrained unified CIG resulting from the conciliation of a CIG for the Peptic Ulcer and a CIG for Venous Thrombosis. Figure 2b represents the CIGs at a different level of abstraction. Legend: blue nodes are work actions, green query actions, yellow decisions and red composite actions. (Color figure online)

patient's condition in that specific point of the execution. Thus, all the paths starting from it have to be considered "necessary" for the execution. As a second experiment, we changed the management of the interactions choosing an option that added the actions in the middle rectangle of Fig. 2a. Notice that, even if we added an "interaction free" path, interactions can still occur before the conciliation (in case the output of H. pylori evaluation is positive and the physician selects path a after Anticoagulant decision). Differently from previous case, the conciliation returned a result, adding the elements with red labels. In particular, the new constraints point that, if H. pylori evaluation has a "positive" output, the only acceptable path is the one containing the (alternative) action Acetylsalicylic acid therapy. Given that, CIGs can be executed without that the (detected) interactions happen.

4   Conclusions

The treatment of comorbidities constitutes a challenging problem, since it involves the "merge" of two or more CIGs. Such a goal can be achieved through three steps: (i) interaction detection (to analyse possible interactions between actions from different CIGs), (ii) interaction management (to propose a "local" treatment of each single interaction), and (iii) integration, to finally put together the original CIGs with the proposed interaction managements, achieving a unifying CIG to be executed on the comorbid patient. In this paper, we focus on the third step. Two main requirements must be satisfied by such a step, in order to make our approach applicable in the medical application domain: (i) to maintain as much as possible the evidence-based character of CIGs, the output CIG must adhere as much as possible to the original ones, and (ii) to support physicians without limiting them in the choice of possible treatment options, the integration must not return an unique solution, but a constrained unified CIG that represent, in a compact way, all the possible alternative treatments. To comply the above requirements, the use of "traditional" AI techniques is not sufficient: in this paper, we propose a new methodology, "conciliation", based on the *Constraint Satisfaction Problem* framework. In literature, [21] already faced the problem through an approach based on constraint programming (thus similar to CSP). However our methodology significantly differs from it, since (i) we support the possibility of interactively analyzing and changing the management options (to cope with interactions), and (ii) the output of our framework is not a CSP, but a new "conciliated" CIG, which can then be executed in GLARE.

From the technical point of view, a crucial aspect for our approach is scalability. The propagation process discussed in Subsect. 3.2 is, in the worst case, exponential in time and space. In addition, obtaining a CSP allowing a backtrack-free search is not always possible. For these reasons, in our future work we aim at investigating the possibility of (i) optimizing the propagation process and (ii) providing "alternative" tools to support physicians in case the (defined) propagation cannot be performed. Regarding point (i), we are experimenting the adoption of "hybrid" constraint propagators, or approximate techniques, in which CSPprop represents only a subset of the possible solutions (where the level of approximation can be chosen, see, e.g., [2]). Regarding point (ii), we are planning to develop a "what-if" tool that, instead of trying to achieve a backtrack-free CSP, exploits CSP *solving* techniques and some partial propagation (e.g., arc consistency) allowing physicians to make hypotheses on the next action(s) to be executed and to check whether such hypotheses can produce a consistent CSP.

References

1. Anselma, L., Terenziani, P., Montani, S., Bottrighi, A.: Towards a comprehensive treatment of repetitions, periodicity and temporal constraints in clinical guidelines. Artif. Intell. Med. **38**(2), 171–195 (2006)
2. Beck, J.C., Carchrae, T., Freuder, E.C., Ringwelski, G.: A space-efficient backtrack-free representation for constraint satisfaction problems. Intl. J. Artif. Intell. Tools **17**(4), 703–730 (2008)
3. Bottrighi, A., Giordano, L., Molino, G., Montani, S., Terenziani, P., Torchio, M.: Adopting model checking techniques for clinical guidelines verification. Artif. Intell. Med. **48**(1), 1–19 (2010)
4. López-Vallverdú, J.A., Riaño, D., Collado, A.: Rule-based combination of comorbid treatments for chronic diseases applied to hypertension, diabetes mellitus and heart failure. In: Lenz, R., Miksch, S., Peleg, M., Reichert, M., Riaño, D., ten Teije, A. (eds.) ProHealth 2012 and KR4HC 2012. LNCS, vol. 7738, pp. 30–41. Springer, Heidelberg (2013)

5. Committee to Advise the Public Health Service on Clinical Practice Guidelines, Institute of Medicine: Clinical Practice Guidelines Directions for a New Program. National Academy Press, Washington, D.C. (1990)
6. Fraccaro, P., Arguello Castelerio, M., Ainsworth, J., Buchan, I.: Adoption of clinical decision support in multimorbidity: a systematic review. JMIR Med. Inform. **3**(1), e4 (2015)
7. Freuder, E.C.: Synthesizing constraint expressions. Commun. ACM **21**(11), 958–966 (1978)
8. Freuder, E.C.: A sufficient condition for backtrack-free search. J. ACM **29**(1), 24–32 (1982)
9. Jafarpour, B., Abidi, S.S.R.: Merging disease-specific clinical guidelines to handle comorbidities in a clinical decision support setting. In: Peek, N., Marín Morales, R., Peleg, M. (eds.) AIME 2013. LNCS, vol. 7885, pp. 28–32. Springer, Heidelberg (2013)
10. Michalowski, M., Wilk, S., Michalowski, W., Lin, D., Farion, K., Mohapatra, S.: Using constraint logic programming to implement iterative actions and numerical measures during mitigation of concurrently applied clinical practice guidelines. In: Peek, N., Marín Morales, R., Peleg, M. (eds.) AIME 2013. LNCS, vol. 7885, pp. 17–22. Springer, Heidelberg (2013)
11. Montanari, U.: Networks of constraints: fundamental properties and applications to picture processing. Inf. Sci. **7**, 95–132 (1974)
12. Montani, S., Terenziani, P.: Exploiting decision theory concepts within clinical guideline systems: toward a general approach. Int. J. Intell. Syst. **21**(6), 585–599 (2006)
13. Peleg, M.: Computer-interpretable clinical guidelines: a methodological review. J. Biomed. Inform. **46**(4), 744–763 (2013)
14. Piovesan, L., Anselma, L., Terenziani, P.: Temporal detection of guideline interactions. In: Proceedings of the International Conference on Health Informatics (HEALTHINF-2015), pp. 40–50. Scitepress (2015)
15. Piovesan, L., Molino, G., Terenziani, P.: An ontological knowledge and multiple abstraction level decision support system in healthcare. Decis. Analytics **1**(8), 1–24 (2014)
16. Piovesan, L., Terenziani, P.: A mixed-initiative approach to the conciliation of clinical guidelines for comorbid patients. In: Riaño, D., et al. (eds.) KR4HC/ProHealth 2015. LNCS, vol. 9485, pp. 95–108. Springer, Heidelberg (2015). doi:10.1007/978-3-319-26585-8_7
17. Sánchez-Garzón, I., Fernández-Olivares, J., Onaindía, E., Milla, G., Jordán, J., Castejón, P.: A multi-agent planning approach for the generation of personalized treatment plans of comorbid patients. In: Peek, N., Marín Morales, R., Peleg, M. (eds.) AIME 2013. LNCS, vol. 7885, pp. 23–27. Springer, Heidelberg (2013)
18. Ten Teije, A., Miksch, S., Lucas, P. (eds.): Computer-Based Medical Guidelines and Protocols: A Primer and Current Trends, Studies in Health Technology and Informatics, vol. 139. IOS Press, Amsterdam (2008)
19. Terenziani, P., Molino, G., Torchio, M.: A modular approach for representing and executing clinical guidelines. Artif. Intell. Med. **23**(3), 249–276 (2001)
20. Terenziani, P., Montani, S., Bottrighi, A., Torchio, M., Molino, G., Correndo, G.: A context-adaptable approach to clinical guidelines. Stud. Health Technol. Inf. **107**(Pt 1), 169–173 (2004)
21. Wilk, S., Michalowski, W., Michalowski, M., Farion, K., Hing, M.M., Mohapatra, S.: Mitigation of adverse interactions in pairs of clinical practice guidelines using constraint logic programming. J. Biomed. Inform. **46**(2), 341–353 (2013)

Intelligence Amplification Framework for Enhancing Scheduling Processes

Andrej Dobrkovic$^{(\boxtimes)}$, Luyao Liu, Maria-Eugenia Iacob,
and Jos van Hillegersberg

Industrial Engineering and Business Information Systems,
University of Twente, Enschede, The Netherlands
a.dobrkovic@utwente.nl

Abstract. The scheduling process in a typical business environment consists of predominantly repetitive tasks that have to be completed in limited time and often containing some form of uncertainty. The intelligence amplification is a symbiotic relationship between a human and an intelligent agent. This partnership is organized to emphasize the strength of both entities, with the human taking the central role of the objective setter and supervisor, and the machine focusing on executing the repetitive tasks. The output efficiency and effectiveness increase as each partner can focus on its native tasks. We propose the intelligence amplification framework that is applicable in typical scheduling problems encountered in the business domain. Using this framework we build an artifact to enhance scheduling processes in synchromodal logistics, showing that a symbiotic decision maker performs better in terms of efficiency and effectiveness.

Keywords: Intelligence amplification · Intelligent agents · Synchromodal logistics · Scheduling

1   Introduction

The concept of intelligence amplification (IA) aims to enhance human decision making abilities through a symbiotic relationship between a human and an intelligent agent. Opposed to the concept of artificial intelligence (AI), where the machine is made to mimic and replace the cognitive abilities of a human brain, the IA concept puts the human in the central role, augmenting natural brain processes with enhanced input information, and output extension handled by a software component.

The challenge of synchromodal logistics involves decision making in a complex environment requiring input from various sources, getting insight into the estimated state of a logistics network and selecting the most appropriate multi-criteria optimization to formulate the shipping policy. We aim to tackle the core challenge of scheduling container transport by applying symbiotic decision makers based on the concept of intelligence amplification. By creating a symbiotic bond between a logistics planner and a supporting system, we can organize the scheduling process in such a way that the human part gets the role of a manager that defines goals and strategies, supervising intelligent agents that carry out routine tasks. The process benefits from the

© Springer International Publishing AG 2016
M. Montes-y-Gómez et al. (Eds.): IBERAMIA 2016, LNAI 10022, pp. 89–100, 2016.
DOI: 10.1007/978-3-319-47955-2_8

inclusion of intelligent agents, as they can process large volumes of data quickly, and cannot be fatigued as opposed to a human planner. On the other hand, humans excel when dealing with uncertainties and other loosely structured tasks requiring experience from different fields.

In this paper we present how the concept of intelligence amplification can be implemented in logistics. Then we use a serious game to simulate typical weekly planning tasks of synchromodal planners. After incorporating intelligent agents into the game, we evaluate the performance of each entity separately, as well as the score when the symbiotic decision maker is used.

Our goal is to show that in a typical scheduling scenario containing uncertainties and limited time to make decision, collaborative human-machine effort produces superior results. With development of IA-based artifact, we aim to advance the IA concept towards industry acceptance.

We streamline the process of fitting the IA framework to one specific industrial domain (logistics) and show how to use it to build the artifact, but also how this artifact enhances the scheduling process and the new division of labor that follows. Although we focus on logistics domain, the concept is applicable in any other business field characterized with rapid decision making under uncertainties, time constraints, existence of large data volumes, and decisions requiring unavoidable human expertise.

The research methodology followed in this study is the design science research methodology [1].

The remainder of this paper is organized as follows. In Sect. 2 we present the problem background and related work on this topic. Section 3 is dedicated to the solution design. Here we present our IA framework and the artifact implementation in a serious game environment. The validation results are given in Sect. 4. Section 5 summarizes the work done, and discusses the future work and impact.

2 Problem Background

2.1 Human-in-the-Loop and Intelligent Systems

Licklider presented a vision on tightly coupling human brains and computing machines, creating a mutually interdependent relationship, capable of processing data, and making decisions in a way not even conceivable by each entity separately [2]. With both human decision maker, and intelligent agent exhibiting different strengths and weaknesses, the goal is to organize a partnership in such a way, that they complement each other. This human-in-the-loop approach allows the human to stay on top of the process, making strategic decisions that influence its outcome, while the role of the machine is to help with the task by processing large volumes of data, identifying alternatives, simulating outcomes, and automating repetitive operations.

Griffit and Greitzer re-address Licklider's vision as the emerging field of neo-symbiosis, arguing that it is essential to have humans incorporated into computer architectures [3]. The same requirement of having human-in-the-loop is recognized in [4] by Williams et al. They use capabilities of a human to help overcome the problems of target recognition algorithms characterized by the excessive amount of false positives, and their inability to adapt to different environmental conditions.

According to Garcia, the knowledge keeps evolving with time, is fragmented, and scattered across many resources [5]. She presents the AGUIA model for amplifying human intelligence based on the agent's technology for the task-oriented contexts using domain ontology. Casini *et al.* recognize in [6] that automating processing without human intervention is error prone, while the human's contribution in a massively distributed, and high volume information environment can dramatically increase the system's accuracy through the prevention of error propagation. Human insights also help to find solutions significantly faster, and of lower complexity, according to Woolley and Stanley [7].

Delibašić *et al.* design an open source data mining platform for a white-box algorithm design, where users assemble a data mining algorithm from algorithm design components. They prove that results achieved by students using the white-box approach exceeds the black-box approach, containing inflexible inputs and outputs [8].

Ahmed and Hasan research human-agent teamwork in data mining for early detection of the breast cancer [9]. They assign human to be the supervisor, and control the autonomous agent. When left unattended, the agent acts upon predefined rules, and priorities. The human can override the agent if he is not satisfied with its decision.

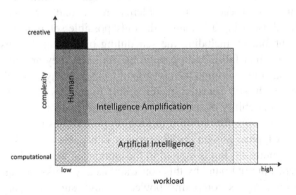

Fig. 1. The decision making domain according to problem complexity and workload

2.2 Synchromodal Logistics

Synchromodal logistics is a form of multimodal logistic, in which the most efficient combination of transport modes is selected for every transport order, aiming at an optimal resource utilization, and improved service levels, while reducing road congestion and CO_2 emissions [10]. Implementing the synchromodal concept in a real-life industry environment, requires the development of a specific synchromodal platform, which facilitates the increased cooperation between logistics service providers [11]. This platform must also be able to provide decision support for the planners by suggesting optimal policies, while taking into account the current, and predicted future state of the logistics network [12].

3 Solution Design

3.1 Symbiotic Man-Machine Relationship

Key Assumptions. In order to define the foundations for man-machine symbiosis we start by making the following assumptions:

- Every problem can be decomposed in two or more sub-problems, and the process recursively repeated up to the atomic level,
- Every problem and sub-problem can be characterized by two factors: complexity, and workload,
- High complexity, low workload problems are ideally suited for humans,
- Low complexity, high workload problems are ideally suited for intelligent agents.

If we consider a sub-problem at an atomic level, and such a problem requires knowledge from other areas (experience) and an innovative approach in finding a solution (creativity), then the problem can only be solved by a human entity. Alternatively, if a solution to a sub-problem relies only on logic, and computational power, then a machine is a much better candidate, as it can process large data volumes, in a short time. Therefore, if a top-problem is a set of exclusively computational sub-problems, then we have an automation case, and we should assign a machine to solve it. At the other extreme, a set of creative only problems are beyond the logic of intelligent machines, and for those using a human is the only possible solution. However, typical decision making problems, including the scheduling ones from the business domain, contain sub-elements that fall in both categories. As such, they are neither ideal for a human, nor for a machine. Through the division of tasks in a way that utilizes the best of both worlds, the new hybrid decision maker is able to address the problem in a novel way that is beyond the ability of these two entities independently.

In Fig. 1 we show how the decision making domain relates to the type of problem being addressed. A typical human decision maker is capable of solving any problem, regardless of its complexity, but is limited in the amount of workload it can handle. An artificial intelligence in not bound by this constraint as any workload can be handled, if necessary, by increasing its computing power, through supplementary memory and processor cores. Still, the AI is limited as far as complexity is concerned, as the machine cannot handle tasks that require creativity.

Intelligence amplification has the potential to expand the decision making into a completely new area, as it can handle task requiring both high creativity, and high workload. However, the fact that IA involves two different entities that constantly exchange information, plus the fact that their native languages and information processing differ significantly, we can expect "friction" (obstacles in information exchange). The friction level depends on the quality of the interface for human-computer interaction and although it should be minimized, an ideal communication is not possible. Hence, the maximum value for the IA workload will be lower in comparison to the AI and the complexity value will also be lower in comparison to the human.

Intelligence Amplification Framework. Building upon Licklider's vision [2], we propose to arrange the symbiotic man-machine relationship along the following lines:

- The human entity is given the master role, and it oversees the artificial intelligence,
- The artificial intelligence is given the assistant role,
- The human is responsible for the strategic decision making,
- AI is responsible for the tactical/operational tasks,
- The human is also responsible for the creative tasks that AI cannot handle,
- The AI is pre-processing the data, and brings awareness to the human component,
- The AI acts upon meta instruction given by the human,
- The AI analyses the human output in context using the available input, and learns to recognize and adapt to the human's behavior,
- Depending on the level of autonomy of the AI, the machine will either automatically complete all computational tasks that conform with the strategic goals set by the human, or will suggest a solution for the human to verify, executing only the tasks that the human has approved,
- If the AI neither can understand the input, nor can process the task, it will ask for human assistance,
- The human can overrule the AI.

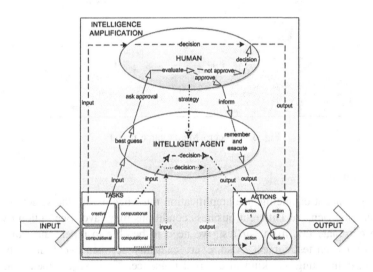

Fig. 2. Hierarchical organization of the human-machine partnership depicting the information flow and the task division

We illustrate the flow of information and the corresponding action of the decision maker built upon intelligence amplification concept in Fig. 2. The decision maker receives external stimuli from the environment. The input is broken down into a set of tasks which can be classified as either creative or computational. The intelligent agent will search through the set, selecting tasks it can identify, and for which it has enough information in its knowledge base to be able to handle successfully. The creative tasks will be ignored by the agent and left to the human to handle.

Depending on the level of complexity of the computational tasks, the intelligent agent will have a different response. If the task is purely computational, the machine will automatically handle it. If the task involves decision, the agent will strive towards the optimal solution, provided it conforms to the strategy set by the human. If no solution is possible, or the input contains a level of ambiguity beyond a desired threshold, the agent should make the best guess, and ask human's approval. Should the human respond positively, the agent will learn from the user's response and proceed to execute an action. Alternatively, the task will be left to the human to process, along with the tasks initially labeled as creative.

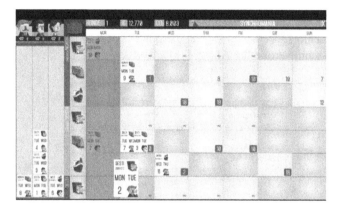

Fig. 3. Synchro Mania game round

3.2 Environment

Proving the concept of intelligence amplification must begin with the selection of the appropriate environment. Different business conditions may favor one option over the other. To eliminate this from our test, we need to be able to repeat the exact same conditions for each test subject. Also, every causality between events must be fully understood, eliminating any kind of external influence. These requirements direct us towards business simulation. We searched for serious games for logistics and found the Synchro Mania [13] to be the best option as it focuses on synchromodal planning.

Figure 3 shows one round from Synchro Mania that represents one planner's week. On the left part of the screen the player receives orders from clients, each with a specific destination, transport mode, availability date, and a number of containers. In the larger window on the right, the player has an overview of the available capacity for each destination and each possible mode. New orders come in real-time, and the player has to make a choice how to schedule orders by assigning them to appropriate transport slots (on the right side of the game canvas) with drag and drop. The choices made have impact on the total costs, CO_2 emissions, and the client satisfaction. Using Synchro

Mania allows us to analyze decisions of each subject, quantify results as score, and compare from the score of different subjects.

3.3 Framework Design for Enhanced Scheduling in Synchromodal Logistics

The design of the IA framework revolves around answering the question of what is the best way to utilize the collaboration between the man and the machine? Finding an effective allocation of tasks can be viewed as an assignment problem from operations research. The assignment problem is the fundamental of combinatorial optimization. It assumes that there are several tasks that can be performed by several agents. Any agent can perform any of those tasks but with varying efficiency, which is modeled as the cost. When assigning specific tasks to specific agents the objective is to minimize the total cost while performing all tasks [14]. In case of IA, we have only two actors: the human and the machine. By assuming that each agent is better suited for specific tasks than the other, the framework will consist of a variant of the assignment problem, aiming at an optimal division of tasks.

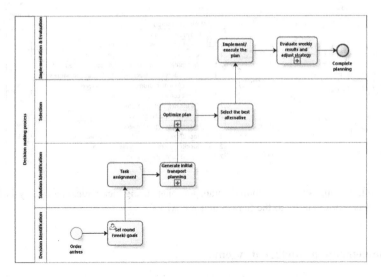

Fig. 4. Decomposing decision making process

Task Decomposition and Task Assignment. We invite two players familiar with the Synchro Mania to play the game. We observe and record their actions to identify the top-level processes. After that, we use hierarchical task analysis [15] to decompose the problem space into sub-tasks. Due to page limitation and result similarity we present the top level process (Fig. 4) and zoom to the sub-task involving the optimization of the transport plan (Fig. 5).

Using the task list obtained through hierarchical task analysis, we evaluate each task according to its complexity level. The evaluation is done by the specialist in industrial engineering, working as full time synchromodal logistics consultant. Those that require external knowledge, creative thinking or contain a strategic component are assigned to the human. The computational ones are assigned to the machine.

The table on the right in Fig. 5 gives the assignment result for the "optimize transport plan" sub-task. We use 'x' in the appropriate column to indicate if an atomic task is assigned to either the human or the AI. The first sub-task "2.1 minimize transport costs" contains four atomic tasks which are all suited for the AI, hence this task will be completely handled by the machine. The last one "2.3 process unassigned orders" requires a creative approach to identify the causes that prevented the order to be included in the automatic planning, and how to deal with them. Since these task typically involve choosing the "lesser evil" from the possible options, the result varies upon human's style and priorities.

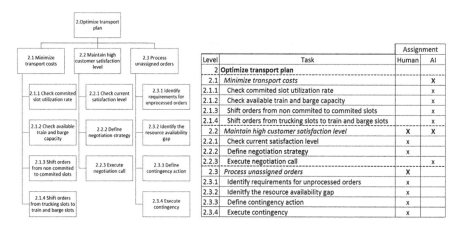

Fig. 5. Left: decomposing task "optimize transport plan" using hierarchical task analysis. Right: assigning decomposed tasks to the human and the AI

3.4   Implementing Intelligent Agents

Based on the results from the task decomposition and assignment, we build the intelligent agents to handle all atomic tasks assigned to the machine. Depending on the manner in which they are used, we distinguish between two types: agents that execute when invoked by the user, and agents that continuously run in the background. A complete list with all agents we add to the game, and their functionality is given in Table 1.

Table 1. Implemented agents, functionalities and execution mode

No.	Agent name	Functionality/Description	Execution mode
1	Urgent order identifier	Search through the list of received orders. Identify the urgent ones and leave visual cue for the human planner	Continuous
2	Auto assigner	Assign specific or all incoming orders to transport slots in a manner that minimize expenses and maximizes resource utilization	On call
3	Quick assigner	Find any available transport slot for the given order to ensure its deliverence	On call
4	Negotiator	Send change request to clients asking for order parameter change (due date, mode, destination,....). Use learning algorithm to adapt to the player's strategy	Continuous
5	Protector	Monitor current time. If an order is about o expire and the capacity is available, ensure it is being shipped.	Continuous

4 Testing and Validation

For the testing phase we use two game scenarios, A and B. Ten students of industrial engineering program at Univeristy of Twente were invited to play the game, which we divided into two groups of five. The first group was playing scenario A without aid. Following that, they played scenario B using intelligent agents. The second group played scenario A with the intelligent agents and scenario B without. We record four values after every game: the service level (service), the average transportation cost per container (costTEU), the average CO_2 emissions per container (emmisionCO$_2$), and the average customer satisfaction level (satisfaction).

To enable comparison, first we quantify the user satisfaction. Then we normalize all four values on a scale from 1 to 10. Finally, we add weights to the variables calculating the score with the following formula:

$$score_i = 0.25 * service_i + 0.31 * costTEU_i + 0.13 * emmisionCO2_i + 0.31 * satisfaction_i$$

We also play both scenarios without any human intervention using only artificial intelligence. The scores for all players playing the rounds independently, with the aid of the intelligent agent, and the fully automated planner are shown in Fig. 6.

When analyzing the scores of human decision makers playing the game without any aid, we can identify a high variation. The median score in this group is 5.78. In general, players with more experience in planning did considerably better than players with no background in logistics scheduling. The scores for 75 % of the group range between 1.77 and 7.77. The best player managed to reach the score of 8.56. Next, we took a look at the score achieved by the fully automated solution (AI). The machine scored 4.6, which is below the median score of the human group. This is expected due

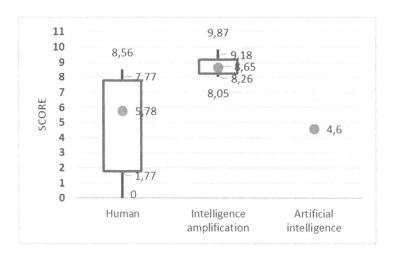

Fig. 6. Synchro Mania score comparison for different decision makers

to the fact that the machine has no external knowledge that could be used to anticipate clients' behaviors. Also, the experienced players were able to handle uncertainties by anticipating potential future bottlenecks, and by freeing capacity in advance. The AI was going for cheaper slots. However, when late orders unexpectedly arrived, the AI was unable to complete several shipments due to no available capacity. Experienced players were able to overcome this, while beginners performed less than the automated solution.

Finally, we took a look at the hybrid approach based on intelligence amplification concept. The results of this groups are concentrated, and balanced around the median 8.65, while the range for 75 % of the group is in the interval 8.26–9.18. The most significant benefit of synergetic approach is that the average player performs within the range of the best human player. This becomes even more evident when looking at the lower performing participants, whose results dramatically improve. For the best player, there is an improvement from 8.56 to 9.87. We conclude from these results that IA benefits every player, yet the improvement is greater for the less experienced players in comparison to the more skilled ones.

Regarding the results we have to note that our scenarios were allowing the player 30 s per day to make decisions. This duration is the standard game time and we found that it gives the experience player just enough time to make the decisions. The less experience players found this time constraint limiting. Furthermore, although experience players were able to deal with all orders, they stated that they felt the pressure.

For the better players, the main advantage that IA brings is that it relieves them of routine work and helps them focus on urgent orders, thus having more time to identify potential bottlenecks in advance, and to contact the clients with change requests. Those players dealing with planning for the first time relied completely on automation to make the initial order placement, only intervening to redistribute the orders when the machine would decide too early in the game to wait until the later in the week to send shipments.

5 Conclusion and Future Work

In this paper we proposed an intelligence amplification framework that would be of value for the business. Starting with the scheduling problem, this concept has the potential for application in diverse decision making processes. After explaining our view on how to organize a symbiotic man-machine partnership, and how to apply it to solve scheduling problems in synchromodal logistics, we implemented and tested the proposed solution in a serious game environment. We analyzed the typical planner behavior, we broke down the scheduling process into tasks and sub-tasks and used the above task decomposition to do the assignment of responsibilities to either the intelligent agent, or to the human player. In the end, we tested the performance of the decision makers with and without the aid of the artificial agent and concluded that the human-machine partnership is capable of improving significantly efficiency, and effectiveness of logistics scheduling processes.

Our conclusion is that humans and AIs have different strengths and weaknesses, and, should a suitable collaborating framework be designed, the best results can be achieved through the human-in-the-loop partnership, where the human will set goals, oversee the system and deal with special cases, while the machine will fast process routine tasks.

In our future work we aim to focus on two objectives. One is to run additional tests in a real-life logistics environment, and speed up the acceptance of intelligence amplification concept in industrial decision making. The second one is to develop learning mechanisms, allowing AI to observe, understand and adopt its behavior in relation to the human's actions. We expect this to decrease the development time and further improve AI autonomy. Also it could potentially increase the decision making performance even further in a repetitive environment with unavoidable uncertainty.

References

1. Peffers, K., Tuunanen, T., Rothenberger, M.A., Chatterjee, S.: A design science research methodology for information systems research. J. Manag. Inf. Syst. **24**, 45–77 (2007)
2. Licklider, J.C.: Man-computer symbiosis. In: IRE Transactions on Human Factors Electron, pp. 4–11 (1960)
3. Griffith, D., Greitzer, F.L.: Neo-symbiosis: the next stage in the evolution of human information interaction. Int. J. Cogn. Inf. Nat. Intell. (IJCINI) **1**, 39–52 (2007)
4. Williams, D.P., Couillard, M., Dugelay, S.: On human perception and automatic target recognition: strategies for human-computer cooperation. In: 2014 22nd International Conference on Pattern Recognition (ICPR), pp. 4690–4695 (2014)
5. Garcia, A.C.B.: AGUIA: agents guidance for intelligence amplification in goal oriented tasks. In: 2010 International Conference on P2P, Parallel, Grid, Cloud and Internet Computing (3PGCIC), pp. 338–344 (2010)
6. Casini, E., Depree, J., Suri, N., Bradshaw, J.M., Nieten, T.: Enhancing decision-making by leveraging human intervention in large-scale sensor networks. In: 2015 IEEE International Inter-Disciplinary Conference on Cognitive Methods in Situation Awareness and Decision Support (CogSIMA), pp. 200–205 (2015)

7. Woolley, B.G., Stanley, K.O.: A novel human-computer collaboration: combining novelty search with interactive evolution. In: Proceedings of the 2014 Conference on Genetic and Evolutionary Computation, pp. 233–240 (2014)

8. Delibasic, B., Vukicevic, M., Jovanovic, M.: White-box decision tree algorithms: a pilot study on perceived usefulness, perceived ease of use, and perceived understanding. Int. J. Eng. Edu. **293**, 674–687 (2013)

9. Ahmed, A.-I., Hasan, M.M.: A hybrid approach for decision making to detect breast cancer using data mining and autonomous agent based on human agent teamwork. In: 2014 17th International Conference on Computer and Information Technology (ICCIT), pp. 320–325 (2014)

10. Mes, M.R., Iacob, M.E.: Synchromodal transport planning at a logistics service provider. In: Zijm, H., Klumpp, M., Clausen, U., ten Hompel, M. (eds.) Logistics and Supply Chain Innovation, pp. 23–36. Springer, Heidelberg (2016)

11. Singh, P.M.: Developing a service oriented IT platform for synchromodal transportation. In: On the Move to Meaning Ful Inrternet Systems OTM 2014 Workshop, pp. 30–36 (2016)

12. Dobrkovic, A., Iacob, M.-E., van Hillegersberg, J., Mes, M., Glandrup, M.: Towards an approach for long term AIS-based prediction of vessel arrival times. In: ten Hompel, M., Clausen, U., Klumpp, M., Zijm, H. (eds.) Logistics and Supply Chain Innovation, pp. 281–294. Springer, Cham (2016)

13. Buiel, E., Visschedijk, G., Lebesque, L., Lucassen, I., Riessen, B.v., Rijn, A.v., et al.: Synchro mania-design and evaluation of a serious game creating a mind shift in transport planning, In: 46th International Simulation and Gaming Association Conference, ISAGA 2015, Kyoto, Japan, 18–25 July 2015, pp. 1–12 (2015)

14. Munkres, J.: Algorithms for the assignment and transportation problems. J. Soc. Ind. Appl. Math. **5**, 32–38 (1957)

15. Annett, J.: Hierarchical task analysis. Handbook Cogn. Task Des. **2**, 17–35 (2003)

A Column Generation Approach for Solving a Green Bi-objective Inventory Routing Problem

Carlos Franco[1(✉)], Eduyn Ramiro López-Santana[2],
and Germán Méndez-Giraldo[2]

[1] Universidad del Rosario, Bogotá, Colombia
carlosa.franco@urosario.edu.co
[2] Universidad Distrital Francisco José de Caldas, Bogotá, Colombia
{erlopezs,gmendez}@udistrital.edu.co

Abstract. The aim of this paper is present a multi-objective algorithm embedded with column generation to solve a green bi-objective inventory routing problem. In contrast with the classic Inventory Routing Problem where the main objective is to minimize the total cost overall supply chain network, in the green logistics besides this objective a minimization of the CO_2 emisions is included. For solving the bi-objective problem, we proposed the use of NISE (Noninferior Set Estimation) algorithm combined with column generation for reduce the amount of variables in the problem.

Keywords: Green logistics · Inventory routing problem · Column generation · Multi-objective optimization · NISE

1 Introduction

Nowadays many companies have to deal with the problem of coordinate inventory policies and transportation management. For this problem a variety of strategies are used for minimizing logistics costs. One of these strategies is the Vendor Managed Inventory System (VMI) where the suppliers decide the quantity that will be sent to customers, the period of time that will be made the delivery and how the product will be distributed. When these decisions are made by retailers the logistics cost are reduced and the customer can reduce the ordering cost [1].

The VMI can be modeled as an Inventory Routing Problem (IRP), which is an attractive problem due to its complexity and for its applicability. this problem is NP-hard because the TSP, which is NP-complete, can be reduced to it [1].

The IRP and its extensions usually looks for minimizing the overall logistics cost (routing and inventory). This traditional model does not take into account the emissions of greenhouse gases. The green logistics refers to the design of models that consider the emissions and other factor that impact negatively the environment.

Policy makers have been interested in contributing to the global environment, for this reason, supply chain management is not excluded of design distribution networks that help to reduce the environmental impact.

© Springer International Publishing AG 2016
M. Montes-y-Gómez et al. (Eds.): IBERAMIA 2016, LNAI 10022, pp. 101–112, 2016.
DOI: 10.1007/978-3-319-47955-2_9

Transportation is one of the most significant contributor to the global warming in the world because of the uses of fossil fuels for energy [2], also during the storage of products are produced several emissions.

In this paper, we model the problem of determining the set of routes that satisfy a set of requirements by minimizing the inventory and distribution costs and minimizing the emission of greenhouse gases. The objective of the model is to design a network of transportation over a planning horizon, determining the inventory levels at the customers and the routes but taking into account that each route an each unit of inventory has emissions of greenhouse gasses.

The paper is organized as follows: Sect. 2 contains a literature overview of the IRP, Multi-objective IRP and Green IRP. Section 3 presents the mathematical model that is used and Sect. 4 shows the methodology approach for solving the problem. The computational results are presented in Sect. 5. Finally, Sect. 6 states the conclusions and future works.

2   Overview of Related Literature

This section introduces an overview of related literature with green bi-objective inventory routing problem. Firstly, we describe the IRP problem. Secondly we describe as the IRP is a multi-objective problem and shows some examples. Then we describe the green IRP and to sum up we present a classification of the literature with the Mono-objective, multi-objective and green factors objectives.

2.1   Inventory Routing Problem (IRP)

The IRP was introduced in 1983 [2], the main idea was to combine decisions of inventory, vehicle scheduling and distribution in a gas company. Also in 1983 a problem of coordinating inventory and routing decisions was presented by Zipkin and Federguen [3].

The problem shifts responsibilities for customers and retailers [4]. Bertazzi and Speranza [5] present a tutorial to introduce the inventory routing problem. Customers transfer the inventory control to the retailer. The retailers monitor the inventory level and decide when to deliver, how much to deliver and how to deliver. This change has advantages like less resources required for inventory management for the customers, and the freedom for the retailer to decide the routes, when, how and how much to deliver.

This problem is not only attractive for its computational complexity but also for its industrial applicability. This model can be used in wholesales, where they have to decide the distribution of products to their stores; another application is the distribution of drugs between centers of attention. We can also see this problem in distribution of blood between hospitals [6]), animals between farms [7], applications in the maritime transportation [8], bulk gas distribution [9]. A wide variety of distribution problem can be modeled using this problem.

There are several algorithms proposed to solve mono-objective deterministic problem [1, 10–13]. In these papers, authors develop several mathematical models and solutions methods based on heuristics and metaheuristics using a wide number of objective functions such as minimization of inventory and routing costs, minimization of the cost penalty of time windows, minimization of the production costs, minimization of travel time, among others.

2.2 Multi-objective IRP

There has not been several approaches for the multi-objective IRP, in fact, the classic IRP is a multi-objective problem due to the inventory costs and the routing costs are conflicting. An application of multi-objective IRP was develop in medicinal drug distribution [14]. The author proposed the use of three different objectives: minimize the holding, transportation and shortage costs, minimize the amount of errors in demand forecasting and minimize the amount of emissions. The mathematical model presented in this paper is Non Linear Mixed-Integer but some linearization is presented. For solving the problem the authors use two phases, the first one consists in use possibilistic methods for generate the auxiliary crisp model. In the second phase, a fuzzy multi-objective optimization method is used.

Some developments of multi-objective IRP are developed in the literature. One of them include in the objective functions the minimization of the inventory levels and the minimizations of the inventory and routing costs [15]. The proposed solution includes the development of an Iterated Local Search (ILS) saving the non-dominated functions. Some applications of IRP in the waste collection were developed including social and health issues [16, 17] and uncertainty demand. Also application for perishable products taking into account the objectives of minimizing the inventory and routing costs and maximizing customer level satisfaction [18].

2.3 Green IRP

Nowadays green logistics and green economics are concepts that have gained a significant importance in the concept of vehicle routing problems. It has been included as the carbon emissions (CO_2) [19]. A survey of Green Vehicle Routing Problems was presented by [20]. The authors provide a classification of problems categorizing the green feature as Green, Pollution and Reverse Logistics, also they suggest research gaps between its state and richer models describing the complexity in real-world cases.

The IRP has several studies including the green logistics factor. The green factor is considered as the carbon emissions for the routing and inventory [21], for solving the problem, a mathematical model is proposed including a demand uncertainty. Other models are proposed taking into account the emissions in the objective function as [3, 22–25] and solving it by various metaheuristics.

2.4 Classification of IRP Models and Green IRP

Table 1 summarizes the literature overview of the Inventory Routing Problem and the Green Inventory Routing Problem including the Green Vehicle Routing Problem.

As we can see there are only the paper proposed by Malekly [25] works the Multi-objective Inventory Routing Problem including green factors. Table 2 presents the environmental aspects present in the different IRP models. All the authors consider the emissions produced by transportation and only one considers additionally the inventory emissions, the model contemplate the emissions as a monetary cost.

As a preliminary conclusion, we can observe that no one has worked the Multi-objective Inventory Routing Problem considering the emissions produced by transportation and inventory, only it has been approximated as the monetary cost produce by both emissions.

Table 1. Summary of literature about the IRP

Type of objective	References
Mono-objective	[2] [4] [1] [6] [7] [8] [9] [10] [11] [12] [14] [13] [17] [23] [22] [3] [24] [26]
Multi-objective	[15] [16] [18] [19] [25]
Green Factors	[21] [25] [23] [22] [3] [24] [26]

Table 2. Components of the green IRP

Type of emission	References
Transportation	[21] [25] [23] [22] [3] [24] [26]
Inventory	[25]
As cost	[25] [22] [3] [24]

3   Problem Definition

To develop our proposal we use the mathematical model presented by Coelho et al. [11] and include the green logistics costs. The problem is modeled using a graph $\mathcal{G} = \{\mathcal{N}, \mathcal{A}\}$ consisting of a set of arcs $\mathcal{A} = \{(i,j), i,j \in \mathcal{N}, \ i \neq j\}$ and a set of nodes $\mathcal{N} = \{0, \ldots, n\}$, where the node 0 is the depot, and the subset $\mathcal{N}' = \mathcal{N} \backslash \{0\}$ represents the customers. At each time, $t \in \mathcal{T} = \{1, \ldots, p\}$ a quantity d_{it} of product is generated by customer $i \in \mathcal{N}'$. There is an initial inventory level at the customers $i \in \mathcal{N}'$ defined by S_i. For each customer $i \in \mathcal{N}'$ there is a maximum level of inventory denoted by U_i. The set $\mathcal{K} = \{1, \ldots, k\}$ defines the set of vehicles available at the depot, each one with a capacity denoted by Q_k. The cost of traveling on the arc $i, j \in \mathcal{A}$ is c_{ij} and the inventory holding cost is denoted by h_i. Finally, $p_{ij} \ \forall i, j \in \mathcal{A}$ denotes the emissions produced by routing the vehicles and pp denotes the emission produced by the products in inventory.

3.1   Mathematical Model

The following are the variables used in the mathematical model:

- I_{it} : denotes the inventory level at node $i \in \mathcal{N}'$ at the end of time $t \in \mathcal{T}$,
- q_{ikt} : represents the quantity that vehicle $\mathcal{K} = \{1, .., k\}$ collected of product from customer $i \in \mathcal{N}'$ at period time $t \in \mathcal{T}$
- x_{ij}^{kt}; is the binary variable that takes the value of one if a vehicle $k \in \mathcal{K}$ traveling on a specific arc (i, j) in a period time $t \in \mathcal{T}$
- w_{ikt} : auxiliary variable used for sub tour elimination constraints

The mathematical formulation can be stated as mixed integer programming model as follows:

$$\min z_1 = \sum_{i \in \mathcal{N}} \sum_{j \in \mathcal{N}} \sum_{k \in \mathcal{K}} \sum_{t \in \mathcal{T}} c_{ij} x_{ij}^{kt} + \sum_{i \in \mathcal{N}'} \sum_{t \in \mathcal{T}} h_i I_{it} + \sum_{t \in \mathcal{T}} h_0 I_{0t} \quad (1)$$

$$\min z_2 = \sum_{i \in \mathcal{N}} \sum_{j \in \mathcal{N}} \sum_{k \in \mathcal{K}} \sum_{t \in \mathcal{T}} p_{ij} x_{ij}^{kt} + \sum_{i \in \mathcal{N}'} \sum_{t \in \mathcal{T}} pp I_{it} \quad (2)$$

subject to,

$$I_{0t} = I_{0t-1} + r_t + \sum_{i \in \mathcal{N}'} \sum_{k \in \mathcal{K}} q_{ikt} \quad \forall t \in \mathcal{T} \quad (3)$$

$$I_{it} = I_{it-1} + \sum_{k \in \mathcal{K}} q_{ikt} - d_{it} \quad \forall i \in \mathcal{N}', \quad \forall t \in \mathcal{T} \quad (4)$$

$$I_{it} \leq U_i \quad \forall i \in \mathcal{N}', \quad \forall t \in \mathcal{T} \quad (5)$$

$$\sum_{i \in \mathcal{N}'} q_{ikt} \leq Q_k \quad \forall k \in \mathcal{K} \quad (6)$$

$$\sum_{j \in \mathcal{N}} x_{ij}^{kt} = \sum_{j \in \mathcal{N}} x_{ji}^{kt} \quad \forall i \in \mathcal{N}', \quad \forall t \in \mathcal{T}, \quad \forall k \in \mathcal{K} \quad (7)$$

$$\sum_{j \in \mathcal{N}'} x_{0j}^{kt} \leq 1 \quad \forall t \in \mathcal{T}, \quad \forall k \in \mathcal{K} \quad (8)$$

$$w_{ikt} - w_{jkt} + Q_k x_{ij}^{kt} \leq Q_k - q_{ikt} \quad \forall i, j \in \mathcal{N}', \quad \forall t \in \mathcal{T}, \quad \forall k \in \mathcal{K} \quad (9)$$

$$q_{ikt} \leq w_{ikt} \leq Q_k \quad \forall i \in \mathcal{N}', \quad \forall t \in \mathcal{T}, \quad \forall k \in \mathcal{K} \quad (10)$$

$$q_{ikt} \geq 0 \quad \forall i \in \mathcal{N}', \quad \forall t \in \mathcal{T}, \quad \forall k \in \mathcal{K} \quad (11)$$

$$I_{it} \geq 0 \quad \forall i \in \mathcal{N}', \quad \forall t \in \mathcal{T} \quad (12)$$

$$x_{ij}^{kt} \in \{0, 1\} \quad \forall i, j \in \mathcal{N}', \quad \forall t \in \mathcal{T}, \quad 3 \forall k \in \mathcal{K} \quad (13)$$

$$y_{it} \in \{0, 1\} \quad \forall i \in \mathcal{N}', \quad \forall t \in \mathcal{T} \quad (14)$$

In this problem, we have two related objective functions. The first one (1) is the sum of the routing and inventory costs and the second one (2) is the sum of emissions of the inventory levels and the routing emissions over the planning horizon. Both objectives look for minimizing its values. Constraints (3) and (4) define the inventory levels at customer $i \in \mathcal{N}'$ at period time $t \in \mathcal{T}$. Constraints (5) ensure that the inventory level cannot exceed the maximum level allowed. In constraints (6), the capacity of each vehicle is guaranteed. The flow conservation constraints are presented in (7). Constraints (8) state that each vehicle can only make one route for each period. Constraints (9) and (10) are the classical sub-tour elimination. Finally, the type of variable used are presented in Eqs. (11), (12), (13) and (14).

4 Solution Approach

Figure 1 presents the strategy for solving the bi-objective IRP using an embedded column generation method. Our solution strategy consist in solving the bi-objective green IRP only using a reduced number of routes due to the problem has an exponential number of routes. For solving the routes that must be generated, we will use column generation technique for create attractive routes based on the dual information then only those routes with an interesting reduced costs are added to the problem. For use column generation we will use a new mathematical formulation that allow us to use this technique.

The methodology for obtain the Pareto's frontier begins solving the NISE algorithm. Due to NISE algorithm requires to solve the MIP with different weights of the objectives function, the column generation method is used several times and it can use extra computational time. For save time we save the basis generated in each iteration in order to avoid routes that have been obtained in previous iterations so only routes that have not been previously generated and given its reduced costs, can help to improve the solution of the problem will be generated.

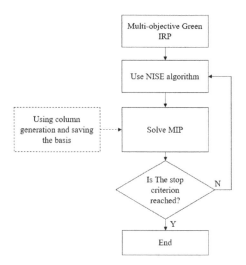

Fig. 1. Flow diagram of column generation methodology for the multi-objective Green IRP

Finally when the criterion stop is reached the algorithm finished and report the Pareto's frontier.

4.1 Column Generation Formulation

For solving the problem via column generation, we reformulate the problem changing a variable as follows:

- R represent the set of all possible routes, where each route consist in a vector of 1 and 0 if the customer below to the route
- I_{it} : denotes the inventory level at node $i \in \mathcal{N}'$ at the end of time $t \in T$
- q_{ikt} : represents the quantity that vehicle $\mathcal{K} = \{1,..,k\}$ collected of product from customer $i \in \mathcal{N}'$ at period time $t \in T$
- x_{rkt}; is the binary variable that takes the value of one if a vehicle $k \in \mathcal{K}$ use a route $r \in R$ on a period time $t \in T$.

In the new model, introduce a variable that select a route r of a set of routes R. For generate only attractive routes, the subproblem is formulated as a shortest path problem using the dual information and the objective functions as is described in [27]. The mathematical model can be described as follows where p_r indicates the total emission of a route:

$$\min z_1 = \sum_{r \in R} \sum_{k \in \mathcal{K}} \sum_{t \in T} c_r x_r^{kt} + \sum_{i \in \mathcal{N}'} \sum_{t \in T} h_i I_{it} + \sum_{t \in T} h_0 I_{0t} \qquad (15)$$

$$\min z_2 = \sum_{r \in R} \sum_{k \in \mathcal{K}} \sum_{t \in T} p_r x_r^{kt} + \sum_{i \in \mathcal{N}'} \sum_{t \in T} p p I_{it} \qquad (16)$$

subject to,

$$\sum_{r \in R} x_{rkt} \leq 1 \quad \forall k \in \mathcal{K}, t \in T \qquad (17)$$

$$\sum_{r \in R} \sum_{k \in \mathcal{K}} a_{ir} x_{rkt} \leq 1 \quad \forall t \in T, i \in \mathcal{N}' \qquad (18)$$

$$q_{ikt} \leq Q \sum_{r \in R} x_{rkt} a_{ir} \quad \forall k \in \mathcal{K}, t \in T, i \in \mathcal{N}' \qquad (19)$$

$$\sum_{i \in \mathcal{N}'} q_{ikt} \leq Q \quad \forall k \in \mathcal{K}, t \in T \qquad (20)$$

$$I_{0t} = I_{0t-1} + r_{t-1} - \sum_{i \in \mathcal{N}} \sum_{k \in K} q_{ikt-1} \quad \forall t \in T' \backslash \{1\} \qquad (21)$$

$$I_{it} = I_{it-1} + \sum_{k \in K} q_{ikt-1} - d_{it-1} \quad \forall i \in \mathcal{N}', t \in T' \backslash \{1\} \qquad (22)$$

$$I_{it} \geq L_i \quad \forall i \in \mathcal{N}', t \in T' \qquad (23)$$

$$I_{it} \leq U_i \quad \forall i \in \mathcal{N}', t \in \mathcal{T}' \tag{24}$$

$$I_{it} \geq 0 \quad \forall k \in \mathcal{K}, t \in \mathcal{T}', \, i \in \mathcal{N} \tag{25}$$

$$q_{ikt} \geq 0 \quad \forall k \in \mathcal{K}, t \in \mathcal{T}, \, i \in \mathcal{N} \tag{26}$$

$$x_{rkt} \in \{0, 1\} \quad \forall k \in \mathcal{K}, \, t \in \mathcal{T}, \, r \in \mathcal{R}. \tag{27}$$

For this formulation, we also have two related objective functions. The first one (15) is to minimize the sum of the routing and inventory costs and the second one (16) is to minimize the sum of emissions of the inventory levels and the routing emissions over the planning horizon. Constraints (17) ensure that maximum one route is selected for each vehicle and period. Constraints (18) state that a customer is visited maximum once. Constraints (19) and (20) define the feasible capacity for each vehicle. Constraints (21) and (22) relate the inventory levels at customer $i \in \mathcal{N}'$ at period time $t \in \mathcal{T}$. Constraints (23) and (24) ensure that the inventory level must be fulfill the lower and upper bounds, respectively. Finally, the type of variable used are presented in constraints (25), (26) and (27).

4.2 Multi-objective Algorithm

For solving the problem, we select the Non Inferior Set Estimation (NISE) algorithm. Cohon et al. [28] developed this algorithm in 1978. This algorithm has been used in several multi-objective problems like regional forest planning, game theory but it has not been applied in vehicle routing and its extensions. The mainly idea of the algorithm is presented in Table 3.

The algorithm begins with solving the two objectives individually obtaining two no dominated solutions. After that, a weighted problem is solved for find the weights of the new objective functions. Once the new weights are defined a new MIP is solved and

Table 3. NISE algorithm description

Resolve z1 and z2 singly
Calculate uncertainty
n=2
while uncertainty<=max error
if uncertainty<=max error **then** stop
else
find a new point with the adyacent points
if the new point improve a objective function **then**
calculate the new uncertainty and new points
end if
end if
end while

the solution obtained is classified into dominated solution or non-dominated solution and then is calculated the new uncertainty and prove if meet the stop criterion where the uncertainty is the altitude of a triangle containing a possible non dominated solutions.

5   Computational Experiments and Results

For proving the algorithm, we use the instances provided by Coelho et al. [11] and use for tested the mono objective Inventory Routing Problem with several algorithms [2, 12]. In order to made a comparison we will use first the pure NISE algorithm over a number of instances and then we will provide the comparison using the column generation method.

In Fig. 2, an example is presented. By solving the two objectives separately, we obtain the following coordinates:

$$z^1 = (1870.88, 729.88), \tag{28}$$

$$z^2 = (3842.66, 703.66). \tag{29}$$

After obtain the two points we use the algorithm and obtain the Pareto frontier as is presented in the example.

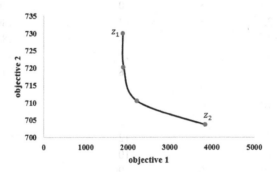

Fig. 2. Example of Pareto frontier

We run the NISE algorithm on five instances with various number of customers. The results are provided in Table 4. The table is organized as follows: the name of the instance is provided in the first column, the second one contain the number of customers while the third contain the number of periods. The third column presents the number of points in the Pareto frontier obtain with the algorithm and in the last column the total amount of time spend in seconds is presented.

As we can see in Table 4, the number of points does not depend of the size of the instance, but the time increases when the number of customers is bigger. This happen because the algorithm requires to solve several times the linear model presented in the previous section and the problem is classified as NP-Hard problem.

Table 4. Results of the algorithm

Instance	# Customers	# periods time	# pareto frontier points	Time (s)
abs1n5	5	3	5	1,9
abs1n10	10	3	6	31243
abs1n15	15	3	9	335
abs1n20	20	3	7	66234
abs1n25	25	3	8	58460

Finally, we present the results of combining column generation with NISE algorithm in Table 5. We can conclude that using column generation approach the computational time can be reduced. In fact, we can solve instances with a mayor number of time periods and customers.

Table 5. Results of column generation with NISE algorithm

Instance	# Customers	# periods time	# pareto frontier points	Time (s)
abs1n5	5	3	5	0,50
abs1n10	10	3	6	30,67
abs1n15	15	3	9	58,25
abs1n20	20	3	7	402,62
abs1n25	25	3	8	918,99
abs1n30	30	3	10	700,89
abs1n35	35	3	12	1869,72
abs1n40	40	3	9	2878,21
abs1n45	45	3	11	4569,99
abs1n50	50	3	8	7429,43
abs1n5	5	6	8	1,74
abs1n10	10	6	9	20,30
abs1n15	15	6	5	40,04
abs1n20	20	6	12	1240,66
abs1n25	25	6	10	4566,54
abs1n30	30	6	9	9432,57

6   Conclusions

In this paper, we have presented a green Inventory Routing Problem with two objective functions related with the monetary costs and the emissions. This the first addressed problem related in the literature considering the emissions produced by transportation and inventory, only it has been approximated as the monetary cost produce by both emissions.

For solving the multi-objective green IRP, we have used the NISE algorithm embedded with column generation for reduce the computational time. Using a reformulation of the mathematical model, we can use column generation and generate attractive routes that contribute to improve the objective function. Our results show that the number of points does not depend of the size of the instance, but the computational time increases when the number of customers is bigger, since it requires solving several times a linear model.

Future work should focus on the extension of the problem, for example consider the locations decisions or several sources of the emissions. In addition, for larger instances and real-world problems, a metaheuristic solution could be developed to solve the problem within a reasonable execution time. Another opportunity to improve the execution time lies in the development of a parallelizable algorithm, which could be distributed among several computers (or multiple cores) thus reducing the computation time.

Acknowledgments. We thank Fair Isaac Corporation (FICO) for providing us with Xpress-MP licenses under the Academic Partner Program subscribed with Universidad Distrital Francisco Jose de Caldas (Colombia). Last, but not least, the authors would like to thank the comments of the anonymous referees that significantly improved our paper.

References

1. Federgruen, A., Zipkin, P.: A combined vehicle routing and inventory allocation problem. Oper. Res. **32**, 1019–1037 (1984)
2. Archetti, C., Bertazzi, L., Laporte, G., Speranza, M.G.: A branch-and-cut algorithm for a vendor-managed inventory-routing problem. Transp. Sci. **41**, 382–391 (2007)
3. Wakeland, W., Cholette, S., Venkat, K.: Food transportation issues and reducing carbon footprint. In: Boye, J.I., Arcand, Y. (eds.) Green Technologies in Food Production and Processing, pp. 211–236. Springer, Boston (2012)
4. Bell, W.J., Dalberto, L.M., Fisher, M.L., Greenfield, A.J., Jaikumar, R., Kedia, P., Mack, R.G., Prutzman, P.J.: Improving the distribution of industrial gases with an on-line computerized routing and scheduling optimizer. Interfaces. **13**, 4–23 (1983)
5. Bertazzi, L., Speranza, M.G.: Inventory routing problems: an introduction. EURO J. Transp. Logist. **1**, 307–326 (2012)
6. Savelsbergh, M., Song, J.-H.: An optimization algorithm for the inventory routing problem with continuous moves. Comput. Oper. Res. **35**, 2266–2282 (2008)
7. Hemmelmayr, V.C., Doerner, K.F., Hartl, R.F.: A variable neighborhood search heuristic for periodic routing problems. Eur. J. Oper. Res. **195**, 791–802 (2009)
8. Aghezzaf, E.-H., Raa, B., Van Landeghem, H.: Modeling inventory routing problems in supply chains of high consumption products. Eur. J. Oper. Res. **169**, 1048–1063 (2006)
9. Andersson, H., Hoff, A., Christiansen, M., Hasle, G., Løkketangen, A.: Industrial aspects and literature survey: combined inventory management and routing. Comput. Oper. Res. **37**, 1515–1536 (2010)
10. Dubedout, H., Dejax, P., Neagu, N., Yeung, T.: A GRASP for real life inventory routing problem: application to bulk gas distribution. In: 9th International Conference on Modeling, Optimization & SIMulation, pp. 1–11 (2012)
11. Coelho, L.C., Cordeau, J.-F., Laporte, G.: Consistency in multi-vehicle inventory-routing. Transp. Res. Part C: Emerg. Technol. **24**, 270–287 (2012)
12. Coelho, L.C., Laporte, G.: The exact solution of several classes of inventory-routing problems. Comput. Oper. Res. **40**, 558–565 (2013)

13. Solyalı, O., Süral, H.: A branch-and-cut algorithm using a strong formulation and an a priori tour-based heuristic for an inventory-routing problem. Transp. Sci. **45**, 335–345 (2011)
14. Coelho, L.C., Laporte, G.: Optimal joint replenishment, delivery and inventory management policies for perishable products. Comput. Oper. Res. **47**, 42–52 (2014)
15. Niakan, F., Rahimi, M.: A multi-objective healthcare inventory routing problem; a fuzzy possibilistic approach. Transp. Res. Part E: Logist. Transp. Rev. **80**, 74–94 (2015)
16. Geiger, M.J., Sevaux, M.: The biobjective inventory routing problem – problem solution and decision support. In: Pahl, J., Reiners, T., Voß, S. (eds.) INOC 2011. LNCS, vol. 6701, pp. 365–378. Springer, Heidelberg (2011). doi:10.1007/978-3-642-21527-8_41
17. Nolz, P.C., Absi, N., Feillet, D.: Optimization of infectious medical waste collection using RFID. In: Paias, A., Ruthmair, M., Voß, S. (eds.) ICCL 2016. LNCS, vol. 9855, pp. 86–100. Springer, Heidelberg (2011). doi:10.1007/978-3-642-24264-9_7
18. Nolz, P.C., Absi, N., Feillet, D.: A bi-objective inventory routing problem for sustainable waste management under uncertainty. J. Multi-Criteria Decis. Anal. **21**, 299–314 (2014)
19. Rahimi, M., Baboli, A., Rekik, Y.: A bi-objective inventory routing problem by considering customer satisfaction level in context of perishable product. In: 2014 IEEE Symposium on Computational Intelligence in Production and Logistics Systems (CIPLS), pp. 91–97. IEEE (2014)
20. Lin, C., Choy, K.L., Ho, G.T.S., Chung, S.H., Lam, H.Y.: Survey of green vehicle routing problem: past and future trends. Expert Syst. Appl. **41**, 1118–1138 (2014)
21. Jabir, E., Panicker, V.V., Sridharan, R.: Multi-objective optimization model for a green vehicle routing problem. Procedia – Soc. Behav. Sci. **189**, 33–39 (2015)
22. Al Dhaheri, N., Diabat, A.: A mathematical programming approach to reducing carbon dioxide emissions in the petroleum refining industry. In: 2010 Second International Conference on Engineering Systems Management and Its Applications (ICESMA), pp. 1–5. IEEE (2010)
23. Alkawaleet, N., Hsieh, Y.-F., Wang, Y.: Inventory routing problem with CO2 emissions consideration. In: Golinska, P. (ed.) Logistics Operations, Supply Chain Management and Sustainability, pp. 611–619. Springer International Publishing, Cham (2014)
24. Mirzapour Al-e-hashem, S.M.J., Rekik, Y.: Multi-product multi-period Inventory Routing Problem with a transshipment option: a green approach. Int. J. Prod. Econ. **157**, 80–88 (2014)
25. Malekly, H.: The inventory pollution-routing problem under uncertainty. In: Fahimnia, B., Bell, M.G.H., Hensher, D.A., Sarkis, J. (eds.) Green Logistics and Transportation, pp. 83–117. Springer International Publishing, Cham (2015)
26. Treitl, S., Nolz, P.C., Jammernegg, W.: Incorporating environmental aspects in an inventory routing problem. A case study from the petrochemical industry. Flex. Serv. Manuf. J. **26**, 143–169 (2014)
27. Toth, P., Vigo, D.: The Vehicle Routing Problem. SIAM, Philadelphia (2002)
28. Cohon, J.L.: Multiobjective Programming and Planning. Academic Press, New York (1978)

Natural Language Processing

Enhancing Semi-supevised Text Classification Using Document Summaries

Esaú Villatoro-Tello[2(✉)], Emmanuel Anguiano[1], Manuel Montes-y-Gómez[1], Luis Villaseñor-Pineda[1], and Gabriela Ramírez-de-la-Rosa[2]

[1] Language Technologies Laboratory, Computational Sciences Department, Instituto Nacional de Astrofísica Óptica y Electrónica (INAOE), Puebla, Mexico
{eanguiano,mmontesg,villasen}@ccc.inaoep.mx
[2] Language and Reasoning Research Group, Information Technologies Department, Universidad Autónoma Metropolitana (UAM) Unidad Cuajimalpa, Mexico City, Mexico
{evillatoro,gramirez}@correo.cua.uam.mx

Abstract. The vast amount of electronic documents available on the Internet demands for automatic tools that help people finding, organizing and easily accessing to all this information. Although current text classification methods have alleviated some of the above problems, such strategies depend on having a large and reliable set of labeled data. In order to overcome such limitation, this work proposes an alternative approach for *semi-supervised* text classification, which is based on a new strategy for diminishing the sensitivity to the noise contained on labeled data by means of automatic *text summarization*. Experimental results showed that our proposed approach outperforms traditional semi-supervised text classification techniques; additionally, our results also indicate that our approach is suitable for learning from only one labeled example per category.

Keywords: Text classification · Text summarization · Semi-supervised learning · Self-training · Feature selection

1 Introduction

Nowadays there are millions of digital texts available on the Internet, which rapidly increase. This situation has produced a growing need for tools that help users to find, organize, and analyze all these resources in short periods of time. Particularly, Text Classification (TC) [1], *i.e.* the task of automatic assignment of free text documents to one or more predefined categories or topics, has emerged as a very important component in many information management tasks.

Traditionally, TC is approached by means of supervised techniques, *i.e.*, automatically constructing a classifier from a large set of labeled (*i.e.*, already categorized) documents. In addition, a common practice considered in the TC pipeline is applying feature selection methods, which use statistical information from the

© Springer International Publishing AG 2016
M. Montes-y-Gómez et al. (Eds.): IBERAMIA 2016, LNAI 10022, pp. 115–126, 2016.
DOI: 10.1007/978-3-319-47955-2_10

training set in order to identify those attributes that better describe the documents among different categories [2]. However, under this paradigm, a major problem is the high cost involved in collecting enough labeled data for building an effective classification model as well as for performing an appropriate feature selection. For example, in classification tasks such as deceptive opinion identification [3] and author profiling [4] it is very difficult to effectively collect or even validate enough labeled data.

In order to overcome the above mentioned problems, semi-supervised learning techniques have gained popularity among the scientific community. For instance, the *self-training* algorithm represents a learning strategy that aims at building an accurate classification model through iteratively increasing the training set by means of labeling and selecting the most confident recently classified instances from the unlabeled data. Some research works have demonstrated the pertinence of the *self-training* method in some TC tasks [5–8]. However, some of the difficulties faced by these approaches is the sensitivity of the algorithm to the noise contained in the labeled data as well as how to effectively select useful new labeled instances in every iteration.

In this paper we propose a modification to the self-training approach for diminishing the sensitivity of the learning algorithm to the noise contained on the labeled data, which consists in using document summaries. This idea was motivated by the work described in [9,10], where the benefits of using document summaries in a supervised TC approach are described. Contrastingly, we evaluate the pertinence of text summaries in a semi-supervised TC scenario, *i.e.,* when very few labeled data are available for building the classification model. As an additional contribution, we propose a new strategy for performing the instance selection process within the self-training algorithm, which helps in the process of preserving high homogeneity values among classes. To sum up, we aim to determine the usefulness of summarization as a noise filtering technique for improving *semi-supervised* text classification. Our main hypothesis establishes that if we incorporate a highly-confident summary, instead of a full document, to the set of labeled data during the training process of a TC model, we will be able to avoid dramatic changes on the behavior of our self-training algorithm, leading to a correct learning of the true target function.

Particularly, we are interested in evaluating the proposed method under exceptional circumstances, *i.e.,* having only one labeled document. Under this scenario, known as *one-shot learning* in the pattern recognition field [11]; our results indicate that having just one labeled summary per category provides enough information to our text classification system, outperforming the traditional semi-supervised configuration (*i.e.,* using full documents). Furthermore, obtained results also demonstrate that our proposed method for selecting highly-confident labeled documents tends to get better performance across iterations when short summaries are added to the set of labeled data rather than full documents.

The rest of this document is organized as follows. Section 2 presents some related work concerning the use of *text summarization* in the task of TC.

Section 3 describes our proposed method; particularly it details the automatic *text summarization* process as well as our proposed self-training method. Then, Sect. 4 describes the used datasets, the experimental configuration and shows the results achieved by the proposed approach as well as some baseline results corresponding to the application of traditional text classification techniques under a semi-supervised paradigm. Finally, Sect. 5 depicts our conclusions and some future work ideas.

2   Related Work

Although there are several works that proposed solving the problem of TC using different strategies of Text Summarization (TS), to the best of our knowledge there is no prior work that has explored the importance of using automatic generated summaries as noise filtering strategies under a *semi-supervised* text classification configuration, *i.e.*, having very few labeled data for training a classifier.

Ideally, a summary contains only the most relevant information from a document and given that such summary represents a significantly shorter document than the original, there are several approaches that prefer using summaries instead of full documents for improving the performance of a supervised TC system. For example in [12], authors proposed a new form of weighting terms by taking into account their frequency and their position within documents; whereas in [13], it is considered a weighting scheme that rewards terms from those phrases selected by a summarization method. Similar methods are described in [14], where relevant sentences are used for training a supervised classifier.

Works described in [10,15] explicitly proposed using summaries as a feature selection strategy. Authors applied different summarization techniques and compared their achieved TC results against those obtained when a statistical feature selection techniques are employed, *e.g.* information gain. Both papers conclude that using document summaries as a feature selection method, represents a competitive strategy against traditional statistical techniques. As a consequence of summarizing documents prior to a TC process, the dimensionality required for representing such documents is considerably reduced. The advantages obtained from the dimensionality reduction has been widely discussed in [9,13].

As we mentioned before, to the best of our knowledge there is no prior work related to the use of TS under a semi-supervised paradigm, hence we describe only those that we consider are the more related to the proposed approach. In [5], authors introduce a self-trainig algorithm that employs an ensemble of classifiers, namely *Ordered Classification*. This method allows selecting highly confident documents (in every iteration) to be included in the training data. In [7] authors describe a method for discovering the constant common knowledge in both, training and test sets by means of semi-supervised strategies. One of the closest works is the proposed by [6] which proposes a semi-supervised method for TC, which considers the extraction of unlabeled examples from the Web and the

application of a enriched self-training approach for the construction of the classification model. And finally, in [8] authors propose a novel self-training approach
for sentiment classification, their method uses multiple feature subspace-based
classifiers for exploring a set of good features for better classification decision
and to select the informative samples for automatically labeling data. Although
these are related works, all of them employ complete documents during training
and classification stages. In this work, we incorporate the advantages of text
summaries as a noise filtering strategy for improving semi-supervised text classification. In addition, we propose a novel strategy for selecting high-confident
instances during self-training, which allows to obtain high homogeneity among
classes. Following sections describe into detail the proposed approach.

3   Proposed Method

Our proposed method represents a modification of the traditional self-training
algorithm. This algorithm assumes that, at the beginning, there are very few
labeled data (D_L) and a very large set of unlabeled data (D_U). The goal is to
obtain and use relevant information, extracted from D_U in order to improve the
initial classifier (Φ_0) which was trained over D_L. To obtain such information,
Φ_0 is used to classify the elements of D_U, then, by means of a specific selection
criteria, some elements of D_U are considered for augmenting the set D_L. Once
D_L has been updated, a new training process is performed to construct the
classifier Φ_1. As expected, self-training algorithm represents an iterative process
that is repeated until some stop criteria is reached.

A general view of the proposed self-training method is shown in Fig. 1. Generally speaking, our proposed method starts by automatically constructing summaries from the set of labeled data D_L (*Text Summarization* module). Then,
while some stop criteria is not achieved, such summaries are employed to construct a classification model Φ_0 (*Classifier Construction* module), which is used
to classify all unlabeled data D_U. Next, we evaluate and preserve, through our
Instance Selection module, those documents that represent the most confident
labeled instances. The exact same number of documents for each category c_j
are preserved, and selected documents are removed from D_U. The following step
consists in creating their respective summaries and incorporating them into the
original training data set. Then, we retrain our classifier to create the Φ_i classification model.

The *instance selection* module represents an important contribution since
contrary to the traditional self-training algorithm, we perform this step separately from the classification stage *i.e.,* the classifier's confidence is not considered for this process. By means of this we favour high homogeneity among
classes and avoid the bias introduced by a classifier trained with very few labeled
data. Following sections describe in detail each one of the main modules from
our proposed method (see Fig. 1).

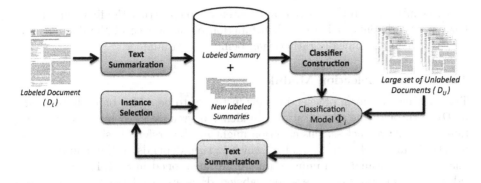

Fig. 1. General architecture of the proposed self-training text classification method.

3.1   Text Summarization Module

For the implementation of the TS module, we used an unsupervised strategy that has demonstrated being able to construct high quality summaries, particularly we employed a graph based technique as explained in [9].

The underlying idea of this method is that every sentence can be represented as the vertex of some graph. By means of similarity measures it is possible to assign an specific rank to each sentence (*i.e.*, an importance value to each element on the graph). Finally, assuming an ideal ranking, this strategy preserve the top n better ranked sentences to construct the final summary.

Similarly to [9], we employed a measure of ranking, originally proposed for Web pages which is called *Hyperlinked Induced Topic Search* [16] to determine the associated value of ranking for each sentence from a document. First, the similarity of each sentence against the rest is computed in order to assign a weight value to all edges[1]. Once all edges have their corresponding weight, two different coefficients are iteratively computed as follows:

$$HITS_A(V_i) = \sum_{V_j \in In(V_i)} w_{ji} HITS_H(V_j) \tag{1}$$

$$HITS_H(V_i) = \sum_{V_j \in Out(V_i)} w_{ij} HITS_A(V_j) \tag{2}$$

where V_i represent a vertex on the graph $G = (V, E)$; $In(V_i)$ represent the number of incoming links to V_i, whereas $Out(Vi)$ represent the out-coming links from vertex V_i. Consequently, $HITS_A$ represent the "authority value" (vertex with a large number of incoming links), while $HITS_H$ the "hub value" (vertex with a large number of outgoing links).

Once the HITS value of each vertex has been computed, the assigned value represents the importance of each sentence within the document. From here we

---

[1] Normally the direction of the edges is determined by the order of the sentences in the original document.

preserve only the top n most relevant sentences to construct the final summary. For our experiments, n is defined in function of the length of the document[2], *i.e.*, is defined dynamically rather than being a fixed number.

3.2 Instance Selection Module

The most natural form of selecting new instances to be added on each iteration to D_L is through the classification confidence degree assigned by the classification model. Nevertheless, this criterion might not be preferable since it depends directly from the classifier's quality, which in turn depends on the quantity and quality of the available training data. As we mentioned in Sect. 1, we are considering a scenario where very few labeled data are available (*e.g.*, one-shot learning), thus, traditional instance selection criteria are not suitable for this type of situations[3].

To overcome the above problem, we propose a novel method for assigning a confidence value to recently classified documents, which is independent from the employed learning algorithm as well as independent from the quantity and the quality of the labeled data. Our proposed instance selection criteria allows to preserve high homogeneity values among classes, and represents a distance based approach, which is computed as follows:

$$dist(d_U, C_j) = \frac{\sum_{d_L \in C_j} dist(d_U, d_L)}{|C_j|} \tag{3}$$

where d_L represents the summary of a labeled document such that $d_L \in D_L$. Hence, after applying Formula 3, those documents from D_U that were assigned to the label c_j and with the minor average distance to the corresponding class C_j are considered as highly confident.

During our experimental phase, we defined $dist(d_U, C_j)$ by means of an euclidean distance. Then, we preserve the top k documents most similar to the class c_j to include them into the labeled set of summaries.

3.3 Classification Module

As explained before, determining the confidence degree of the classified instances does not depend on any particular algorithm, thus, any learning algorithm can be employed in the classification module.

In particular, we use the Support Vector Machine (SVM) method given that is especially suited to work with datasets with high dimensionality. For performing our experiments we employed the SVM implementation included in the Weka[4] toolkit with the default parameters.

---

[2] The parameter that defines the length of a summary is also known as the *compression rate parameter*, and represents a number that indicates the percentage of the information that we are requiring to preserve from the original document.

[3] One disadvantage of self-training is that mistakes reinforce/strengthen themselves; it is well known that accuracies lower than random at the beginning tend to conduct to worst results in subsequent iterations.

[4] http://www.cs.waikato.ac.nz/ml/weka/.

4 Experimental Results

4.1 Data Set

For validating our hypothesis, we performed experiments with the R8 data set. This collection is formed by the eight largest categories from the Reuters-21578 corpus, which documents belong to only one class. In order to know a more detailed description of this data set refer to [17]. Table 1 shows some basic statistics (*e.g.*, number of training/test documents, vocabulary size, etc.) from the employed data set. As can be seen, it is a highly unbalanced data set.

Table 1. Statistics from the employed document collection: number of documents per category, average documents size (in tokens) and average vocabulary size

Categories names	Training documents			Test documents		
	Num. docs	Docs size	Vocab. size	Num. docs	Docs size	Vocab. size
Earn	2701	49.99	31.00	1040	45.01	26.42
Acq	1515	74.74	50.36	661	71.25	48.16
Trade	241	121.78	81.46	72	125.12	83.01
Crude	231	110.57	71.87	112	102.21	66.66
Money-fx	191	97.46	65.14	76	95.55	65.92
Interest	171	88.43	57.35	73	91.43	59.04
Ship	98	82.56	59.01	32	79.25	57.71
Grain	41	115.89	75.40	9	85.66	50.33
Total:8	*5189*	*67.36*	*43.80*	*2075*	*63.42*	*40.66*

For performing our experiments we randomly select *one* document from the training set as the labeled training document, *i.e.*, our algorithm always begin iterating with only one single document and the rest are considered as the unlabeled data. It is important to mention that our method does not require knowing if the dataset is unbalanced or not, since given the nature of the proposed algorithm, this will converge to an accurate classification model.

4.2 Method Configuration

As mentioned in Sect. 3, our method requires an user defined parameter k, which represents the number of documents to be added for each category into the labeled data. For our experimental phase we considered adding $k = 1$ and $k = 5$ documents in every iteration. It is worth to remember that at the beginning, the self-training algorithm starts by having only one labeled document, replicating a *one-shot learning* scenario. Our reported experiments represent the average performance obtained when randomly varying five times the initial labeled document. As stop-criteria we defined the following: *(i)* when a top of 20 iterations is reached, and *(ii)* when there are no more unlabeled documents in D_U.

Finally, it is worth mentioning that for all the experiments, documents were represented following the traditional *vector model* from the Information Retrieval field, specifically a *tf* (term-frequency) weighting scheme.

4.3 Baseline Definition

Since we aim at demonstrating that using summaries allows to improve the performance of the proposed self-training method, we defined as our *baseline* approach the exact same algorithm but using complete documents instead of summaries. In other words, this baseline depicts the traditional self-training algorithm for text classification without summarizing the considered documents. Similarly, documents were represented by means of a vector model using a *tf* weighting scheme.

As an additional baseline, we consider the result obtained by the classifier when training only with a single labeled data. In our experiment this result corresponds to iteration number 0. This baseline aims at demonstrating that using summaries for training a TC model allows to obtain, from the beginning, better classification results.

4.4 Evaluation Metrics

The effectiveness of the proposed method was measured by means of the macro-averaged F_1 evaluation measure. Using this type of measure is very useful since it allows obtaining a confident perspective of the system's performance, particularly for cases where classes are highly unbalanced.

4.5 Results

Figure 2 shows the obtained results for the performed experiments having different compression rate summaries (30 %, 50 % and 70 %). The graph on the left represents the behaviour of the proposed method when one document ($k = 1$) is added in every iteration, whereas the graph on the right depicts the results when five documents ($k = 5$) are added in every iteration. As we mentioned, all the experiments started with only one document per class on the D_L set.

It is important to remark that our proposal of using summaries for training rather than full documents allows to improve the performance of the classification system from iteration 0, *i.e.*, when our proposed method still has not started iterating yet. Notice that for the baseline configuration, *i.e.*, when training with a full document the $F_1 = .34$ at iteration 0, and for the same case we get a $F_1 = .43$ when using a summary of 30 % compression rate, which confirms the ability of *text summarization* as feature selection method.

These results also indicate that the baseline configuration hardly improves its performance even when more documents are added to the training data. Particularly, it goes from $F_1 = .34$ on iteration 0 to $F_1 = .38$ on iteration 20 when $k = 1$, and from $F_1 = .34$ on iteration 0 to $F_1 = .37$ on iteration 20 when

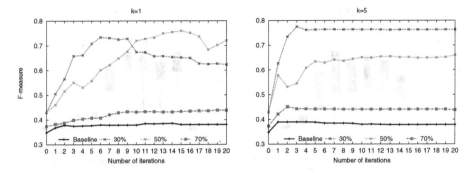

Fig. 2. Obtained results on the R8 collection. On the left 1 document is added every iteration ($k = 1$), whilst on the right 5 documents are added each iteration ($k = 5$).

$k = 5$. On the contrary, our proposed method obtains a significant improvement as more iterations are performed. Particularly, when we add only 1 document per iteration per category (Fig. 2, $k = 1$), and using summaries that represent only 30 % of the size of the original document, our system goes from a $F_1 = .43$ at iteration 0 to $F_1 = .65$ at iteration 3. A similar situation occurs for the case of adding 5 documents on each iteration (Fig. 2, $k = 5$), where using summaries of 30 % size allow to our method to go from $F_1 = .43$ at iteration 0 to $F_1 = .73$ at iteration 3. This behavior indicates that our *instance selection* criteria, in combination with the use of summaries allows to maintain high homogeneity among classes, reduces the noise contained in labeled documents and converges to the true target function.

Finally, in order to validate the importance of the compression rate parameter from the *text summarization* module, we performed a series of experiments varying the size of the produced summary across the 20 iterations that are executed by the self-training algorithm. Figure 3 show the statistical variance of the F-score for $k = 1$ and $k = 5$, respectively.

In general, we can observe that the compression rate is an important parameter of the proposed method, however, it is clear that using summaries is in fact a better strategy than using full documents (*i.e.*, the baseline configuration). In addition, it is possible to notice that using summaries of 30 %–50 % compression rate allow obtaining the best results.

4.6   Discussion

What to summarize: With the intention of evaluating the pertinence of using summaries in both training and test phases, we carried out some experiments considering a supervised classification scenario. Table 2 shows the results from these experiments. The conclusion is clear: summarizing both training and test documents (*Sum-Sum*) worsen the classifier performance; on the contrary, the best performance was obtained when only training documents were summarized

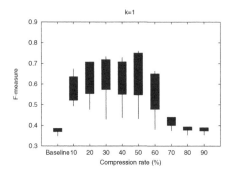 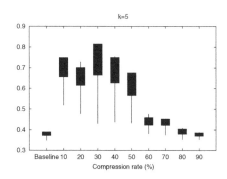

Fig. 3. Variance of the F-score performance for the proposed method when $k = 1$ and $k = 5$. Reported results are for 20 iterations at different compression rate values (10 %–90 %); first column depicts the baseline performance

(*Sum-Doc*). These results supported the design of the proposed system architecture, where summaries were employed only on the training (see Fig. 1).

Table 2. Results of a supervised TC method varying the TS module location. Notice that when no summarization is applied neither to training or test documents, the obtained performance is $F = 0.84$

Compression rate	Method configuration		
	Sum-sum	Doc-sum	Sum-doc
30 %	0.87	0.77	**0.89**
50 %	0.86	0.82	**0.87**
70 %	0.86	0.83	**0.86**

Some Learned Lessons: Despite obtained results, experiments indicate that the success of the proposed method depends to some extent on the following factors: (*i*) the quality of the initial labeled data, *i.e.*, the initial labeled documents must be representative of their categories; (*ii*) the number of unlabeled data: for those categories having a small number of examples in the unlabeled set, the selection of confident instances at every iteration becomes into a very hard task; (*iii*) a good selection of the compression rate parameter: our method showed outstanding results in news classification when using compression rates from 10 to 60 %, however, for noisier documents, such as social media texts, the compression rate definition could be complex and critical.

5   Conclusions

In this paper we have proposed a modification to the *self-training* algorithm for improving text classification when very few labeled documents are available. Our proposal considers using text summaries instead of full documents as a strategy for diminishing the sensitivity of the learning algorithm to the noise contained on the labeled data. Additionally, as a second contribution of this work, we have proposed a novel criterion for selecting highly confident elements to be included in the set of labeled data. The proposed criteria is performed separately from the classification process, which makes it independent from the learning algorithm, allowing to preserve high homogeneity values among labeled documents.

The performed experiments showed that our proposed algorithm is able to incorporate information from the unlabeled data for improving the performance of the classifier. By means of using automatic text summaries we are able to discard noisy information during the *self-training* process. Particularly, experimental results showed that shorter summaries are the best choice (30 % to 50 % compression rate). Further more, we demonstrated that the proposed approach is very suitable for collections with very few labeled data. Particularly, we evaluated our proposed method under a *one-shot learning* scenario, *i.e.*, having only one labeled document. Obtained results are promising and represent an initial effort towards the problem of one-shot text classification.

As future work we are interested in evaluating different unsupervised summarization techniques [18], aiming at determining the sensitivity of the proposed method towards the TS module. We are also interested in evaluating the performance of the proposed method using other semi-supervised strategies, *e.g.*, co-training and multi-view approaches, as well as in other larger datasets. Additionally, we intent to determine the pertinence of the proposed algorithm for solving non-thematic text classification tasks, such as author profiling problems (*e.g.*, age, gender, and personality recognition), where not enough/reliable labeled data are available.

Acknowledgments. This work was partially funded by CONACyT, project number 247870 and 258588. We appreciate the support provided by the Thematic Networks program (Language Technologies Thematic Network projects 260178 and 271622). We thank to UAM Cuajimalpa and SNI for their support.

References

1. Sebastiani, F.: Machine learning in automated text categorization. ACM Comput. Surv. (CSUR) **34**(1), 1–47 (2002)
2. Villuendas-Rey, Y., Garcia-Lorenzo, M.M.: Attribute and case selection for nn classifier through rough sets and naturally inspired algorithms. Computación y Sistemas **18**(2), 295–311 (2014)
3. Fusilier, D.H., Montes-y-Gómez, M., Rosso, P., Cabrera, R.G.: Detecting positive and negative deceptive opinions using PU-learning. Inf. Process. Manag. **51**(4), 433–443 (2015)
4. López-Monroy, A.P., Montes-y-Gómez, M., Escalante, H.J., Villaseñor-Pineda, L., Stamatatos, E.: Discriminative subprofile-specific representations for author profiling in social media. Knowl.-Based Syst. **89**, 134–147 (2015)
5. Solorio, T.: Using unlabeled data to improve classifier accuracy. M. Sc. Degree thesis, Computer Science Department, Inaoe, Mexico (2002)

6. Guzmán-Cabrera, R., Montes-y-Gómez, M., Rosso, P., Villaseñor-Pineda, L.: Using the web as corpus for self-training text categorization. Inf. Retrieval **12**(3), 400–415 (2009)

7. Zheng, Y., Teng, S., Liu, Z., Sun, M.: Text classification based on transfer learning and self-training. In: 2008 Fourth International Conference on Natural Computation, vol. 3, pp. 363–367, October 2008

8. Gao, W., Li, S., Xue, Y., Wang, M., Zhou, G.: Semi-supervised sentiment classification with self-training on feature subspaces. In: Su, X., He, T. (eds.) CLSW 2014. LNCS, vol. 8922, pp. 231–239. Springer, Heidelberg (2014)

9. Mihalcea, R., Hassan, S.: Using the essence of texts to improve document classification. In: Proceedings of the Recent Advances in Natural Language Processing (RANLP-2005) (2005)

10. Anguiano-Hernández, E., Villaseñor-Pineda, L., Montes-y-Gómez, M., Rosso, P.: Summarization as feature selection for document categorization on small datasets. In: Loftsson, H., Rögnvaldsson, E., Helgadóttir, S. (eds.) IceTAL 2010. LNCS, vol. 6233, pp. 39–44. Springer, Heidelberg (2010)

11. Fei-Fei, L., Fergus, R., Perona, P.: One-shot learning of object categories. IEEE Trans. Pattern Anal. Mach. Intell. **28**, 594–611 (2006)

12. Ker, S.J., Chen, J.-N.: A text categorization based on summarization technique. In: Proceedings of the ACL-2000 Workshop on Recent Advances in Natural Language Processing and Information Retrieval: Held in Conjunction with the 38th Annual Meeting of the Association for Computational Linguistics, vol. 11, pp. 79–83. Association for Computational Linguistics (2000)

13. Ko, Y., Park, J., Seo, J.: Automatic text categorization using the importance of sentences. In: Proceedings of the 19th International Conference on Computational linguistics, vol. 1, pp. 1–7. Association for Computational Linguistics (2002)

14. Xiao-Yu, J., Xiao-Zhong, F., Zhi-Fei, W., Ke-Liang, J.: Improving the performance of text categorization using automatic summarization. In: International Conference on Computer Modeling and Simulation, ICCMS 2009, pp. 347–351. IEEE (2009)

15. Kolcz, A., Prabakarmurthi, V., Kalita, J.: Summarization as feature selection for text categorization. In: Proceedings of the Tenth International Conference on Information and Knowledge Management, pp. 365–370. ACM (2001)

16. Kleinberg, J.M.: Authoritative sources in a hyperlinked environment. J. ACM (JACM) **46**(5), 604–632 (1999)

17. Cachopo, A.M.D.J.C.: Improving methods for single-label text categorization. Ph.D. thesis, Universidade Técnica de Lisboa (2007)

18. Litvak, M., Vanetik, N.: Multi-document summarization using tensor decomposition. Computación y Sistemas **18**(3), 581–589 (2014)

A Comparison Between Two Spanish Sentiment Lexicons in the Twitter Sentiment Analysis Task

Omar Juárez Gambino[1,2][✉] and Hiram Calvo[2]

[1] Instituto Politécnico Nacional, ESCOM,
Av. Juan de Dios Bátiz esq. Av. Miguel Othón de Mendizábal,
Col. Lindavista, Del. Gustavo A. Madero, 07738 Mexico City, Mexico
b150697@sagitario.cic.ipn.mx
[2] Instituto Politécnico Nacional, CIC,
Av. Juan de Dios Bátiz esq. Miguel Othón de Mendizábal,
Col. Nueva Industrial Vallejo, Del. Gustavo A. Madero,
07738 Mexico City, Mexico
hcalvo@cic.ipn.mx

Abstract. Sentiment analysis aims to determine people's opinions towards certain entities (e.g., products, movies, people, etc.). In this paper we describe experiments performed to determine sentiment polarity on tweets of the Spanish corpus used in the TASS workshop. We explore the use of two Spanish sentiment lexicons to find out the effect of these resources in the Twitter sentiment analysis task. Rule based and supervised classification methods were implemented and several variations over those approaches were performed. The results show that the information of both lexicons improve the accuracy when is provided as a feature to a Naïve Bayes classifier. Despite the simplicity of the proposed strategy, the supervised approach obtained better results than several participant teams of the TASS workshop and even the rule based approach overpass the accuracy of one team which used a supervised algorithm.

1 Introduction

Sentiment analysis can be tackled as a classification problem, where the classes are the polarity of the expressed opinions (i.e., positive or negative opinion). There are usually two approaches for sentiment analysis classification, the supervised and the unsupervised approach.

The unsupervised approach tries to determine the polarity of the sentiments without using prior knowledge of the data. Such methods are usually based on lexical resources like sentiment lexicons, which are a list of words with a sentimental attachment. In [1] dictionaries of words annotated with their semantic orientation or polarity were created and used for classifying the polarity of different user's reviews. Every word of the reviews was compared to the words on the dictionaries, in order to find a match; if they matched, the polarity of the words was used to determine the global sentiment polarity of the review.

© Springer International Publishing AG 2016
M. Montes-y-Gómez et al. (Eds.): IBERAMIA 2016, LNAI 10022, pp. 127–138, 2016.
DOI: 10.1007/978-3-319-47955-2_11

For English, numerous lexicons have been created over the years, for instance: SentiWordnet [2], the Harvard inquirer [3] and LIWC [4]. Many English lexicons have been translated to Spanish and have been used for sentiment analysis in Spanish.

On the other hand, most of the supervised approaches for sentiment analysis have used machine learning algorithms to train with data examples and then apply the learned model to unseen data. In [5] Naïve Bayes, Maximum Entropy and Support Vector Machine algorithms were used to classify sentiment polarity on a movie reviews English corpus. Even though the experiments obtained 82 % of accuracy, the authors point out that the applied algorithms were not able to achieve accuracies comparable to those reported for standard topic-based categorization, concluding that sentiment analysis is a more difficult problem than text categorization. There have been also some efforts to apply supervised methods to classify sentiment polarity for Spanish texts [6,7].

One of the most remarkable efforts for Spanish sentiment analysis is done in TASS (sentiment analysis workshop) organized by SEPLN (Spanish society for natural language processing). This workshop has created a general Corpus of Twitter posts [8], known as tweets, annotated with their global sentiment polarity. The workshop has been celebrated annually since 2012 and recurrently has included a task for determining the goblal polarity of every tweet in the corpus. The participants have tried different approaches for the task, from unsupervised [9,10] and supervised methods [11,12] to assembled systems [13].

In this paper we describe several experiments for classifying sentiment polarity on tweets performed on the Spanish corpus used in TASS workshop, in order to find out the contribution of the two selected lexicons. In the following sections we describe the resources we used (Sect. 2); the experiments performed and the results obtained (Sect. 3); and finally our conclusions and future work (Sect. 4).

2   Resources

The sentiment analysis task has attracted the scientific community attention during the last years. Thanks to the social networks, a lot of opinions are publicly available. Twitter is a social network in which users post messages (tweets) with a maximum length of 140 characters. Due to the importance of this social network and the difficulty to automatically determine the associated sentiment polarity of the tweets, we are interested in exploring different linguistic resources that could help in this task for Spanish.

Workshop
First of all, we have selected the general corpus used in TASS workshop. This corpus contains 68,017 tweets written in Spanish, posted between November 2011 and March 2012. The corpus is divided into a training set of 7,219 tweets (10 %) and a test set of 60,798 tweets (90 %). Every tweet is annotated with its global sentiment polarity at five levels: positive (P), strong positive (P+), negative (N), strong negative (N+), neutral (NEU) and an extra tag (NONE) for those tweets with no sentiment at all. Because we want to classify the tweets

according to their polarity instead of their intensity, we consider only 3 levels and NONE. The tweets from classes P and P+ are joined in a new positive class, and the same treatment is done to the tweets from classes N and N+, which are joined in a new negative class. Besides the difficulty related to the short length of the tweets, the corpus has the problem of being unbalanced. In Table 1 we show the frequency and the total number of tweets for every class in the training and testing sets.

Table 1. Frequency and tweets of the general corpus.

Sentiment	Train		Test	
	Frequency	Tweets	Frequency	Tweets
Positive	39.94 %	2,884	36.57 %	22,233
Negative	30.22 %	2,182	26.06 %	15,844
None	20.54 %	1,483	35.22 %	21,416
Neutral	9.28 %	670	2.15 %	1,305

Lexicons

Sentiment lexicons are created for determining the attachment of a word to a sentiment or sentiments. For our research we have selected the Spanish Emotion Lexicon (SEL) [14] and LIWC [15] with a special list of Spanish words. SEL has already been used in [16] for Twitter sentiment analysis, while to our knowledge LIWC has been used for opinion mining in Spanish in [17], but not for Twitter sentiment analysis. Performance comparison with other lexicons such as those translated from other languages [2], [3] has been left as future research.

SEL is composed by 2,036 words. For every word SEL includes the probability factor for affective use (FPA for its acronym in Spanish). This value indicates how often a word is used to express some of the six different sentiments considered in the lexicon. Some words can be used to express more than one sentiment. In Table 2 we show the total number of words classified into every sentiment. Because this lexicon does not include an explicit reference to positive or negative sentiments, we have considered the words classified into the Joy and Surprise sentiments as positives, and the words classified into the rest of the sentiments as negatives.

Table 2. Sentiments and words of SEL.

Sentiment	Joy	Surprise	Anger	Fear	Disgust	Sadness
Words	668	175	382	211	209	391

LIWC is a lexicon composed by 12,656 words and stems. Words and stems are classified into four groups and every group is composed by several categories

Table 3. Groups and some categories of LIWC.

Group	Categories
Standard linguistic dimension	Personal pronouns, impersonal pronouns, articles, verbs, adverbs, prepositions, conjunctions, negations, quantifiers, numbers
Psychological process	Social process, **affective process**, cognitive process, perceptual process, biological process, relativity
Personal concerns	Work, achievement, leisure, home, money, religion, death
Spoken categories	Assent, nonfluencies, fillers

(a total of 464), some of them are shown in Table 3. This lexicon has a special category (marked in bold face) which indicates whether a word is used for a positive or negative sentiment.

3   Method

In order to measure the contribution of different lexicons and text character-istics, several experiments were ran. In this section we describe the performed experiments and the obtained results.

Rule based approach
Two experiments were carried out using the lexicons described in the previous section, following a rule based approach.

The algorithm to determine the sentiment polarity with the SEL lexicon uses the next steps for every tweet:

1. Tokenize
2. Lemmatize[1]
3. For every word in the tweet:
    (a) Compare the word with the words in the lexicon
    (b) Get the FPA value and the related sentiment (positive or negative) of the matched word in the lexicon
    (c) Accumulate the FPA value into its corresponding sentiment
4. Calculate the difference (df) between the positive and negative sentiment values
5. Get the overall sentiment polarity using the selected threshold shown in Table 7.

The algorithm used with LIWC follows the next steps:

1. Tokenize
2. Compare the word with the words in the lexicon
3. Count the words marked in the lexicon with the affective process category, according to its corresponding sentiment (positive or negative)
4. Calculate the difference (df) between the positive and negative counting

---

[1] All the words in the SEL lexicon are lemmatized.

5. Get the overall sentiment polarity using the selected threshold shown in Table 7

TASS corpus has extra information like date, user and topic. Topics has proved to be useful for polarity determination [18], but we did not take into account because Twitter does not include this information automatically (topics were manually annotated) and date and user seem irrevalevant for our task. For both algorithms the none class is assigned when no match is found between the words of the tweet and the words of the lexicon. For example the tweet *Medir las palabras en 140 caracteres:* http://t.co/s41kO7jt (Measure the words in 140 characters) has not any sentiment related word and therefore not of them would be found in the lexicons.

Supervised approach
We used a Multinomial Naïve Bayes classifier trained with the train set described in Table 1. Different features were used in order to find out the best performance. All the following representations used a lemmatized version of the corpus, except for the LIWC representation. Besides, for the first two representations a vocabulary of the full corpus is calculated.

1. Bag of words (BOW). Every tweet is represented as a vector of lemmatized word frequency.
2. Binarized BOW (BBOW). Instead of the word frequency, the vector contains the values 1 or 0 depending on the existence of the word in the vocabulary.
3. FPA. The FPA values of the six sentiments (see Table 2) are obtained for every word in the tweet, and the FPA values are used in the vector of characteristics.
4. LIWC. The vector contains the frequency of every[2] category (see Table 3) marked in the lexicon for the words in tweet.

The following representations are combinations of the above described. For all the representations, the values are concatenated to each other, resulting in an augmented vector. The plus symbol (+) is used to indicate the concatenation.

5. BBOW + FPA.
6. BBOW + LIWC.
7. BBOW + FPA + LIWC.
8. BBOW + FPA + FPAG + LIWC. The new feature FPAG is obtained by accumulating the FPA values of the six sentiments and grouping them into positive and negative words, as it was done in the rule based approach.

Given the original tweet *Habia prometido responder a todos, pero me ha sido imposible. Y hoy no doy para mas. MUCHAS GRACIAS A TODOS* (sic) (I had promised to respond to everyone, but it has been impossible. And today I cannot give more. THANK YOU VERY MUCH TO ALL) the lemmatization process (using Freeling [19]) generates the following output *habia prometer responder a todo, pero me haber ser imposible. y hoy no dar para mas. mucho gracia a todo*

---

[2] Experimentally better results were achivied when using all the categories instead of using only the affective process category like in the rule based approach.

(i have promise to respond to everyone, but it have be impossible. and today i can not give more. thank you very much to all), after that the above mentioned variations operate over the obtained tokens (20 in the Spanish version) of the lemmatized tweet. In Table 4 we show the generated feature vectors.

Table 4. Variations on feature vector representation.

	BOW features					
	w_1	w_2	w_3	w_4	...	w_n
(1)	1	0	1	2	...	1
(2)	1	0	1	1	...	1
	FPA features					
	joy	anger	fear	disgust	surprise	sadness
(3)	0.597	0	0	0	0	0
	LIWC features					
	c_1	c_2	c_3	c_4	...	c_{464}
(4)	2	8	0	3	...	5
(5)	(2) + (3)					
(6)	(2) + (4)					
(7)	(2) + (3) + (4)					
(8)	(2) + (3) + FPAG + (4)					

4   Results and Discussion

Table 5 shows the accuracy obtained by all the performed experiments over the test set. As can be seen, the accuracy obtained on the rule based approach shows a little improvement when LIWC is used. We consider that this improvement occurs because LIWC lexicon has over 10,000 words more than SEL, so the probability to find a match is higher. In addition to that, LIWC provides a specific indicator for positive or negative words while SEL does not. Nevertheless, the idea of considering the Joy and Surprise categories as positive and the rest as negatives and the use of FPA values gives similar results than LIWC. Table 6 shows the best results obtained by the participants on the TASS 2015 workshop [8]. All systems used a supervised approach, so a fair comparison is not possible. Even though, our rule based algorithms obtained better results than the GAS-UCR team.

Results using the supervised approach were better than those obtained with the rule based experiments even though the simplicity of the features used. The improvement when using LIWC instead of SEL is preserved with this approach. Moreover, the best result is obtained when the information provided by both

Table 5. Experiments on TASS corpus with different lexicons and configurations.

Rule based experiments	Accuracy
SEL	0.529
LIWC	0.531
Supervised experiments	Accuracy
BOW	0.585
BBOW	0.586
FPA	0.425
LIWC	0.516
BBOW + FPA	0.596
BBOW + LIWC	0.599
BBOW + FPA + LIWC	0.605
BBOW + FPA + FPAG +LIWC	0.608

Table 6. Best results on TASS 2015.

Team	Accuracy
LIF	0.726
ELiRF	0.721
GTI-GRAD	0.695
GSI	0.690
LyS	0.664
DLSI	0.655
DT	0.625
SINAI_wd2v	0.619
INGEOTEC	0.613
UCSP	0.613
ITAINNOVA	0.610
Us Supervised	*0.608*
BittenPotato	0.602
CU	0.597
TID-spark	0.594
Us Rule based	*0.531*
GAS-UCR	0.446

lexicons is added to the feature vector. This suggests that each lexicon provides complementary information to the classifier allowing a better accuracy. In Table 6 we compare our results with other TASS workshop participants. Most of the participants used elaborated preprocessing steps [12, 20, 21] like spelling correction, emoticon handling and special treatment for Twitter elements (hashtags, URLs, users). Besides, the classifiers used like SVM and Logistic Regression are more complex and some participants used more than one [13]. Despite using only a lemmatizer during preprocessing and a Multinomial Naïve Bayes classifier with the information provided by both lexicons, the reached accuracy is better than the obtained by four participant teams.

4.1 Analysis of Results

(a) Threshold selection for rule based approach. Several experiments were run to determine suitable threshold values. The first experiment consider an initial range of 0.66 diference value to determine the neutral polarity, the following experiments decrease this range knowing that they are less frequent that other classes. We selected the threshold values of experiment 3 because no improvement was found when decreasing the range. In Table 7 we show the variations over the threshold values.

Table 7. Threshold values variations.

Experiments			
No.	Sentiment polarity	Threshold	Accuracy
1	Negative	$df < -0.33$	0.501
	Neutral	$-0.33 \leq df \leq 0.33$	
	Positive	$df > 0.33$	
2	Negative	$df < -0.23$	0.517
	Neutral	$-0.23 \leq df \leq 0.23$	
	Positive	$df > 0.23$	
3	Negative	$df < -0.1$	0.529
	Neutral	$-0.1 \leq df \leq 0.1$	
	Positive	$df > 0.1$	
4	Negative	$df < -0.05$	0.529
	Neutral	$-0.05 \leq df \leq 0.05$	
	Positive	$df > 0.05$	

(b) Confusion matrix. In order to identify the misclassification problems we show a confusion matrix in Table 8. None class is mostly confused with negative and positive class, while all the actual neutral tweets were misclassified. We consider that the unbalance in corpus and the few examples available for the neutral class generate these errors.

Table 8. Confusion matrix obtained in the experiment with best results in supervised approach.

Actual	Predicted	Total
N	P	1688
N	N	13959
N	NONE	192
N	NEU	5
NEU	N	940
NEU	P	362
NEU	NONE	3
NONE	P	8815
NONE	N	8337
NONE	NONE	4249
NONE	NEU	15
P	P	18756
P	N	2880
P	NONE	594
P	NEU	3

(c) Surprise as negative emotion. SEL lexicon does not include an explicit reference to positive or negative polarity, therefore we grouped the joy and surprise sentiments as positive while the rest of the sentiments listed in Table 2 were considered negative. Nevertheless some words marked in the corpus as surprise can be considered negative, for example *confuse* and *scare*. An additional experiment was run to determine if change of class would improve the performance, obtaining an accuracy 0.526. The result shows a decrease of 0.003 in comparison with the original experiment, leading us to think that the words related to surprise can have both positive and negative polarity.

4.2   Corpus Analysis

The results shown in Table 5 were obtained following the specifications of TASS workshop, which consider 10 % of the corpus for training and the remaining 90 % for testing. In order to explore how the learning rate is affected when the training size is increased, we run the experiments shown in Table 9 using the eighth variation from Table 4. For these experiments, we joined the training and testing corpora in a single new corpus and the tweets were shuffled. After that, we selected a fixed 20 % of the new corpus for testing and the remaining 80 % was left for training purposes. For the first experiment 1/8 part of the training corpus was used, for the second one the training corpus was incremented with other 1/8 (2/8 in total) and so on, until the eighth experiment that covers the whole training corpus. Our results show that for the first fourth experiments the

Table 9. Learning rate and vocabularies intersection.

Vocabulary size (considering types)		
V_{SEL}	V_{LIWC}	V_{test}
1,891	12,552	41,173
Intersections		
$V_{test} \cap V_{SEL}$	$V_{test} \cap V_{LIWC}$	
662	4466	

Experiments					
No.	Training size	Accuracy	V_{train}	$V_{train} \cap V_{SEL}$	$V_{train} \cap V_{LIWC}$
1	1/8	0.637	25,180	539	3,615
2	2/8	0.649	41,394	661	4,491
3	3/8	0.654	54,747	740	4,992
4	4/8	0.663	66,920	798	5,358
5	5/8	0.665	78,052	846	5,641
6	6/8	0.666	88,718	888	5,912
7	7/8	0.667	98,817	929	6,136
8	8/8	0.669	108,305	964	6,295

accuracy improve in 0.01, but from the fifth the improvement is around 0.001. We can conclude that using 50 % of the corpus for training is enough to improve the initial results. Another important information showed in Table 9 is the size of the intersection between the vocabulary of test corpus and the vocabulary of the lexicons as well as the size of the intersection between the vocabulary of training corpora and the vocabulary of the lexicons. Results show very low intersection between corpora's vocabulary (testing and training) and lexicons' vocabulary (SEL and LIWC). We consider that to improve the overall accuracy both aspects must be taken into account, therefore the size of the training corpus and the intersection between vocabularies need to be increased.

5   Conclusions and Future Work

Sentiment analysis is an interesting but difficult task, especially with the characteristics of tweets. A significant effort has been done for Spanish sentiment analysis and the TASS workshop has been an important platform. In this work two different Spanish lexicons were used with rule based and supervised approaches. To our knowledge the LIWC lexicon had not been used for determining the sentiment polarity of Spanish tweets, and according to our experimentation, this lexicon got better results than SEL lexicon. In addition to that, when the information of both lexicons was used with the Multinomial Naïve Bayes the best results were obtained. Despite the fact that our results are not the best when compared with the obtained by the participant teams on TASS 2015, the simple proposed strategy was able to overpass the results of four teams, which shows the potential of combining the information of the selected lexicons with a supervised approach. By doing a deeper analysis, we created a confusion matrix showing that the neutral and none classes has many misclassification issues, we think that this is because there are fewer examples of these classes in the training corpus and the classifier can not learn from them. In addition to that, several experiments were run to determine how the size of the training corpus and vocabularies intersection impact the performance. We conclude that when the size is increased to 50 % better results are obtained, but without an increment in vocabularies's intersection the following increments has no significant impact in the accuracy. For future work we would improve the preprocessing phase to increase the intersection between vocabularies as well as use other lexicons. More complex machine learning algorithms must be tried and other text representations can be explored.

Acknowledgments. We thank the support of Instituto Politécnico Nacional (IPN), ESCOM-IPN, CIC-IPN, SIP-IPN projects number 20160815, 20162058, COFAA-IPN, and EDI-IPN.

References

1. Taboada, M., Brooke, J., Tofiloski, M., Voll, K.D., Stede, M.: Lexicon-based methods for sentiment analysis. Comput. Linguist. **37**, 267–307 (2011)
2. Baccianella, S., Esuli, A., Sebastiani, F.: SentiWordNet 3.0: an enhanced lexical resource for sentiment analysis and opinion mining. In: Calzolari, N., Choukri, K., Maegaard, B., Mariani, J., Odijk, J., Piperidis, S., Rosner, M., Tapias, D. (eds.) LREC. European Language Resources Association (2010)
3. Stone, P.J.: The General Inquirer: A Computer Approach to Content Analysis. The MIT Press, Cambridge (1966)
4. Tausczik, Y.R., Pennebaker, J.W.: The psychological meaning of words: LIWC and computerized text analysis methods. J. Lang. Soc. Psychol. **29**, 24–54 (2010)
5. Pang, B., Lee, L., Vaithyanathan, S.: Thumbs up? Sentiment classification using machine learning techniques. In: EMNLP 2002, Philadelphia, Pennsylvania pp. 79–86 (2002)
6. Urizar, X.S., Roncal, I.S.V.: Detecting sentiments in Spanish tweets. TASS 2012 Working Notes (2012)
7. Sidorov, G., et al.: Empirical study of machine learning based approach for opinion mining in tweets. In: Batyrshin, I., González Mendoza, M. (eds.) MICAI 2012, Part I. LNCS, vol. 7629, pp. 1–14. Springer, Heidelberg (2013)
8. Villena-Román, J., García-Morera, J., Cumbreras, M., Martínez-Cámara, E., Martín-Valdivia, M.T., López, L.A.U.: Overview of TASS 2015. In: Villena-Román, J., García-Morera, J., Cumbreras, M.Á.G., Martínez-Cámara, E., Martín-Valdivia, M.T., López, L.A.U. (eds.) TASS@SEPLN, CEUR Workshop Proceedings, vol. 1397, pp. 13–21 (2015)
9. Garcıa, D., Thelwall, M.: Political alignment and emotional expression in Spanish Tweets. In: Proceedings of the TASS Workshop at SEPLN, pp. 151–159 (2013)
10. Moreno-Ortiz, A., Pérez Hernández, C.: Lexicon-based sentiment analysis of twitter messages in Spanish. Procesamiento del Lenguaje Natural **50**, 93–100 (2013)

11. Urizar, J., San Vicente Roncal, I.: Elhuyar at TASS 2013. In: Proceedings of the TASS Workshop at SEPLN (2013)
12. Araque, O., Corcuera, I., Román, C., Iglesias, C.A., Sánchez-Rada, J.F.: Aspect based sentiment analysis of Spanish tweets. In: Villena-Román, J., García-Morera, J., Cumbreras, M.Á.G., Martínez-Cámara, E., Martín-Valdivia, M.T., López, L.A.U. (eds.): TASS@SEPLN, CEUR Workshop Proceedings, vol. 1397, pp. 29–34 (2015). CEUR-WS.org
13. Valverde, T.J., Tejada, C.J.: Comparing supervised learning methods for classifying Spanish tweets. In: Villena-Román, J., García-Morera, J., Cumbreras, M.Á.G., Martínez-Cámara, E., Martín-Valdivia, M.T., López, L.A.U. (eds.) TASS@SEPLN, CEUR Workshop Proceedings, vol. 1397, pp. 87–92 (2015). CEUR-WS.org
14. Rangel, I.D., Guerra, S.S., Sidorov, G.: Creación y evaluación de un diccionario marcado con emociones y ponderado para el español. Onomazein **29**, 31–46 (2014)
15. Pennebaker, J.W., Francis, M.E., Booth, R.J.: Linguistic Inquiry and Word Count. Lawerence Erlbaum Associates, Mahwah (2001)
16. Cámara, E.M., Cumbreras, M., Martín-Valdivia, M.T., López, L.A.U.: SINAI-EMMA: Vectores de Palabras para el Análisis de Opiniones en Twitter. In: Villena-Román, J., García-Morera, J., Cumbreras, M.Á.G., Martínez-Cámara, E., Martín-Valdivia, M.T., López, L.A.U. (eds.) TASS@SEPLN, CEUR Workshop Proceedings, vol. 1397, pp. 41–46 (2015). CEUR-WS.org
17. del Pilar Salas-Zárate, M., López-López, E., Valencia-García, R., Aussenac-Gilles, N., Almela, Á., Alor-Hernández, G.: A study on LIWC categories for opinion mining in Spanish reviews. J. Inf. Sci. **40**, 749–760 (2014)
18. Vázquez, S., Bel, N.: A classification of adjectives for polarity lexicons enhancement. In: Calzolari, N., Choukri, K., Declerck, T., Dogan, M.U., Maegaard, B., Mariani, J., Odijk, J., Piperidis, S. (eds.) LREC, pp. 3557–3561. European Language Resources Association (ELRA) (2012)
19. Padró, L., Stanilovsky, E.: FreeLing 3.0: towards wider multilinguality. In: Proceedings of the Language Resources and Evaluation Conference (LREC 2012). ELRA, Istanbul (2012)
20. Hurtado, L.F., Plà, F., Buscaldi, D.: ELiRF-UPV en TASS 2015: Análisis de Sentimientos en Twitter. In: Villena-Román, J., García-Morera, J., Cumbreras, M.Á.G., Martínez-Cámara, E., Martín-Valdivia, M.T., López, L.A.U. (eds.) TASS@SEPLN, CEUR Workshop Proceedings, vol. 1397, pp. 75–79 (2015). CEUR-WS.org
21. Álvarez-López, T., Juncal-Martínez, J., Gavilanes, M.F., Costa-Montenegro, E., González-Castaño, F.J., Cerezo-Costas, H., Celix-Salgado, D.: GTI-Gradient at TASS 2015: a hybrid approach for sentiment analysis in twitter. In: Villena-Román, J., García-Morera, J., Cumbreras, M.Á.G., Martínez-Cámara, E., Martín-Valdivia, M.T., López, L.A.U. (eds.) TASS@SEPLN, CEUR Workshop Proceedings, vol. 1397, pp. 35–40 (2015). CEUR-WS.org

Is This a Joke? Detecting Humor
in Spanish Tweets

Santiago Castro[✉], Matías Cubero, Diego Garat, and Guillermo Moncecchi

Universidad de la República, Montevideo, Uruguay
{sacastro,mcubero,dgarat,gmonce}@fing.edu.uy

Abstract. While humor has been historically studied from a psychological, cognitive and linguistic standpoint, its study from a computational
perspective is an area yet to be explored in Computational Linguistics.
There exist some previous works, but a characterization of humor that
allows its automatic recognition and generation is far from being specified. In this work we build a crowdsourced corpus of labeled tweets,
annotated according to its humor value, letting the annotators subjec-
tively decide which are humorous. A humor classifier for Spanish tweets
is assembled based on supervised learning, reaching a precision of 84 %
and a recall of 69 %.

Keywords: Humor · Computational humor · Humor recognition ·
Machine learning · Natural language processing

1 Introduction

The human being as a species is characterized by laughter. Humor, which is a
potential cause of laughter, is an essential component of human communication.
Not only does it allow people to feel comfortable, but also produces a cozier
environment. While humor has been studied from a psychological, cognitive [1]
and even linguistic [2] standpoint, its study from a computational viewpoint
is still an area to be explored within Computational Linguistics. There exist
some previous works [3]; however, a humor characterization that allows its automatic recognition and generation is far from being specified, particularly for the
Spanish language.

Identifying humor in a text can be seen as an intermediate step for the
resolution of more complex tasks. It would be interesting to generate jokes, or
humor in general, based on the knowledge of which attributes enrich texts in
a better way. Another appealing use case is to exploit the outcome of a humor
detector to decide automatically if a text span can be taken seriously or not. On
the other hand, by way of a more direct use, humor identification can be used
to find jokes on Twitter, to search for potentially funny tweets about certain
trending topic or to search for humorous answers to comments on the social
network.

© Springer International Publishing AG 2016
M. Montes-y-Gómez et al. (Eds.): IBERAMIA 2016, LNAI 10022, pp. 139–150, 2016.
DOI: 10.1007/978-3-319-47955-2_12

We address herein the problem of detecting humor in Spanish tweets. It should be noted that this is different from trying to recognize humor in arbitrary texts, due to tweets' length. Here it could be assumed that tweets are either humorous or not, but not both, because they are brief (up to 140 characters). This is not always the case in others texts, as jokes could only exist in some parts but not on the whole text. Another advantage considered is that there are plenty of tweets available to analyze.

Since there is no clear definition of what humor is, how can we detect something that is in principle vaguely stated? We explore different ideas, and we finally decide to let people define it themselves by voting tweets from a web page and an Android app, in which they can label a tweet as humorous or not humorous. Once we have defined which tweets are humorous, we tackle the problem of humor detection using a supervised learning approach. In other words, we infer a function that identifies humor from labeled data. We use several techniques such as Support Vector Machine, Nearest Neighbors, Decision Trees and Naive Bayes. In order to build a set of features, we first study the state of the art of the Computational Humor area, focused on recognition and in Spanish.

In Sect. 2 we present the humor detection problem and its state of the art, including features studied in previous works. In Subsect. 3.1 we show the corpus built for this purpose and in Subsect. 3.2 we describe the classifier used. Afterwards, we present an experimental evaluation in Sect. 4 and finally the conclusions in Sect. 5.

2   Computational Humor

Computational Humor is a recent field of study about recognizing and generating humor through automatic processing. The task of language understanding is rather hard, and so are tasks related to humor. Furthermore, humor entails the usage of figurative language, which obviously makes language handling harder.

Humor by itself is not a clearly determined concept. According to Real Academia Española[1], humor is defined as a way of presenting reality, highlighting the comic or ridiculous side. As for comedy, it is a kind of drama meant to cause laughter. However, what causes laughter? There are several theories which try to answer this question, and consequently attempt to find what humor is. A report on the state of the art about Humor and Computational Humor [3] enumerates some of them. The main ideas of these theories are described hereinafter. Readers will notice that these ideas are similar, in spite of putting the focus on different attributes.

Gruner [4] develops a theory which claims that humor is related to superiority feelings, asserting that there is always a winner in every joke. Freud and Strachey [5] and Minsky [6] state that humor is about relieving repressed feelings. In this case, laughter relieves the stress caused by taboo topics, such as death, marriage or sex. The Theory of the Incongruity Resolution [7] claims that two

---

[1] http://dle.rae.es/.

objects are presented under the same concept, with details applying to both and with similarities, but as narration progresses it turns out that only one is possible. Furthermore, we have The Semantic Script Theory of Humor and The General Theory of Verbal Humor [8,9]. They state that humor is about two scripts which come into conflict with each other, where there are two opposed subjects contrasted, such as big vs small, death vs life, normal vs abnormal, among others.

Let us introduce an example[2]:

— Nada es imposible.
— A ver, tocate la espalda con la rodilla, mente positivista.

— Nothing is impossible.
— Seriously? Touch your back with your knee, you positivist mind.

Following the Superiority Theory, the reader is the winner when he laughs at the positive person, feeling superior as the latter lose the dispute. According to the Relief Theory, we laugh with the purpose of releasing tension, which in this case can be provoked by talking about the limits of life, such as when saying "nothing is impossible". The Theory of the Incongruity Resolution also applies here due to the fact that there is ambiguity; with "nothing is impossible" the example implies that all your dreams may come true, but the person is answered as if the statement was literal.

2.1   Humor Detection

The concrete goal of this research is to classify tweets written in Spanish as humorous or not humorous. In order to accomplish this, jokes need to be completely expressed within the text, and no further information must be required (apart from contextual information). Since Twitter allows only brief publications — no more than 140 characters — we freely assume the text to be a unit: either the whole tweet is humorous, or it is not.

2.2   State of the Art

We did not find any attempt to automatically recognize humor for Spanish. Notwithstanding, Mihalcea and Strapparava [10] and Mulder and Nijholt [3] built humor detectors for English making use of one-liners, i.e., texts of approximately fifteen words. Supervised learning was used to produce an outcome — humorous or not humorous content — based on features which might reflect certain properties that humor should satisfy. Furthermore, Reyes et al. [11] and Reyes et al. [12] have gathered and studied features specific to humor, without having the objective of creating a recognizer.

---

[2] Taken from https://twitter.com/chistetipico/status/430549009812291584. It has been slightly adapted to maintain an appropriate language.

A concise compilation of the features presented in these works is shown below:

Adult Slang: According to Mihalcea and Strapparava [10], adult slang is popular in jokes. Let us remember that the Relief Theory states that laughter releases stress caused by taboo subjects, and adult slang could be one. WordNet Domains [13] can be used to search for words tagged with the domain "Sexuality" in potentially humorous texts.

Alliteration: This is about the repetition of phonemes in a text. It is a generalization of the rhyme. As stated in [10], structural and phonetic properties of jokes are at least as important as their content.

Ambiguity: It may be explained by the Incongruity Resolution Theory that ambiguity plays an important role, as it gives more than one interpretation to texts. Sjöbergh and Araki [14], Basili and Zanzotto [15], and Reyes et al. [11] mention different ways to measure it, such as counting the number of meanings of the words that appear or counting the number of possible syntax trees.

Antonymy: Following the Semantic Script Theory of Humor, we could look for opposed terms in texts, and that is how this feature is supported. The idea is to take into account pairs of antonym words mentioned in texts. Wordnet [16] is useful since it is a lexical database which contains antonyms for English words, among other relations.

Keywords: There are certain words that are more used in humorous contexts than in normal situations [14]. An example of these are words related to animal contexts, lawyers, etc.

Language model perplexity: In Reyes et al. [11] a language model is built from narrative texts, and perplexity[3] is used as a feature. Humorous texts have a higher perplexity than those which are not humorous.

Negativity: There is a certain kind of humor which tends to have negative connotations [12,17]. It can be about denying, such as when saying "no", "don't" or "never", when talking about subjects with negative polarity such as "bad", "illegal" or "wrong" or when it is related to words referring to stressful subjects, such as "alcohol" or "lie".

People-centered words: Humorous texts are constantly referring to scenarios related to people, with dialogues and references such as "you", "I", "woman" and "my". This is supported by Mihalcea and Pulman [17], Mihalcea and Strapparava [18].

Mihalcea and Strapparava [10] used the features Adult Slang, Alliteration and Antonymy, while Sjöbergh and Araki [14] focused on Alliteration, Ambiguity, Keywords and People-centered words. Both studies collected humorous one-liners from the Internet. Sjöbergh and Araki [14] employed only the British National Corpus (BNC) as negative samples whereas Mihalcea and Strapparava

---

[3] Perplexity is a measurement of how well a probability model predicts a sample. Low perplexity indicates the probability model is good at predicting the sample. It is defined as $2^{-\frac{1}{n}\sum_{i=1}^{n}\log_2 p(x_i)}$, where $x_1, \ldots, x_n$ are the sample data and $p(x_i)$ is the probability assigned to each one.

Table 1. Comparison of the approach of both works. The results are not directly comparable as they use different corpora.

	Mihalcea and Strapparava [10]	Sjöbergh and Araki [14]
Negative samples	BNC sentences, news headlines and proverbs	Other sentences from BNC
Accuracy	96.95 % with headlines, 79.15 % with the BNC and 84.82 % with the proverbs	85.40 %
Features	Adult Slang, Alliteration and Antonymy	Alliteration, Ambiguity, Keywords and People-centered words

[10] additionally used proverbs and news headlines from Reuters. In both works they tried with Naïve Bayes and Support Vector Machine classifiers, resulting in no significant difference between these techniques. On one hand, Mihalcea and Strapparava [10] achieved their best accuracy with headlines: 96.85 %, while they reached 84.82 % with proverbs and 79.15 % with the BNC. Alliteration proved to be the most accurate feature. On the other hand, Sjöbergh and Araki [14] achieved an accuracy of 85.40 %, with Keywords being the most useful. Table 1 summarizes the main differences and compares both studies.

3 Proposal

3.1 Corpus

Our first goal is to build a corpus with samples of humorous and non-humorous tweets. Based on Mihalcea and Strapparava [10], we choose to use non-humorous sample tweets that fall into the following topics: news, reflections and curious facts. For humorous samples, we extracted tweets from accounts which appeared after having searched for the keyword "chistes" ("jokes" in Spanish). In total, 16,488 tweets were extracted from humorous accounts and 22,875 from non-humorous. The two groups are composed of 9 Twitter accounts[4] each, with the non-humorous containing 3 of each topic. The amount of tweets in each topic is similar.

We tagged all tweets from news, reflections and curious facts as non-humorous, as random sampling showed that there was no humor in them. Conversely, not all tweets that were extracted from a humorous account were in fact humorous. Many of them were used to increase their number of followers, to express their opinion about a fact or to support a cause through retweets.

A crowdsourced web[5] and a mobile[6] annotation was carried out in order to tag all tweets from humorous accounts. In order to obtain as many annotations

---

[4] These tweets show traits of different variaties of the Spanish language among them. At least three countries were identified: Colombia, Spain and Uruguay.

[5] http://clasificahumor.com.

[6] https://play.google.com/store/apps/details?id=com.clasificahumor.android.

as possible, we wanted to keep it simple. Therefore, we showed random tweets to annotators (avoiding duplicates), providing no instructions, and let them implicitly define what humor is. In addition, the user interface was simple, as shown in Fig. 1. The users could either provide a ranking of humor between one and five, express that the tweet was not humorous or skip it.

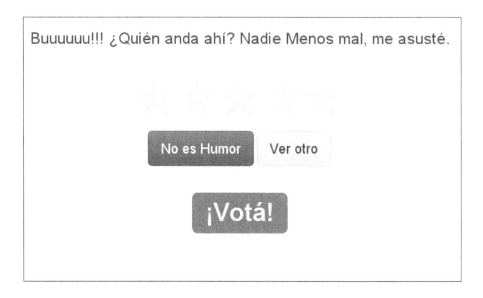

Fig. 1. Page used to annotate tweets, with an example tweet on screen.

In total, 33,531 annotations were achieved, after filtering some of them that occurred in a short time lapse in the same session and with the same tag. About half of the labels were non-humorous, while the other half was divided approximately between the five rankings. A histogram of the annotations is shown in Fig. 2. Regarding the agreement among annotators, the Fleiss' Kappa measurement for tweets with 2 annotations[7] is 0.416 and for those with 6 annotations it is 0.325.

Based on this analysis, we have to decide which tweets are considered humorous. Let us define the tweets considered humorous as *positives* and the ones considered as non-humorous as *negatives*. The decision consisted in marking as positives those tweets whose ratio of humorous annotations is greater than or equal to 0.6 and as negatives those lower than or equal to 0.3. The rest are considered as *doubtful*. The criterion of giving a 0.1 handicap to the positives was thereby performed, as they are obtained from humorous accounts. This may be seen as if the source is giving its opinion too. Additionally, those tweets with

---

[7] Note that Kappa assumes a fixed number of annotators. For this reason, we measure it with 2 and 6, in order to give an idea of the agreement having a value with many tweets but few annotators, and other value with few tweets but many annotators.

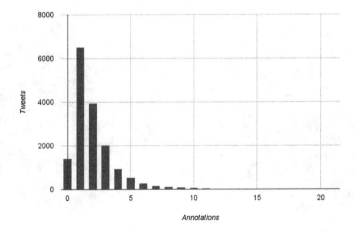

Fig. 2. Histogram of annotations. Here the tweets are grouped in bins according to the number of annotations each one received. Note that most tweets have few annotations.

no annotations fall into the category of doubtful. Figure 3a illustrates the proportions of each category. To sum up, 5,952 tweets are considered positive. The rest of the tweets obtained from humorous accounts are not taken into account, even though the negatives can also be used. The corpus composition is shown in Fig. 3b.

3.2   Classifier

Firstly, we split data into 80 % for training and 20 % for later evaluation. The former were used to fit the model and also to test it by means of cross-validation before carrying out the final evaluation with fresh unseen data. Similarly to the works mentioned in this document, we built a humor classifier but for the Spanish language. Such works used Support Vector Machine (SVM) and a Multinomial version of Naïve Bayes (MNB). However, more machine learning techniques are tried here: Decision Trees (DT), k Nearest Neighbors (kNN) and a Gaussian version of Naïve Bayes (GNB). Tweets are tokenized using Freeling [19]. Also, a higher quantity of features was implemented, which is described below.[8]

Adult slang: Here we count the relative number of tokens in the tweets which appeared in a previously built dictionary about adult slang. This dictionary contains 132 words, and it was built using bootstrapping, in a similar manner Mihalcea and Strapparava to [20], with a seed of 21 words. Dictionary-lookup features are computed with this formula (where the multiset intersection is used):

---

[8] The codebase for the classifier and the corpus built can be found in https://github.com/pln-fing-udelar/pghumor.

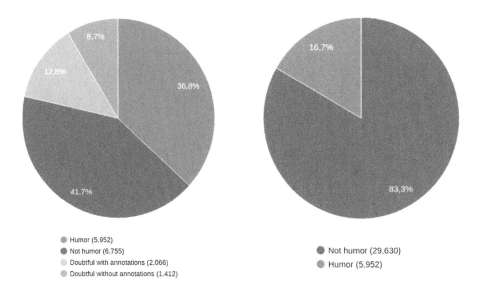

Humor (5,952)
Not humor (6,755)
Doubtful with annotations (2,066)
Doubtful without annotations (1,412)

Not humor (29,630)
Humor (5,952)

(a) Graph showing the percentage of tweets from humorous accounts in each category.

(b) Pie displaying the ratio between positives and negatives in the corpus, after the decision was made.

Fig. 3. Charts showing information about the data.

$$featureValue(tweet) = \frac{|tweet \cap dictionary|}{\sqrt{|tweet|}}$$

Animal presence: In this case we compare against a handcrafted dictionary about animals. This dictionary contains 103 names, including typical typographic misspellings and grammatical mistakes.

Antonyms: Given a tweet, this feature counts the relative number of pairs of antonyms existing in it. WordNet [16] antonymy relationship and Spanish language enrichment provided by the Multilingual Central Repository [21] are used for this. This feature was discarded since after performing Recursive Feature Elimination [22] (RFE) we found out the classification worsened.

Dialog: This feature only establishes if a tweet is a dialog. Its value is 1 if it starts with a dash or hyphen (or another similar character) and 0 otherwise.

Exclamations: The relative number of exclamation marks are counted.

First and Second person: These two features try to capture verbs conjugated in the first and second persons and nouns and adjectives which agree with such conjugations (in Spanish, nouns and adjectives express gender and number at the end of the word).

Hashtags: The amount of hashtags in the tweet is counted. It is suspected that the higher this amount is, the more informal the tweet is. Thus, it is more likely to be humorous.

Keywords: An intuitively handmade dictionary of 43 common words found in jokes was built for this, and it was used for checking purposes.

Links: This feature counts the number of links contained in a tweet.

Negation: Here we count the relative quantity of times the word "no" appears in the tweet. It was removed after running RFE.

Non-Spanish words: The relative number of words with characters not present in the Spanish alphabet is counted. It was discarded after running RFE.

Out of vocabulary: The idea behind this is to keep record of the relative count of words not found in dictionaries. These are four features based on the combination of the dictionaries used: Freeling, Freeling-Google[9], Freeling-Wiktionary[10] and Wiktionary.

Questions-answers: One interesting attribute for tweets is to count how many questions and answers are present, one after another.

Topic distance: The idea is to check if a tweet is somewhat near to a joke category in *Chistes.com*, or whether it is closer to a Wikipedia's sentence, from Wikicorpus [23]. This is carried out using a Multinomial Naïve Bayes classifier together with the Bag of Words technique.

Uppercase words: The relative amount of words completely in uppercase is counted.

4 Experimental Evaluation

Provided that our work is the only one using this corpus, and even the only one with the goal of classifying humor in Spanish, we cannot directly compare it with any other work. Hence, we developed two baselines to compare it with, aiming them to be simple ideas which could be crafted to face this task. The first one (BL1) is a Multinomial Naïve Bayes classifier combined with Bag of Words similarly to the Topic Distance feature. The second one (BL2) is a classifier which predicts all tweets with the most likely outcome, *non-humorous*, having a frequency of almost 83 %.

A comparison using mainly the F_1 score is intended. We want to pay attention to the positives (the humorous) but also granting the same degree of importance to false positives and false negatives. Nonetheless, we take advantage of the runs in order to also pay attention to other measurements. The results are shown in Table 2.

The best results are obtained with SVM, even in terms of accuracy. Also, kNN shows satisfactory output. These two approaches outperform the baselines, with the former clearly surpassing the latter. Meanwhile, GNB and DT have poor precision, although GNB certainly does a better job among these two and has the best recall. The confusion matrix for SVM is shown in Table 3.

---

[9] https://www.google.com.

[10] https://www.wiktionary.org.

Table 2. Results obtained with the different techniques over the test set. NPV, TNR and Neg. F1 refer to Precision, Recall and F_1 score, respectively, when reversing the roles positive-negative.

	Precision	Recall	F_1	NPV	TNR	Neg. F_1	Accuracy
BL1	0.617	0.846	0.714	0.966	0.892	0.714	0.885
BL2	N/A	0.000	N/A	0.830	1.000	0.907	0.830
SVM	0.836	0.689	**0.755**	0.938	0.972	**0.955**	**0.925**
DT	0.665	0.675	0.670	0.933	0.930	0.932	0.889
GNB	0.575	**0.782**	0.663	**0.952**	0.882	0.915	0.865
MNB	**0.848**	0.600	0.703	0.923	**0.978**	0.950	0.914
kNN	0.813	0.663	0.730	0.934	0.969	0.951	0.917

Table 3. Confusion matrix for SVM classifier with respect to the test set

	Positive	Negative
Positive	842	381
Negative	165	5805

5   Conclusions

A crowdsourced corpus has been assembled, which serves the purpose of this work and could be useful for future research. It contains over 30,000 annotations for 16,488 tweets, coming from humorous accounts, and it also counts with 22,875 sourced from non-humorous accounts. Uses of such corpus include analyzing its data, as well as performing tasks similar to the work described herein.

We have built a classifier which outperforms the baselines outlined. Support Vector Machine proved to be the best technique. It has a precision of 83.6 %, a recall of 68.9 %, a F_1 score of 75.5 % and an accuracy of 92.5 %. Nevertheless, it must be highlighted that the corpus built does not depict a great variety of humor. Hence, some features perform well in this work but might not perform so well in another context.

As a future work, more complex features could be crafted, such as trying to detect wordplay and puns, ambiguity, perplexity against some language model, inter alia. Other Machine Learning techniques could also be tried. It would be interesting if we take advantage of the star ranking people provided; maybe this can also suggest how funny a joke is. As a harder task, humor generation could be tackled. Finally, it could be studied how the influence of humor varies between different social contexts, depending on gender, age, interest areas, mood, etc.

References

1. International Journal of Humor Research: HUMOR (1988). http://www.degruyter. com/view/j/humr. Visited May 2015
2. Raskin, V.: Semantic Mechanisms of Humor. Springer, Heidelberg (1985)
3. Mulder, .M.P., Nijholt, A.: Humour Research: State of Art. Technical report TR-CTIT-02-34, Enschede: Centre for Telematics and Information Technology University of Twente (2002)
4. Gruner, C.: The Game of Humor: A Comprehensive Theory of Why We Laugh. Transaction Publishers, Piscataway (2000)
5. Freud, S., Strachey, J.: Jokes and Their Relation to the Unconscious (1905)
6. Minsky, M.: Jokes and the logic of the cognitive unconscious. In: Vaina, L., Hintikka, J. (eds.) Cognitive Constraints on Communication, vol. 18, pp. 175–200. Springer, Heidelberg (1980)
7. Rutter, J.: Stand-up as interaction: performance and audience in comedy venues. Citeseer (1997)
8. Attardo, S., Raskin, V.: Script theory revis(it)ed: joke similarity and joke representation model. Humor: Int. J. Humor Res. 4, 293–347 (1991)
9. Ruch, W., Attardo, S., Raskin, V.: Toward an empirical verification of the general theory of verbal humor. HUMOR: Int. J. Humor Res. 6(2), 123–136 (1993)
10. Mihalcea, R., Strapparava, C.: Making computers laugh: investigations in automatic humor recognition. In: Proceedings of the Conference on Human Language Technology and Empirical Methods in Natural Language Processing, HLT 2005, pp. 531–538. Association for Computational Linguistics, Vancouver (2005)
11. Reyes, A., Buscaldi, D., Rosso, P.: An analysis of the impact of ambiguity on automatic humour recognition. In: Matoušek, V., Mautner, P. (eds.) TSD 2009. LNCS, vol. 5729, pp. 162–169. Springer, Heidelberg (2009)
12. Reyes, A., Rosso, P., Martí, M.A., Taulé, M.: Características y rasgos afectivos del humor: un estudio de reconocimiento automático del humor en textos escolares en catalán. Procesamiento del Lenguaje Nat. 43, 235–243 (2009)
13. Strapparava, C., Valitutti, A.: WordNet affect: an affective extension of WordNet. In: LREC, pp. 1083–1086 (2004)
14. Sjöbergh, J., Araki, K.: Recognizing humor without recognizing meaning. In: Masulli, F., Mitra, S., Pasi, G. (eds.) WILF 2007. LNCS (LNAI), vol. 4578, pp. 469–476. Springer, Heidelberg (2007)
15. Basili, R., Zanzotto, F.M.: Parsing engineering and empirical robustness. Nat. Lang. Eng. 8(3), 97–120 (2002)
16. Fellbaum, C. (ed.): WordNet: An Electronic Lexical Database. MIT Press, Cambridge (1998)
17. Mihalcea, R.F., Pulman, S.: Characterizing humour: an exploration of features in humorous texts. In: Gelbukh, A. (ed.) CICLing 2007. LNCS, vol. 4394, pp. 337–347. Springer, Heidelberg (2007)
18. Mihalcea, R., Strapparava, C.: Learning to laugh (automatically): computational models for humor recognition. Comput. Intell. 22(2), 126–142 (2006)
19. Padro, L., Stanilovsky, E.: FreeLing 3.0: towards wider multilinguality. In: Proceedings of the Language Resources and Evaluation Conference (LREC 2012), Istanbul, Turkey (2012)
20. Mihalcea, R., Strapparava, C.: Bootstrapping for fun: web-based construction of large data sets for humor recognition. In: Proceedings of the Workshop on Negotiation, Behaviour and Language (FINEXIN 2005), pp. 84–93 (2005)

21. Gonzalez-Agirre, A., Laparra, E., Rigau, G.: Multilingual central repository version 3.0: upgrading a very large lexical knowledge base. In: Proceedings of the 6th Global WordNet Conference (GWC 2012), Matsue (2012)
22. Guyon, I., Weston, J., Barnhill, S., Vapnik, V.: Gene selection for cancer classification using support vector machines. Mach. Learn. **46**(1–3), 389–422 (2002)
23. Reese, S., Boleda, G., Cuadros, M., Padró, L., Rigau, G.: Wikicorpus: a word-sense disambiguated multilingual wikipedia corpus. In: Proceedings of 7th Language Resources and Evaluation Conference (LREC 2010), La Valleta, Malta (2010)

Evaluating Topic-Based Representations for Author Profiling in Social Media

Miguel A. Álvarez-Carmona[1(✉)], A. Pastor López-Monroy[1],
Manuel Montes-y-Gómez[1], Luis Villaseñor-Pineda[1], and Ivan Meza[2]

[1] Computer Science Department, Instituto Nacional de Astrofísica,
Óptica y Eletrónica (INAOE), LabTL,
Luis Enrique Erro No. 1, 72840 Tonantzintla, Puebla, Mexico
miguelangel.alvarezcarmona@ccc.inaoep.mx
[2] Instituto de Investigaciones En Matematicas Aplicadas y En Sistemas (IIMAS),
Universidad Nacional Autonoma de Mexico (UNAM),
Escolar 3000, Ciudad Universitaria, Mexico City D.F., Mexico

Abstract. The Author Profiling (AP) task aims to determine specific demographic characteristics such as gender and age, by analyzing the language usage in groups of authors. Notwithstanding the recent advances in AP, this is still an unsolved problem, especially in the case of social media domains. According to the literature most of the work has been devoted to the analysis of useful textual features. The most prominent ones are those related with *content* and *style*. In spite of the success of using jointly both kinds of features, most of the authors agree in that content features are much more relevant than style, which suggest that some profiling aspects, like age or gender could be determined only by observing the thematic interests, concerns, moods, or others words related to events of daily life. Additionally, most of the research only uses traditional representations such as the BoW, rather than other more sophisticated representations to harness the content features. In this regard, this paper aims at evaluating the usefulness of some *topic-based representations* for the AP task. We mainly consider a representation based on Latent Semantic Analysis (LSA), which automatically discovers the topics from a given document collection, and a simplified version of the Linguistic Inquiry and Word Count (LIWC), which consists of 41 features representing manually predefined thematic categories. We report promising results in several corpora showing the effectiveness of the evaluated topic-based representations for AP in social media.

1 Introduction

The Author Profiling (AP) task aims to analyze written documents to extract relevant demographic information from their authors [14]. The following problems have gained interest recently: gender prediction [2,31], age estimation [23,24], personality detection [33], native language identification [2], and political orientation detection [25]. The AP task has a wide range of practical applications. For example, in marketing, companies may leverage online reviews to improve targeted

M. Montes-y-Gómez et al. (Eds.): IBERAMIA 2016, LNAI 10022, pp. 151–162, 2016.
DOI: 10.1007/978-3-319-47955-2_13

advertising, and in forensics, the linguistic profile of authors could be used as valuable additional evidence. In this paper we are interested in profiling age and gender from authors of social media domains. Social media documents are difficult to analyze by standard text mining methods because of several challenging characteristics such as spelling-grammar errors and out-of-vocabulary terms[1].

The AP task has mainly approached as a single-labeled classification problem, where the different profiles (e.g., *males* vs. *females*, or *teenager* vs. *young* vs. *old*) stand for the target classes. The common processing pipeline is as follows: i) extracting textual features, ii) representing documents by these features, and iii) learning a classification model of documents. The extraction of textual features is the stage that has received more attention. In this direction, two kind of attributes stand out from others: content features (i.e., nouns, verbs and adjectives), and style features (i.e., function words, punctuation marks, emoticons and POS tags) [23,31]. In AP tasks, content and style features are extracted by observing words usage to reveal people interests and writing style. In spite of the success of using jointly both kind of attributes, a number of authors have reported results suggesting that content features are the most valuable for AP [19,27]. This can be explained by the fact that people from the same demographic group tend to share interests, concerns, hobbies and opinions [22,29].

In this work, rather than define a suitable set of features for AP, we focus on studying the informative value of content features. More importantly, unlike other works using standard representations like BoW, in this work we propose using topic-based representations to better exploit the content information. Our hypothesis is that by using content features in conjunction with topic-based representations, it is possible to obtain comparable results than other more elaborated strategies from the state of the art. A second contribution of this paper is the evaluation of two different approaches for computing the topic-based representations. The first approach consists in automatically compute topic-based features by means of Latent Semantic Analysis (LSA) [4]. Although LSA has been preciously used in several text mining problems, to the best of our knowledge this is the first time it is fully evaluated on pure content features for the AP task[2]. The second approach builds the topic-based representation by considering a set of hand-crafted content features. For this, we devise a simplified version of Linguistic Inquiry and Word Count (LIWC) [34], which consists of 41 predefined topic categories. Each LIWC category contain a number of associated words, which were defined by a group of socio-linguistic experts. In particular, the main contribution of this study consists in exposing the strengths and weaknesses of each topic-based approach over different social media domains.

---

[1] It is very hard to accurately apply typical procedures like stemming or extract specific syntactic information from informal documents.

[2] In AP tasks, several authors have used LSA as part of elaborated strategies involving different kinds of features, for example: ensemble strategies, or fusion strategies [21]. Nevertheless, they have not reported experimental results to show the real contribution of LSA features.

The evaluation was done using the data sets from PAN14 [27]. The obtained results showed that the two kinds of topic-based representations outperformed the standard BoW in most social media domains. Furthermore, using only 41 features, manually or automatically defined, they obtained competitive results to state of the art methods.

This paper is organized as follows: Sect. 2 presents some relevant work for this research. Section 3 explains the textual features we used and the considered topic-based representations. Section 4 explains the experimental settings, and then, Sect. 5 shows the evaluation results. Finally, Sect. 6 presents our conclusions and some future work directions.

2   Related Work

The AP task has been approached from different areas, including psychology [26], linguistics [11], socio-linguistics [5], and natural language processing (NLP) [14,31]. In this section we review the related work from the NLP perspective. Mainly, we focus on describing the *content* and *stylistic* features that have been employed.

According to the literature, a wide range of different approaches have been proposed for the AP task. The different methods for learning specific textual patterns range from simple lexical approaches to elaborated strategies requiring syntactic/semantic analysis of the documents. For example, the bag of words (BoW) [14] have been successfully used for gender prediction in formal documents. Another example are Probabilistic Context-Free Grammars (PCFG) [30] and language models, which have been designed for gender detection in scientific articles [3]. Likewise, other authors have gone beyond by exploiting latent biographic attributes (e.g., gender, native language), with the aim of analyzing the discourse style between people of the same/different age-gender [9]. Notwithstanding the usefulness of these features for profile prediction, most of them are only relevant for domains having formal documents (i.e., books, articles, etc.), and they remain unexplored in informal domains, such as the case of social media sources. For example, the building process of a PCFG involves the extraction of part-of-speech (POS) tags, which are difficult to accurately extract from social media texts.

In the case of social media, the majority of the works have focused on using *content* and *stylistic* features [18,27,28]. Moreover, several works suggest that content words usually are much more relevant than style features. For example, an analysis of information gain presented in [31], showed that the most relevant attributes for gender prediction are those related with content words, for example: *linux* and *office* for discriminating males, whereas *love* and *shopping* for discriminating females. Furthermore, Schler et al. (2006) also concluded that syntactic features are less useful than very basic lexical thematic features when analyzing blogs. Other works have also considered interesting stylistic features, namely slang vocabulary and the average sentence length, but in all the cases these features have been used in combination –as a complement– of content features [1,10].

In this work, we attempt to evaluate the relevance of content features for the task of AP in social media. Our main hypothesis is that content features, which capture the topics of interests of users, are the cornerstone to reveal profiling cues in social media domains. In particular, we propose modeling this content features by means of two different topic-based representations: LSA [15], which automatically extracts the topics from the given document collection, and LIWC [34], which is a set of manually defined topics. These two topic-based representations have been previously used in AP [12, 20, 36], but always in combination with other features and strategies, making it impossible to observe its real relevance to the AP task.

3   Features

The main idea behind this paper is that topic-based representations are effective in capturing the content –thematic– information of documents, and therefore that they could be appropriate for the task of AP in social media domains. As mentioned before, we consider two ways of representing the topics from social media profiles. First, we use a set of automatically extracted topics discovered by means of the LSA algorithm [6], and secondly, a set of manually defined topics obtained from the LIWC resource [34]. In the following subsections we describe both approaches.

3.1   LSA

Latent Semantic Analysis (LSA) is a method for representing the contextual-usage meaning of words. It assumes that words close in meaning tend to occur in similar contexts [16], and therefore, uses occurrence and co-occurrence information to associate words and to measure their contribution to automatically generated concepts (topics) [15].

LSA is a method to extract and represent the meaning of the words and documents. LSA is built from a matrix $\mathbf{M}$ where m_{ij} is typically represented by the TFIDF [35] of the word i in document j. LSA uses the Singular Value Decomposition (SVD) to decompose $\mathbf{M}$ as follows:

$$\mathbf{M} = \mathbf{U\Sigma V}^T \tag{1}$$

where the $\mathbf{\Sigma}$ values are called the singular values and $\mathbf{U}$ and $\mathbf{V}$ are the left and right singular vectors respectively. $\mathbf{U}$ and $\mathbf{V}$ contain a reduced dimensional representation of words and documents respectively. $\mathbf{U}$ and $\mathbf{V}$ emphasize the strongest relationships and remove the noise [16]. In other words, it makes the best possible reconstruction of the $\mathbf{M}$ matrix with the less possible information [17]. In this work we compute $\mathbf{U}$ and $\mathbf{V}$ from the given training documents as described in [37].

3.2   LIWC

The way that the Linguistic Inquiry and Word Count (LIWC) works is fairly intuitive. Basically, it reads a given text and counts the percentage of words associated with a set of manually defined categories. Given that LIWC categories were developed by researchers from cognitive psychology, they were created with the aim of capturing people's social and psychological states [13], which have proved to be useful in the AP task [8, 24, 32].

LIWC has two types of categories; the first kind captures the style of the author by considering features like the POS frequency or the length of the used words. The second group captures content information by counting the frequency of words related with some thematic categories such as family, work, friends and others. In this research we focused on the content information, and consequently we decided ignoring the style categories. In particular, we considered the 41 thematic categories, each of them described by a name and a set related words. Table 1 lists the 41 LIWC categories, and Table 2 shows some example words associated to the categories of family, work, body, religion and friends.

Table 1. The 41 LIWC content categories

relativity	feel	money	causation	insight
humans	discrepancy	sad	anger	see
affect	home	work	sexual	negative emotion
death	family	tentative	religion	verbs
quant	achievement	health	body	perception
assent	positive emotion	time	leisure	inhibition
hear	friends	anxiety	cognitive	certainty
space	motion	swear	social	biological
ingestion				

Table 2. Examples of five LIWC categories: name of categories and a subset of associated words

Category		Subset of associated words		
Family	uncle	granddad	mommy	son
Work	sector	commerce	feedback	corps
Body	thigh	flesh	cornea	hands
Religion	amish	pope	rabbi	zen
Friends	comrad	sweetheart	mates	roomate

Table 3. Distribution of the gender classes across the different domains

Class	Blogs	Reviews	Social Media	Twitter
Female	73	2080	3873	153
Male	74	2080	3873	153
Σ	147	4160	7746	306

Table 4. Distribution of the age classes across the different domains

Class	Blogs	Reviews	Social Media	Twitter
18–24	6	360	1550	20
25–34	60	1000	2098	88
35–49	54	1000	2246	130
50–64	23	1000	1838	60
65+	4	800	14	8
Σ	147	4160	7746	306

3.3 Corpora

For the experiments we used the datasets from the PAN 2014 AP task. These corpora were especially built to study the AP in social media domains. They consist of two gender profiles (female vs. male) and five non-overlapping age profiles (18–24, 25–34, 35–49, 50–64, 65-plus). All document collections are in English and they belong to four different domains: Blogs, Social Media, Hotel Reviews, Twitter [27]. Tables 3 and 4 describe the distribution of profiles for the different domains for the gender and age classes respectively. It is important to notice that gender classes are balanced, whereas age classes are highly unbalanced.

4 Experimental Settings

In this section we describe the configuration used in all the experiments.

Preprocessing: First we removed stop words, then we extracted content words and applied stemming on them. Finally, we considered the 5000 most frequent terms for each domain.

Text representation: For building the LIWC representation we considered the 41 thematic categories shown in Table 1. For the LSA representation we set the parameter k to 41 in order to be able to compare its results against those using the LIWC topics.

Classification: In all the experiments we used the LibLINEAR classifier [7] and performed a stratified 10 cross fold validation (10-CFV). As a baseline we used the results from the BoW representation considering the 5000 selected words.

5   Results

The goal of the following experiments is two fold: first, to determine the effectiveness of topic-based representations, namely LSA and LIWC, for AP in social media, and second, to compare their performance with the traditional BoW representation as well as with one state of the art (BSoA) approach. In particular, we used the results reported in [19] as BSoA results. This work uses a combination of content and style features and representation based on automatically discovered subprofiles.

5.1   Age Results

Table 5 shows the obtained results. They indicate that the LSA and LIWC based approaches outperform the BoW results in all social media domains. These results allows to conclude that applying a topic-based reprentation is useful for the task of age predecition.

In these experiments LSA obtained the best results for blogs, reviews and social media domains, whereas LIWC obtained the best result for the twitter collection. We presume this may be explained by the great vaiability of topics communicated by a user in their different tweets, which difficults LSA to discover word relations and to extract discriminative topics. On the contrary, LIWC is based on manually defined topics and it is independent from the data. Summarizing, the experimental results show that for highly diverse domais, such as Twitter, it seems a better option to defined the topic representation based on external knowledge.

Table 5. Accuracy results for age classification in four social media domains

Approach	Blogs	Reviews	Social Media	Twitter
BoW	0.34($\pm$0.10)	0.28($\pm$0.02)	0.32($\pm$0.01)	0.42($\pm$0.05)
LSA	**0.48($\pm$0.09)**	**0.34($\pm$0.02)**	**0.36($\pm$0.01)**	0.39($\pm$0.06)
LIWC	0.42($\pm$0.26)	0.29($\pm$0.02)	0.34($\pm$0.02)	**0.47($\pm$0.05)**
BSoA	0.48	0.34	0.37	0.48

The results from Table 5 also show that the best results from the topic-based representations are comparable to those from the BoSA method. Given that the BoSA method captures both content and style information, these results allows to observe the importance of content features (thematic interests) for the subtask of age classification in social media domains. Table 6 shows the three topics with the greatest information gain for both, LSA and LIWC. In the case of LSA we list the four most important words associated to each topic. It is interesting to notice that for the blogs collection there are only 2 topics and for Twitter only one. As we explained before, the Twitter collection has a wide range of subjects, and it was difficult for LSA to find relations between the words and to build relevant topics for the AP task.

Table 6. The topics with more information gain for age classification

Domain	LSA 1	LSA 2	LSA 3	LIWC 1	LIWC 2	LIWC 3
Blogs	thesis	tutorial				
	memory	bank	-	religion	-	-
	technology	market				
	education	company				
Reviews	fantastic	amazing	beach			
	wonderful	balcony	resort	affect	cognitive	positive emotion
	great	excellent	lovely			
	view	lobby	pool			
Social media	boot	vuitton	smoke			
	coach	louis	cigarett	cognitive	work	quant
	handbag	shoes	dog			
	shoes	handbag	nike			
Twitter	fb					
	ow	-	-	assent	swear	certainty
	sigir					
	gamif					

5.2 Gender Results

In this Section we show the results for gender classification on four different social domains. Table 7 shows the obtained accuracy results.

As we can see, the BoW representation obtained the best result for the blogs collection; LSA outperformed the BoW in the reviews and social media domains, and LIWC was the best approach in the Twitter corpus. In all domains, the BoSA method obtained the best results, and, furthermore, it considerably outperformed the results from the topic-based representations. We consider this is because style information is possible more relevant for gender classification than for age prediction.

Table 8 shows the three topics with the greatest information gain for LSA and LIWC. It is interesting to notice that, such as some previous works have pointed out, the some of the topics that helped mostly to distinguish between women and men are those related to work, home and leisure.

Table 7. Accuracy results for gender classification in four social media domains

Approach	Blogs	Reviews	Social Media	Twitter
BoW	**0.72(±0.13)**	0.62(±0.02)	**0.52(±0.02)**	0.70(±0.08)
LSA	0.70(±0.10)	**0.65(±0.01)**	**0.52(±0.02)**	0.66(±0.11)
LIWC	0.60(±0.13)	0.62(±0.01)	0.50(±0.01)	**0.71(±0.07)**
BSoA	0.78	0.69	0.55	0.71

Table 8. The topics with more information gain for gender classification

Domain	LSA 1	LSA 2	LSA 3	LIWC 1	LIWC 2	LIWC 3
Blogs	love	women	tutorial			
	holiday	diet	media	insight	cognitive	work
	conference	food	social			
	system	eat	inventor			
Reviews	lovely	beach	pool			
	wonder	park	bathroom	sexual	biological	social
	great	place	bed			
	nice	york	resort			
Social media	handbag	jersey				
	vuitton	outlet	_	sad	tentative	negative emotion
	louis	jordan				
	bag	replica				
Twitter	wp	instagram	tumblr			
	seo	sigir	instagram	work	home	leisure
	beso	cikm	linkedin			
	swim	trec	vine			

6   Conclusions

This paper studied the relevance of content features for the author profiling task. It proposed using *topic-based representations* to better capture and exploit the thematic information from the documents. The described experiments mainly focused on evaluating the effectiveness of two topic-based representations, LSA and LIWC, to predict gender and age of users from four different social media domains.

The obtained results provide evidence that topic-based representations outperform the traditional BoW representation. Also, these results are comparable to those from a current state of the art approach, which considers content and style information, indicating that content information is highly informative for the AP task. In particular, content information was very important to predict the age of users from social media domains; in the case of gender classification the results were not as conclusive as in the age classification, showing that style information is possible more relevant for discriminating between men and women.

Regarding the use of LSA and LIWC, the results indicate that topics automatically discovered from the training set are, in most of the cases, a better representation for AP than using a set of manually defined topics. However, for the collections having a small number of training examples and high vocabulary richness, such as Twitter, the best results were obtained using the manually defined topics from LIWC.

Acknowledgments. This work was partially supported by CONACYT under scholarships 401887 and 243957, project 247870, and the Thematic Network in Language Technologies, projects 260178 and 271622.

References

1. Argamon, S., Koppel, M., Pennebaker, J.W., Schler, J.: Mining the blogosphere: age, gender and the varieties of self-expression. First Monday **12**(9) (2007)
2. Argamon, S., Koppel, M., Pennebaker, J.W., Schler, J.: Automatically profiling the author of an anonymous text. Commun. ACM **52**(2), 119–123 (2009)
3. Bergsma, S., Post, M., Yarowsky, D.: Stylometric analysis of scientific articles. In: Proceedings of the 2012 Conference of the North American Chapter of the Association for Computational Linguistics: Human Language Technologies, pp. 327–337. Association for Computational Linguistics (2012)
4. Deerwester, S., Dumais, S.T., Furnas, G.W., Landauer, T.K., Harshman, R.: Indexing by latent semantic analysis. J. Am. Soc. Inf. Sci. **41**(6), 391 (1990)
5. Eckert, P.: Age as a sociolinguistic variable. In: The Handbook of Sociolinguistics, pp. 151–167 (1997)
6. Evangelopoulos, N.E.: Latent semantic analysis. Wiley Interdiscip. Rev.: Cogn. Sci. **4**(6), 683–692 (2013)
7. Fan, R.E., Chang, K.W., Hsieh, C.J., Wang, X.R., Lin, C.J.: Liblinear: a library for large linear classification. J. Mach. Learn. Res. **9**, 1871–1874 (2008)
8. Fink, C., Kopecky, J., Morawski, M.: Inferring gender from the content of tweets: a region specific example. In: ICWSM (2012)
9. Garera, N., Yarowsky, D.: Modeling latent biographic attributes in conversational genres. In: Proceedings of the Joint Conference of the 47th Annual Meeting of the ACL and the 4th International Joint Conference on Natural Language Processing of the AFNLP, vol. 2, pp. 710–718. Association for Computational Linguistics (2009)
10. Goswami, S., Sarkar, S., Rustagi, M.: Stylometric analysis of bloggers age and gender. In: Third International AAAI Conference on Weblogs and Social Media (2009)
11. Holmes, J., Meyerhoff, M.: The Handbook of Language and Gender, vol. 25. Wiley, Hoboken (2008)
12. Iqbal, H.R., Ashraf, M.A., Nawab, R.M.A.: Predicting an author's demographics from text using topic modeling approach (2015)
13. Kahn, J.H., Tobin, R.M., Massey, A.E., Anderson, J.A.: Measuring emotional expression with the linguistic inquiry and word count. Am. J. Psychol. 263–286 (2007)
14. Koppel, M., Argamon, S., Shimoni, A.R.: Automatically categorizing written texts by author gender. Lit. Linguist. Comput. **17**(4), 401–412 (2002)
15. Landauer, T.K., Dumais, S.T.: A solution to plato's problem: the latent semantic analysis theory of acquisition, induction, and representation of knowledge. Psychol. Rev. **104**(2), 211 (1997)
16. Landauer, T.K., Foltz, P.W., Laham, D.: An introduction to latent semantic analysis. Discourse processes **25**(2–3), 259–284 (1998)
17. Landauer, T.K., McNamara, D.S., Dennis, S., Kintsch, W.: Handbook of Latent Semantic Analysis. Psychology Press, Abingdon (2013)

18. López-Monroy, A.P., Montes-y-Gómez, M., Escalante, H.J., Villaseñor-Pineda, L.: Using intra-profile information for author profiling. In: CLEF (Working Notes) (2014)

19. López-Monroy, A.P., y Gómez, M.M., Escalante, H.J., Villaseñor-Pineda, L., Stamatatos, E.: Discriminative subprofile-specific representations for author profiling in social media. Knowl.-Based Syst. **89**, 134–147 (2015)

20. McCollister, C., Huang, S., Luo, B.: Building topic models to predict author attributes from twitter messages (2015)

21. Meina, M., Brodzinska, K., Celmer, B., Czokow, M., Patera, M., Pezacki, J., Wilk, M.: Ensemble-based classification for author profiling using various features notebook for PAN at CLEF 2013. In: CLEF (Working Notes) (2013)

22. Newman, M.L., Groom, C.J., Handelman, L.D., Pennebaker, J.W.: Gender differences in language use: an analysis of 14,000 text samples. Discourse Process. **45**(3), 211–236 (2008)

23. Nguyen, D., Gravel, R., Trieschnigg, D., Meder, T.: How old do you think i am?: A study of language and age in twitter. In: Seventh International AAAI Conference on Weblogs and Social Media (2013)

24. Nguyen, D., Smith, N.A., Rosé, C.P.: Author age prediction from text using linear regression. In: Proceedings of the 5th ACL-HLT Workshop on Language Technology for Cultural Heritage, Social Sciences, and Humanities, pp. 115–123. Association for Computational Linguistics (2011)

25. Pennacchiotti, M., Popescu, A.M.: Democrats, republicans and starbucks afficionados: user classification in twitter. In: Proceedings of the 17th ACM SIGKDD International Conference on Knowledge Discovery and Data Mining, KDD 2011, pp. 430–438. ACM (2011). http://doi.acm.org/10.1145/2020408.2020477

26. Pennebaker, J.W., Stone, L.D.: Words of wisdom: language use over the life span. J. Personal. Soc. Psychol. **85**(2), 291 (2003)

27. Rangel, F., Rosso, P., Chugur, I., Potthast, M., Trenkmann, M., Stein, B., Verhoeven, B., Daelemans, W.: Overview of the author profiling task at PAN 2014. In: CLEF (Online Working Notes/Labs/Workshop), pp. 898–927 (2014)

28. Rangel, F., Rosso, P., Koppel, M., Stamatatos, E., Inches, G.: Overview of the author profiling task at PAN 2013. In: Notebook Papers of CLEF 2013 LABs and Workshops, CLEF-2013, Valencia, Spain, September, pp. 23–26 (2013)

29. Rude, S., Gortner, E.M., Pennebaker, J.: Language use of depressed and depression-vulnerable college students. Cogn. Emot. **18**(8), 1121–1133 (2004)

30. Sarawgi, R., Gajulapalli, K., Choi, Y.: Gender attribution: tracing stylometric evidence beyond topic and genre. In: Proceedings of the Fifteenth Conference on Computational Natural Language Learning, pp. 78–86. Association for Computational Linguistics (2011)

31. Schler, J., Koppel, M., Argamon, S., Pennebaker, J.: Effects of age and gender on blogging. In: Proceedings of 2006 AAAI Spring Symposium on Computational Approaches for Analyzing Weblogs, pp. 199–205 (2006)

32. Schwartz, H.A., Eichstaedt, J.C., Kern, M.L., Dziurzynski, L., Ramones, S.M., Agrawal, M., Shah, A., Kosinski, M., Stillwell, D., Seligman, M.E., et al.: Personality, gender, and age in the language of social media: the open-vocabulary approach. PloS One **8**(9), e73791 (2013)

33. Schwartz, H.A., Eichstaedt, J.C., Dziurzynski, L., Kern, M.L., Blanco, E., Kosinski, M., Stillwell, D., Seligman, M.E., Ungar, L.H.: Toward personality insights from language exploration in social media. In: AAAI Spring Symposium: Analyzing Microtext (2013)

34. Tausczik, Y.R., Pennebaker, J.W.: The psychological meaning of words: LIWC and computerized text analysis methods. J. Lang. Soc. Psychol. **29**(1), 24–54 (2010)
35. Turney, P.: Mining the web for synonyms: Pmi-ir versus lsa on toefl (2001)
36. Weren, E.R., Kauer, A.U., Mizusaki, L., Moreira, V.P., de Oliveira, J.P.M., Wives, L.K.: Examining multiple features for author profiling. J. Inf. Data Manag. **5**(3), 266 (2014)
37. Wiemer-Hastings, P., Wiemer-Hastings, K., Graesser, A.: Latent semantic analysis. In: Proceedings of the 16th International Joint Conference on Artificial Intelligence, pp. 1–14. Citeseer (2004)

Using Robustness to Learn to Order Semantic Properties in Referring Expression Generation

Pablo Ariel Duboue[(⊠)] and Martin Ariel Domínguez

Facultad de Matematica, Astronomía y Física,
Universidad Nacional de Córdoba, Córdoba, Argentina
pablo.duboue@gmail.com

Abstract. A sub-task of Natural Language Generation (NLG) is the generation of referring expressions (REG). REG algorithms aim to select attributes that unambiguously identify an entity with respect to a set of distractors. Previous work has defined a methodology to evaluate REG algorithms using real life examples with naturally occurring alterations in the properties of referring entities. It has been found that REG algorithms have key parameters tuned to exhibit a large degree of robustness. Using this insight, we present here experiments for learning the order of semantic properties used by a high performing REG algorithm. Presenting experiments on two types of entities (people and organizations) and using different versions of DBpedia (a freely available knowledge base containing information extracted from Wikipedia pages) we found that robustness of the tuned algorithm and its parameters do coincide but more work is needed to learn these parameters from data in a generalizable fashion.

1 Introduction

The main research focus in NLG is the creation of computer systems capable of generating human-like language. According to the consensus Natural Language Generation (NLG) architecture [1] the NLG task takes as input non-linguistic data and operates over it as a series of enrichment steps, culminating with fully specified sentences from which output strings can be read out. A step in that process is Referring Expressions Generation (REG), where, given an entity (the **referent**) and a set of competing entities (the **set of distractors**), it involves creating a mention (a Referring Expression, RE) to the referent so that, in the eyes of the reader, it is clearly distinguishable from any other entity in the set of distractors. Therefore REG algorithms are expected to select attributes that unambiguously identify an entity with respect to a set of distractors.

Besides identifying the referent unambiguously, REG algorithms are expected to produce a RE that is easy to understand, for example by avoiding long, complex constructs and by referring to properties easily associated with the entity. For an example of REG, see Fig. 2.

A full-fledged REG algorithm selects properties and might also produce a surface form (for example, a noun phrase). In our case (and for most algorithms

© Springer International Publishing AG 2016
M. Montes-y-Gómez et al. (Eds.): IBERAMIA 2016, LNAI 10022, pp. 163–174, 2016.
DOI: 10.1007/978-3-319-47955-2_14

studied in the literature), the construction of the surface form is left to the rest of the NLG system.

A popular algorithm to perform REG is the Incremental Algorithm which assumes the properties P are ordered according to an established criteria. This parameter is called the Preference Order (PO) and learning it from data is the focus on recent research including this present work. Then the algorithm iterates over P, adding each triple one at a time and removing from the confusor set C all entities ruled out by the new triple. Triples that do not eliminate any new entity from C are ignored. The algorithm terminates when C is empty. This algorithm was created in 1995 [2] as a simplification of previous work on the development of REG algorithms.

The PO is thus key to the Incremental Algorithm: the algorithm can generate minimal or verbose REs depending on this ordering. Moreover, the original authors of the algorithm contend that a psycholinguistic motivation lies behind the ordering (a claim contended by other authors, see next section). More recent research has reinforced the idea that most popular properties should be considered and ascertained first. Our experiments in this present work further agrees with this previous research.

In previous work [4], we found that REG algorithms are very robust in the presences of data evolution: a RE calculated on a past version of the data holds remarkably well when the world to which the data refers to naturally evolves. Using this insight, in the current work we designed and performed experiments using robustness as an extra source of information to learn the PO employed by the Incremental Algorithm.

Our results so far speak of a coincidence between robustness and a target PO written by hand but the effect is not enough to learn the PO in a generalizable manner. We suspect this is due to differences in the nature of the ontological changes between the two types as organizations seldom change their properties, but tend to have more information added to them on Wikipedia.

2   Related Work

Dale and Reiter [2] introduced the Incremental Algorithm and compared it against other state-of-the-art algorithms at that time (Full Brevity and Greedy algorithm). They conclude that, first, Incremental Algorithm produces referring expressions more similar to human ones. Second, Incremental Algorithm is computationally more efficient than these competitors. Van Deemter et al. [3] detail a set of experiments based on the annotated TUNA corpus of reference expressions [6], with the objective of assessing the claims made by Dale and Reiter. They conclude that different POs lead to different referring expressions. And, given that the Incremental Algorithm's competitors performed quite reliably, they speculated that, in some domains with no corpus available, other algorithms may perform better.

In a recent review of the state-of-the-art in REG [7], the authors devote a significant amount of time to the Incremental Algorithm, acknowledging that the frequency of a property in naturally occurring REs gives a preference to have it earlier in the PO [3,5]. However, corpora of REs are hard to come by and building them is also costly. Therefore, a question asked by the authors is how much data is necessary to obtain a reasonable PO. The answer might lie in studying different learning curves for different corpora size and comparing improvements for different quality metrics.

Krahmer and colleagues [8] investigated how difficult it is to find "good" POs for Incremental Algorithm, also the topic of the present work, although ours is in a much larger scale. They decided the PO by counting the frequencies of properties of reference expressions in the TUNA corpus, that is, the frequencies of properties that discriminate the referent against the distractors. The TUNA corpus consists of two parts: the furniture corpus, consisting of stylized pictures of furniture; and the people corpus, consisting of descriptions of photographic portraits. They did the counting over a small part of the corpus. They use a training set (165 type of furnitures and 136 peoples) to determine the PO based on the frequency of properties selected in the annotated corpus. Then, they test the resulting orders by using Dice and PRP scores (see Sect. 4.1), and they found that Incremental Algorithm perform very well, particularly for small examples.

In contrast, in this work we do not use an annotated corpus. In our approach in Sect. 4 using the popularity ordering, we compute the frequencies of properties from the set of entities available for experiments.

3   Data

We have chosen Wikipedia as our source of entities, as it represents one of the biggest freely available knowledge base. Started in January 2001 at present it contains over 37 million articles in 284 languages and it continues to grow thanks to the collaborative creation of content by thousands of users around the globe.

Given that the content in Wikipedia pages is stored in a structured way, it is possible to extract and organize it in an ontology-like manner as implemented in the DBpedia community project. This is accomplished by mapping Wikipedia infoboxes from each page to a curated shared ontology that contains 529 classes and around 2.330 different properties. DBpedia contains the knowledge from 111 different language editions of Wikipedia and, for English the knowledge base consists of more than 400 million facts describing 3.7 million things [11]. A noble feature of this resource is that it is freely available to download in the form of *dumps* or it can be consulted using specific tools developed to query it.

These dumps contain the information coded in a language called Resource Description Framework (RDF) [9]. The WWW Consortium (W3C) has developed RDF to encode the knowledge present in web pages, so that it is comprehensible and exploitable by agents during any information search. RDF is based on the concept of making statements about (web) resources using expressions in the subject-predicate-object form. These expressions are known as triples, where the subject denotes the resource being described, the predicate denotes a characteristic of the subject and describes the relation between the subject and

Example tasks for people:

1. Hillary_Rodham_Clinton   Bill_Clinton   Bob_Rae   Bashar_al-Assad
   Nancy_Pelosi Barack_Obama
2. Stephen_Colbert Inez_Tenenbaum Barack_Obama
3. John_Edwards Howard_Dean Barack_Obama
4. Susan_Sarandon Martin_Sheen Fidel_Castro
5. John_McCain Dan_Rather Viktor_Yushchenko

Example tasks for organizations:

1. Verizon_Communications Kansas_City,_Missouri Verizon_Wireless
2. Los_Angeles_Times DirecTV
3. Egypt Israel CBC_News
4. Wikimedia Foundation Brussels

Fig. 1. Example tasks. See Fig. 2 for an example.

```
<http://dbpedia.org/resource/Barack\_Obama>
<http://www.w3.org/1999/02/22-rdf-syntax-ns#type>
<http://dbpedia.org/ontology/Politician>

<http://dbpedia.org/resource/Stephen\_Colbert>
<http://www.w3.org/1999/02/22-rdf-syntax-ns#type>
<http://dbpedia.org/ontology/Artist>

<http://dbpedia.org/resource/Inez\_Tenenbaum>
<http://www.w3.org/1999/02/22-rdf-syntax-ns#type>
<http://dbpedia.org/ontology/Person>
```

Fig. 2. DBpedia triples for a REG task from Fig. 1. When asking for Stephen Colbert, it can be referred to as the "artist".

the object. A collection of such RDF declarations can be formally represented as a labeled directed multi-graph, naturally appropriate to represent ontologies.

We have chosen to use the dumps of different versions of Wikipedia, namely versions 2014 (09/2014) and 3.6 (01/2011). DBpedia 3.6 ontology encompasses 359 classes and 1775 properties (800 object properties, 859 datatype properties using standard units, 116 datatype properties using specialized units) and DBpedia 2014 ontology encompasses 685 classes and 2795 properties (1079 object properties, 1600 datatype properties using standard units, 116 datatype properties using specialized units)[1] These versions have been specifically selected to be able to compare our results with those obtained by Pacheco et al. [12].

We followed an established approach to extract potential REG tasks from journalistic text: we extracted all people that appear explicitly linked in a

---

[1] Statistics taken from the DBpedia change log available at http://wiki.dbpedia.org/services-resources/datasets/change-log.

```
Former [[New Mexico]] {{w|Governor of New Mexico|governor}} {{w|Gary Johnson}}
ended his campaign for the {{w|Republican Party (United States)|Republican Party}}
(GOP) presidential nomination to seek the backing of the {{w|Libertarian Party
(United States)|Libertarian Party}} (LP).
```

Fig. 3. Wikinews example, from http://en.wikinews.org/wiki/U.S._presidential_candidate_Gary_Johnson_leaves_GOP_to_vie_for_the_LP_nom

given Wikinews article. By using Wikinews, we ensured all the people were disambiguated to their DBpedia URIs by construction (Fig. 3).[2] We selected a Wikinews dump as closer to our target DBpedia (20140901). From there, we defined all URIs for which DBpedia has a birthDate relation (761,830 entities) as "people" and everything with a foundDate as an "organization" (19,694 entities). We extracted all such people and organizations that appear in the same Wikinews article using the provided inter-wiki SQL links file. For each article, we randomly chose a person (or organization) as the referent, turning them into a fully defined REG task. This approach produced 4,741 different REG tasks, over 9,660 different people and 3,062 over 8,539 organizations. Each REG task has an average of 4.89 people per task or 2.78 organizations per task. We then created a subset of the relevant tuples for these people (291,039 tuples on DBpedia 2014 and 129,782 on DBpedia 3.6, a 224 % increase[3] for people, and 468,041 tuples and 216,730 tuples for organizations).

4   Experimental Setup

In our work, we want to use the REG tasks obtained from Wikinews to measure of how well a particular PO is. For that we execute the Incremental Algorithm[4] with the given PO and all the tasks and measure how well the obtained REs perform on the new data. These metrics (discussed next) together with other metrics (average length of the obtained RE) are the basis for a quality metric over POs we then perform a Genetic Algorithm over it (Sect. 4.2).

4.1   Metrics

We have two sets of metrics. First, the metrics we use to measure robustness, which we will explain now and then metrics to measure whether the PO is similar to a hand-written one. To measure robustness, let's say we have entities {A, B, C, D}. We need to select C and we choose properties prop1 and prop2 as RE. In a new version of the data, prop1 and prop2 select a new subset of {A, B, C, D}. We can compute Dice on that. We can also count how many times C is no longer

---

[2] These *potential* REG tasks, but not *actual* REG tasks. We use the news article to extract naturally co-occurring entities.

[3] In DBpedia 2014, there was an average of 30.12 properties per person while in DBpedia 3.6, there was an average of 17.3.

[4] Our publicly available implementation: https://github.com/DrDub/Alusivo.

selected (exclusion errors) and how many time other entities different from C are selected (inclusion errors) and the total number of either error (number of errors). These are our robustness metrics (the four numbers).

To measure whether the learned POs were in any way similar to the hand-written PO (Fig. 4), as we are trying to learn an ordering, we initially computed the Kendall's τ [10]:

$$\tau = 1 - \frac{2(\text{number of inversion})}{N(N-1)/2}$$

where N is the number of objects (atomic values) and inversions is the number of exchanges on consecutive objects required to put them in the order appearing in the hand tagged reference. However, the τ proved to be too strict and we decided to move to a metric that considers the REs being generated rather than the exact ordering.

Instead, we wanted to know whether the new PO selected the same properties as the hand-written one. For that we turned to Dice over the selected properties.

Besides the hand-picked PO from Fig. 4, we also ordered the properties according to their total count on DBpedia. We call that the "popularity" PO (Fig. 5).

TYPE   ORDERINOFFICE   NATIONALITY   COUNTRY   PROFESSION   BIRTHPLACE LEADERNAME^{-1} KEYPERSON^{-1} AUTHOR^{-1} COMMANDER^{-1} OCCUPATION KNOWN-FOR INSTRUMENT SUCCESSOR MONARCH SUCCESSOR^{-1} PRIMEMINISTER^{-1} AC-TIVEYEARSENDDATE PARTY DEATHDATE DEATHPLACE CHILD ALMAMATER AC-TIVEYEARSSTARTDATE RELIGION SPOUSE PRESIDENT^{-1} NOTABLECOMMANDER^{-1} VICEPRESIDENT PRESIDENT PRIMEMINISTER AWARD MILITARYRANK CHILD^{-1} MILI-TARYCOMMAND SERVICESTARTYEAR OFFICE BATTLE SPOUSE^{-1} KNOWNFOR^{-1} PRE-DECESSOR FOUNDATIONPERSON^{-1} MONARCH^{-1} PREDECESSOR^{-1} ACTIVEYEARSSTAR-TYEAR ACTIVEYEARSENDYEAR STARRING^{-1} LIEUTENANT PARENT GOVERNOR^{-1} HOMEPAGE RESIDENCE APPOINTER^{-1} SUBJECT^{-1} PARENT^{-1} OCCUPATION^{-1} REGION STATEOFORIGIN EMPLOYER GENRE HOMETOWN ASSOCIATEDMUSICALARTIST ASSOCI-ATEDBAND GOVERNOR DEPUTY VICEPRESIDENT^{-1} LIEUTENANT^{-1} GOVERNORGENERAL GOVERNORGENERAL^{-1} INFLUENCED^{-1} INFLUENCEDBY TEAM MANAGERCLUB INFLU-ENCED GRAMMYAWARD STATISTICLABEL FORMERTEAM OTHERPARTY ASSOCIATE^{-1} AS-SOCIATE RECORDLABEL MILITARYBRANCH MILITARYUNIT DEPUTY^{-1} BEATIFIEDBY^{-1} ASSOCIATEDBAND^{-1} ASSOCIATEDMUSICALARTIST^{-1} RELATION COLLEGE DRAFTTEAM CHANCELLOR^{-1} INCUMBENT^{-1}

Fig. 4. Target Property Ordering for people. A property with a -1 superscript implies being the receiving end of a property. This ordering was done by hand taking into account the nuisances of the data. For example, ORDERINOFFICE is a string that contains an extended information about an occupation ("Governor of Wisconsin" vs. "Politician").

The hand-picked PO in Fig. 4 was elucidated by an AI expert over the period of several weeks of running the Incremental Algorithm over a development set.

TYPE BIRTHPLACE TEAM COUNTRY GENRE STARRING^{-1} RECORDLABEL HOMEPAGE AC-
TIVEYEARSSTARTYEAR OCCUPATION OCCUPATION^{-1}

Fig. 5. Popularity PO (beginning).

When the Incremental Algorithm algorithm does not have any information
about the entities, it fails. For the tasks extracted from a version of Wikinews
close in time to the new version of DBpedia, the old version misses a significant
number of people, rendering 40 % of the people tasks unusable. The situation is
much better for organizations, as there it fails only on 13 % of the cases.

4.2   A Genetic Algorithm Set-Up for Person Priorities Ontology

This section introduces the search space and the strategy to traverse it in our
optimization procedure. The search space consists of different permutation in the
PO within the people ontology, which are an input parameter for the Incremental
Algorithm implementation. The PO contains 92 properties.

As all possible permutations can be candidate POs the search space is rather
large, all 96! permutations are equally likely. We traverse this search space using
Genetic Algorithms starting from a population of POs slightly modified from
the popularity ordering. Genetic Algorithms require three main components:
(1) Definition of individuals: each individual codifies one candidate PO that is
being optimized. (2) A fitness function defined over individuals: the quality measure is computed by computing the amount of errors obtained for this candidate
PO as described in the previous section. Finally, (3) A strategy for evolution:
we apply two different operations to individuals: ad-hoc crossover and mutation,
defined below; crossover gets 0.8 probability of being applied, while mutation
gets 0.08. We used as the selection strategy the *tournament selection* with 7 as
parameter. In our experiments, in each generation there is a population of 200
individuals; we let the population evolve for 50 generations.

We define an individual as an array of 92 elements, one of type *String* which
represents the PO.

The *mutation* function is easily defined by computing a swap between two
elements randomly selected. The *crossover* operation is defined as follows. Let
$[g_1, \ldots, g_i, \ldots, g_n]$ and $[h_1, \ldots, h_i, \ldots, h_n]$ be the "order of properties" vector of
two different individuals and let i be random number between 1 and n. The
crossover produces two new individuals. The new properties permutation vector
of one of the individuals is defined as follows:

$$sub([g_1, \ldots, g_n], [h_{i+1}, \ldots, h_n]) \cdot [h_{i+1}, \ldots, h_n]$$

where the operator ($\cdot$) appends two arrays, and $sub(x,y)$ deletes the elements
in x that are in y, maintaining the order of elements. The vector of the other
individual is defined similarly by changing g by h and vice versa. In this way,
crossover ensures that the resulting permutation vector do not have repeated
properties.

For the case of organizations, the hand-written PO has 554 properties. Trying to learn a full ordering on them would have been a fool's errand so we focus on the top 92 most frequent ones.

For searching, we can start from a random PO, but then the search process will take a very long time. Instead, as we know popularity PO is a good ordering, we started from a population obtained by doing 9 random swaps away from popularity PO.

5   Results

We performed seven experiments. In the first one, we took 10,000 random POs for people properties and executed them over the 4,741 REG tasks for people described in Sect. 3. We then calculated the average length of the generated REs and the error metrics described in Sect. 4.1 when the RE calculated on the old version of DBpedia is then applied to the new version. We also computed the set overlap of selected properties using Dice between the random PO RE and the RE obtained by the hand-written PO. We also did this computation for the popularity PO. Finally, we computed the Dice overlap between the hand-written PO and the popularity PO, which for the case of people is 0.749. Therefore, we have 10,000 rows of numbers, some are observable (length and robustness metrics), the other is our target class (what we want to learn). Before attempting any learning, we want to see whether the observable variables change similarly to the target. For this purpose we compute Spearman's rho (Table 1, first row). In the table we can see that length has a low rho value but the different metrics for errors are, particularly the inclusion error. This experiment is thus a success.

In the second experiment, we want to learn a function that approximates the target (Dice for the hand-picked PO) using the observable variables. We tried a few ML algorithms but our data is simple enough to be tackled with a straight linear regression:

$$target = -1.608 * length + 15.5279 * inclusion + 0.8787 * exclusion + 1.9403$$

performed well with a Pearson's correlation coefficient of 0.872 (although a high relative absolute error or 49.6 %). We consider this experiment also a success.

In our third experiment, we tested whether the Genetic Algorithm solution described in Sect. 4.2 was capable of learning the target function. The learning curve in Fig. 6 shows that it seems to be the case. The best instance has a Dice property overlap with the hand-written PO of 0.774, which exceeds the Dice for popularity PO of 0.749 (we do not analyze whether this differences are statistically significant as these are overfitted results, we used the hand-written PO to derive the fitness function). This experiment thus validates our Genetic Algorithm system.

Now, in experiment four, we took the trained fitness from Experiment 2 and apply it to the same Genetic Algorithm of Experiment 3 but this time over organization data (reduced to 92 most frequent properties). The results here

Table 1. Spearman's rho between hand-written PO and popularity PO and observable metrics, over 10,000 random POs

Exp/metric	Length	Dice	Exclusion errors	Inclusion errors
People				
Hand-written	−0.018	−0.215	0.185	0.397
Popularity	−0.226	0.232	−0.258	0.394
Organization				
Hand-written	0.059	0.832	−0.834	0.840
Popularity	−0.064	0.864	−0.866	0.841

were a disappointment: after 50 generations, we get a Dice property overlap of only 0.435, when popularity PO achieves 0.93. This is our main negative result.

Trying to shed more light to the situation, we performed Experiment 1 over organizations, yielding the third row of Table 1. We can see strong numbers for Spearman's rho, but very different from people's numbers. We further discuss this situation in the next section. Still, there is indication that robustness should help approximate the hand-picked PO.

In our sixth experiment, we attempted to use only length and inclusion errors and see whether this generalizes (this will be a preliminary result, as we obtained this insight from looking at both tables, which contains the test set). When trained in people, we obtain a Dice in organizations of 0.906. When trained in organizations, we obtain a Dice of 0.608. These numbers are below the popularity PO but show more generalization strength than before.

Finally, we did a Genetic Algorithm on both people and organization using only inclusion errors as fitness function. We obtained a Dice of 0.272 and 0.361, respectively. Therefore robustness alone is not enough, combining it with length is key.

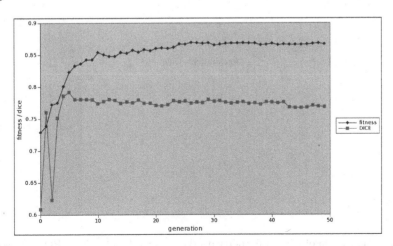

Fig. 6. Learning curve for people using Genetic Algorithms.

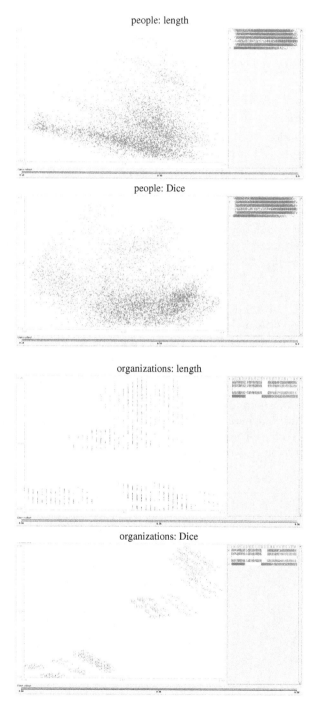

Fig. 7. Visualization of Dice of property overlap against hand-written PO and two key features: length and robustness, both over people and organizations.

6   Discussion

While it is always disappointing when experiments do not turn out the way we expected (the model on people did not generalize well to organizations), we found our results shed light in the complex nature of the REG problem in several levels. First, the Spearman's rho values from Table 1 speak of a clear coincidence between robustness and the hand-written PO. There was no *a priori* reason for this to be the case, both events are very different. These values show the intuitions behind our experiments were in the right track and are the main contribution of this work. The fact that Genetic Algorithms can learn an improved ordering in the case of an overfitted fitness function seems to indicate that Genetic Algorithms are up to the task, when initialized from perturbations around the popularity ordering. While this contribution has not yet realized its full potential, we hope other researchers in the field will continue this work seeing it has shown potential here.

The main disappointment has been in the lack of generalization between people and organizations. Further plotting target Dice vs length and robustness Dice shows that the behavior of the system in each problem is *remarkably* different.

In any learning setting, for generalization to occur, both the train and test sets should be drawn from the same underlining population. From Fig. 7 we can see that this assumption was not the case and that computing REG for people is a different problem altogether than organizations, when taking into account robustness. This difference can be explained by difference in nature of ontological changes on both populations. People properties do change and quite often, while organizations very seldom change properties. Most of the changes between versions of DBpedia relating to properties in organizations lie in contributors completing gaps in the encyclopedic knowledge of the entities. Now, it is impossible to tell apart bonafide changes to an entity from information that it is being added but was true also beforehand. That is a limitation of using this dataset.

From a most practical perspective, given the levels of Dice overlap between hand-picked PO and popularity PO, we are inclined to recommend using popularity PO for most cases (interestingly, the have a very low τ, they look different but behave similarly). This is also a result from the current work and we plan to make popularity POs available alongside with the human curated ones in our open source REG toolkit.

7   Conclusions

Among NLG subproblems, REG has received an unusual amount of attention due to its appeal of being a relatively self-contained problem that encompasses issues in content selection, ontological modeling and psycholinguistics. Using intuitions gathered in previous work, we set ourselves to contribute an extra element in the search for better REG: generating a robust RE should be also a desirable characteristic, akin to generating short REs and REs that are easy to understand. We had hoped that length of the RE, popularity of the selected properties and robustness were enough to learn a reasonable PO for the Incremental Algorithm. That has not been shown by our experiments but we have shown robustnesss coincides enough with a hand-written PO to secure itself as a reasonable source of information in future experiments.

In future work, we want to explore simulating robustness score by producing a simulated "old" version of the data by randomly deleting tuples. We also want to learn POs for subtypes of entities, for example, soccer players and politicians.

Acknowledgments. The authors would like to thank Annie Ying and the three anonymous reviewers for comments and suggestions.

References

1. Cahill, L., Carroll, J., Evans, R., Paiva, D., Power, R., Scott, D., van Deemter, K.: From rags to riches: exploiting the potential of a flexible generation architecture. In: Proceedings of the 39th Annual Meeting on Association for Computational Linguistics, pp. 106–113. Association for Computational Linguistics (2001)
2. Dale, R., Reiter, E.: Computational interpretations of the gricean maxims in the generation of referring expressions. Cogn. Sci. **19**(2), 233–263 (1995)
3. van Deemter, K., Gatt, A., van der Sluis, I., Power, R.: Generation of referring expressions: assessing the incremental algorithm. Cogn. Sci. **36**(5), 799–836 (2012)
4. Duboue, P., Domínguez, M., Estrella, P.: Evaluating robustness of referring expression generation algorithms. In: Proceedings of Mexican International Conference on Artificial Intelligence 2015. IEEE Computer Society (2015)
5. Gatt, A., Belz, A.: Empirical Methods in Natural Language Generation: Data-oriented Methods and Empirical Evaluation. Springer, Heidelberg (2010)
6. Gatt, A., van der Sluis, I., van Deemter, K.: Evaluating algorithms for the generation of referring expressions using a balanced corpus. In: Proceedings of the Eleventh European Workshop on Natural Language Generation, ENLG 2007, pp. 49–56. Association for Computational Linguistics, Stroudsburg, PA, USA (2007). http://dl.acm.org/citation.cfm?id=1610163.1610172
7. Krahmer, E., Deemter, K.V.: Computational generation of referring expressions: a survey. Comput. Linguist. **38**, 173–218 (2009)
8. Krahmer, E., Koolen, R., Theune, D.M.: Is it that difficult to find a good preference order for the incremental algorithm? Cogn. Sci. **36**(5), 837–841 (2012)
9. Lassila, O., Swick, R.R., Wide, W., Consortium, W.: Resource description framework (rdf) model and syntax specification (1998)
10. Lebanon, G., Lafferty, J.: Combining rankings using conditional probability models on permutations. In: Sammut, C., Hoffmann, A. (eds.) Proceedings of the 19th International Conference on Machine Learning. Morgan Kaufmann Publishers, San Francisco, CA (2002)
11. Lehmann, J., Isele, R., Jakob, M., Jentzsch, A., Kontokostas, D., Mendes, P.N., Hellmann, S., Morsey, M., van Kleef, P., Auer, S., Bizer, C.: DBpedia - a large-scale, multilingual knowledge base extracted from wikipedia. Semant. Web J. **6**(2), 167–195 (2015)
12. Pacheco, F., Duboue, P.A., Domínguez, M.A.: On the feasibility of open domain referring expression generation using large scale folksonomies. In: Proceedings of the 2012 Conference of the North American Chapter of the Association for Computational Linguistics: Human Language Technologies, NAACL HLT 2012, pp. 641–645. Association for Computational Linguistics, Stroudsburg, PA, USA (2012). http://dl.acm.org/citation.cfm?id=2382029.2382136

Conditional Random Fields for Spanish Named Entity Recognition Using Unsupervised Features

Jenny Copara[1(✉)], Jose Ochoa[1], Camilo Thorne[2], and Goran Glavaš[2]

[1] Universidad Católica San Pablo, Arequipa, Peru
{jenny.copara,jeochoa}@ucsp.edu.pe
[2] Data and Web Science Group, Universität Mannheim, Mannheim, Germany
{camilo,goran}@informatik.uni-mannheim.de

Abstract. Unsupervised features based on word representations such as word embeddings and word collocations have shown to significantly improve supervised NER for English. In this work we investigate whether such unsupervised features can also boost supervised NER in Spanish. To do so, we use word representations and collocations as additional features in a linear chain Conditional Random Field (CRF) classifier. Experimental results (82.44 % F-score on the CoNLL-2002 corpus) show that our approach is comparable to some state-of-art Deep Learning approaches for Spanish, in particular when using cross-lingual word representations.

Keywords: NER for Spanish · Word representations · Collocations · Conditional random fields

1 Introduction

Supervised Named Entity Recognition (NER) system are typically fed with supervised or manually engineered features [9] such as, e.g., word capitalization or domain-specific lexicons (lists of words related with named entity types) [5,17,19]. The performance of such techniques however depends on the availability, quality and size of annotated data, which can be scarce for NER for languages other than English. More recently, it has been shown that supervised NER can be boosted via unsupervised word features induced from corpora [23], such as **(i)** very large word clusters [3,13], **(ii)** word collocations [13], and **(iii)** very large word embeddings [6,7,14,15]. Such techniques show in particular that it is possible to take advantage of unlabeled data to enrich and boost supervised NER models learned over small gold standards.

For English NER, [11,17] show that (large) word embeddings yield better results than clustering. However, when combined and fed as features to linear chain Conditional Random Field (CRF) sequence classifiers, they yield models comparable to state-of-the-art deep learning models.

In this paper we investigate whether these techniques can be successfully applied to NER in Spanish. In order to do so, we follow Guo's approach in [11], combining probabilistic graphical models in the form of CRFs learned from

© Springer International Publishing AG 2016
M. Montes-y-Gómez et al. (Eds.): IBERAMIA 2016, LNAI 10022, pp. 175–186, 2016.
DOI: 10.1007/978-3-319-47955-2_15

the CoNLL 2002 corpus with word representations learned from large unlabeled Spanish corpora, while exploring the optimal setting and feature combinations that match state-of-the-art algorithms for NER in Spanish.

The paper is organized as follows. In Sect. 2 we provide a review of Spanish NER and unsupervised word features. Section 3 describes the structure of the word representations used. Section 4 shows our experimental setting and discusses our results. Section 5 presents our final remarks.

2   Related Work

2.1   Spanish NER

The first results (CoNLL 2002 shared-task[1]) for supervised Spanish NER were obtained by Carreras et al. [5]. A set of selected word features and lexicons (gazetteers) on an Adaboost learning model were used, obtaining an F-score of 81.39 %. These results remained unbeaten until recently, and the spread of *Deep Learning* (currently achieving an F-score of 85.77 %). The main algorithms that are currently used for NER in Spanish are: Convolutional Neural Networks with word and character embeddings [7], Recurrent Neural Networks (RNNs) with word and character embeddings [12,25], and a character-based RNN with characters encoded as bytes [10].

2.2   Unsupervised Word Features

Unsupervised features based on word representations and word collocations have been successfully used to boost many Natural Language Processing (NLP) tasks (e.g., language modeling [3], English NER [8,11,13,17,23], German NER [8], chunking [23], Chinese word segmentation [13]).

There are two main approaches used to induce word representations. One approach is to compute either clusters [3] or [13] Brown Clustering from unlabeled data and using them as features in NLP models (including NER). Another approach transforms each word into a continuous real-valued vector [6] of n dimensions also known as a "word embedding" [14]. With Brown clustering, words that appear in the same or a similar sentence context are assigned to the same cluster. Whereas in word embeddings similar words occur close to each other in $\mathbb{R}^n$ (the induced n dimensional vector space). Having more data is better for word representations. Cross-lingual datasets can be used to gather data, provided they overlap in vocabulary and domain. In this sense, cross-lingual word representations have been shown to improve several NLP tasks, such as model learning [1,27]. This is because, among other things, they allow to extend the coverage of possibly limited (in the sense of small or sparsely annotated) resources with resources in other languages, such as: using English to enrich Chinese [27], or learning a model in English to solve a text classification task for German (also German-English, English-French and French-English) [1].

---

[1] http://www.cnts.ua.ac.be/conll2002/ner/.

On the other hand, word collocations have also been used as additional word features to solve NLP tasks. In particular, Chinese word segmentation have been significantly improved by using them [13].

3 Unsupervised Word Features for Spanish NER

3.1 Brown Clustering

Brown clustering is a hierarchical clustering of words that takes a sequence $w_1, \ldots, w_n$ of words as input and returns a binary tree as output (a dendogram). The binary tree's leaves are the input words. This clustering method is based on bigram language models [3,13].

3.2 Clustering Embeddings

A clustering method for embeddings based on *k-means* has been proposed by Yu et al. [26]. In this method, different k clusters values convey different clustering granularity levels. The toolkit Sofia-ml [20] [2] was used to compute such k clusters.

3.3 Binarized Embeddings

The idea behind this method is to "reduce" continuous word vectors $\boldsymbol{w}$ in standard word embeddings into discrete $bin(\boldsymbol{w})$ vectors that however preserve the ordering or ranking of the original vectors. To do this, we need to compute two thresholds per dimension (upper and lower) across the whole vocabulary. For each dimension (component) i is computed the *mean* of positives values (C_{i+}, the upper threshold) and negative values (C_{i-}, the lower one). Thereafter, the following function is used over each component C_{ij} of vector $\boldsymbol{w}_j$:

$$\phi(C_{ij}) = \begin{cases} U_+, & if C_{ij} \geq mean(C_{i+}), \\ B_-, & if C_{ij} \leq mean(C_{i-}), \\ 0, & otherwise. \end{cases} \qquad (1)$$

3.4 Distributional Prototypes

This approach, proposed by Guo in [11] is based on the idea that each entity class has a set of words more likely to belong to this class than the other words (i.e., Maria, Jose are more likely to be classified as a *PERSON* entity). Thus, it is useful to identify a group of words that represent each class (*prototypes*) and select *some of them* in order to use them as word features. In order to compute prototypes two steps are necessary:

---

[2] https://code.google.com/archive/p/sofia-ml/.

1. Generate a prototype for each class of an annotated training corpus. This step relies on Normalized Pointwise Mutual Information (NPMI) [2], as word-entity type relations can be modeled as a form of collocation. NPMI is a smoothed version of the Mutual Information measure typically used to detect word associations [24] and collocations [13]. Given an annotated training corpus, the NPMI is computed between labels l and words w using the following two formulas:

$$\lambda_n(l, w) = \frac{\lambda(l, w)}{-\ln p(l, w)}, \quad \lambda(l, w) = \ln \frac{p(l, w)}{p(l)p(w)}.$$

2. Map the prototypes to words in a (large) word embedding. In this step, given a group of prototypes for each class, we find out which prototypes in our set are the most *similar* to each word in the embeddings. *Cosine similarity* is used to do so and those prototypes above a threshold of usually 0.5 are chosen as the prototype features of the word.

3.5 Collocations

A collocation is defined as two or more lexical items that co-occur in a text or in a text corpus, whether or not they form a syntactic pattern [18]. Collocations are induced from unlabeled data by computing bigram counts and Pointwise Mutual Information [13].

4 Experiments and Discussion

Unlike previous approaches, our work focuses on using unsupervised word features in supervised NER for Spanish. We do it within a probabilistic graphical model framework: CRFs. We trained our (enriched) CRF model over the (Spanish) CoNLL 2002 corpus, and built our unsupervised word features over the Spanish Billion Corpus and English Wikipedia.

For Spanish this is a novel approach. The experimental results show it achieves competitive performance w.r.t. the current (Deep learning-driven) state-of-the-art for Spanish NER, in particular when using *cross-* or *multi-lingual* Word Representations.

4.1 NER Model

In order to perform our NER experiments, a linear chain CRF sequence classifier[3] has been used. Our classifier relies on a set of standard baseline features, that we have extended with additional features based on unsupervised word features and collocations. This setup is depicted in Fig. 1. The classifier was implemented using *CRFSuite* [16], due to its simplicity and the ease with which one can add extra features. Additionally, we tried the Stanford CRF classifier for NER [9], for comparison purposes.

---

[3] http://github.com/linetcz/spanish-ner.

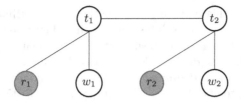

Fig. 1. Linear chain-CRF with word representations as features. The upper nodes are the label sequences, the bottom white nodes are the supervised word features in the model and the filled nodes are the unsupervised word features included in our model.

4.2 Baseline Features

The baseline features were defined over a window of ± 2 *tokens*. The set of features for each word was:

- The word itself.
- Lower-case word.
- Part-of-speech tag.
- Capitalization pattern (e.g. from word "Twitter" we will get ULLLLLL) and type of character in the word(e.g. 'AllUpper', 'AllDigit', 'AllSymbol').
- Characters type information: capitalized, digits, symbols, initial upper case letter, all characters are letters or digits.
- Prefixes and suffixes: four first or latter letters respectively.
- Digit length: whether the current token has 2 or 4 length.
- Digit combination: which digit combination the current token has (alphanumeric, slash, comma, period).
- Whether the current token has just uppercase letter and period mark or contains an uppercase, lowercase, digit, alphanumeric, symbol character.
- Flags for initial letter capitalized, all letter capitalized, all lower case, all digits, all non-alphanumeric characters.

4.3 CoNLL 2002 Spanish Corpus

The CoNLL 2002 shared task [22] gave rise to a training and evaluation standard for supervised NER algorithms used ever since: the CoNLL-2002 Spanish corpus. The CoNLL is tagged using the so-called BIO format for NER gold standards. It covers four entities: *PERSON, ORGANIZATION, LOCATION, MISCELLANEOUS* and nine classes: B-PER, I-PER, B-ORG, I-ORG, B-LOC, I-LOC, B-MISC, I-MISC and O (B for "begin", I for "inside" and O for "outside" an entity mention of any of the four given types).

4.4 Unsupervised Word Features

Spanish Dataset. In order to compute our word representations (viz., the Brown clusters, word embeddings) and word collocations, a large amount of unlabeled data is required. To this end we relied on the Spanish Billion Words (SBW) corpus and embeddings [4]. This dataset was gathered from several public domain resources[4] in Spanish: e.g., a Spanish portion of SenSem, the Ancora Corpus, the Europarl and OPUS Project Corpora, the Tibidabo Treebank and IULA Spanish LSP Treebank and dumps from Spanish Wikipedia, Wikisource and Wikibooks until September 2015 [4]. The corpora cover 3 817 833 *unique* tokens, and the embeddings 1 000 653 *unique* tokens with 300 dimensions per vector.

Cross-Lingual Dataset. Entity names tend to be very similar (often, identical) across languages and domains. This implies that word representation approaches should gain in performance when cross- or multi-lingual datasets are used. To test this hypothesis, we used an English Wikipedia dump from 2012 preprocessed by Guo [11], who removed paragraphs that contained non-roman characters and lowercased words. Additionally they removed very frequent words.

Brown Clustering. The number k of word clusters for Brown clustering was fixed to 1000 following Turian [23]. Sample Brown clusters are shown in Table 1. The cluster is used as feature of each word in the annotated CoNLL 2002. As the reader can see Brown clustering tends to assign entities of same entity type to the same cluster.

Table 1. Brown cluster computed from SBW.

Brown clusters	Word
011100010	Française
011100010	Hamburg
0111100011010	latino
0111100011010	conservador
0111111001111	malogran
0111111001111	paralizaban
011101001010	Facebook
011101001010	Twitter
011101001010	Internet

Binarized Embeddings. Table 2 shows a short view of word "equipo". In the first column we can see each dimension of "equipo" and in the second its continuous value. The third column shows the binarized value. We used the binarized value as feature for each observed word (all dimensions with a *binarized value* different to *zero* will be considered).

---

[4] http://crscardellino.me/SBWCE/.

Table 2. Binarized embeddings from SBW for word "equipo".

Dimension	Value	Binarized
1	−0.008255	0
2	0.145529	U+
3	0.010853	0
⋮	⋮	⋮
298	0.050766	U+
299	−0.066613	B−
300	0.073499	U+

Clustering Embeddings. For cluster embeddings, 500, 1000, 1500, 2000 and 3000 clusters were computed, to model different levels of granularity [11]. As features for each word w, we return the cluster assignments at each granularity level. Table 3 shows the clusters of embeddings computed for word "Maria". The first column denotes the level of granularity. The second column denotes the cluster assigned to "Maria" at each granularity level.

Table 3. Clustering embeddings from SBW for word "Maria".

Granularity k	Cluster
500	31
1000	978
1500	1317
2000	812
3000	812

Distributional Prototypes. We extracted, for each CoNLL BIO label 40 prototypes (the topmost 40 w.r.t. NPMI).

Table 4 shows the top four prototypes per entity class computed from CoNLL-2002 Spanish corpus (training subset). These prototypes are instances of each entity class even non-entity tag(O) and therefore they are compound by entities or entity parts (i.e. *Buenos Aires* is a *LOCATION* so we see the word *Aires* as prototype of I-LOC).

Collocations. Collocations were computed for each word in the the CoNLL 2002 corpus, and added as features. Table 5 depicts collocations for words: Estados" and General".

Table 4. CoNLL-2002 Spanish Prototypes.

Class	Prototypes
B-ORG	EFE, Gobierno, PP, Ayuntamiento
I-ORG	Nacional, Europea, Unidos, Civil
I-MISC	Campeones, Ambiente, Ciudadana, Profesional
B-MISC	Liga, Copa, Juegos, Internet
B-LOC	Madrid, Barcelona, Badajoz, Santander
I-LOC	Janeiro, York, Denis, Aires
B-PER	Francisco, Juan, Fernando, Manuel
I-PER	Alvarez, Lozano, Bosque, Ibarra
O	que, el, en, y

Table 5. Collocations computed for "Estados" and "General"

Word	Collocations	
Estados	los	miembros
	Miembros	Unidos
General	Asamblea	Secretario

4.5   Results

In order to evaluate our proposal the standard `conlleval`[5] script was used. Table 6 shows results achieved on CoNLL-2002 (Spanish), and compares them to Stanford and the state-of-the-art for Spanish NER. The baseline achieved 80.02 % of F-score.

It is worth nothing that *Brown clustering* improves the baseline. The same holds for *clustering embeddings* and *collocations*. By contrast, *binarized embeddings* do worse than the *baseline*. This seems to be due to the fact that the process of binarization apparently discards information quite relevant for Spanish NER. The same holds for *prototypes* which, when taken alone, yield results also below the *baseline*.

Combining the features yields, on the other hand and in all cases, results above the baseline and above Brown clustering or clustering embeddings taken alone.

However, our best results were obtained by using a *cross-lingual combination* of Brown clusters computed from the English Wikipedia dump (2012) with clustered embeddings and prototypes computed from SBW. The same holds when combining Brown clusters, clustered embeddings and prototypes with collocations. The reason why cross-lingual combinations are good in this task is due to the high level of overlap among entities in Spanish and English. Put otherwise,

---

[5] http://www.cnts.ua.ac.be/conll2000/chunking/conlleval.txt.

Table 6. CoNLL2002 Spanish Results. Top: results obtained by us. Middle: results obtained with previous approaches. Down: current Deep Learning-based state-of-the-art for Spanish NER.

Model	F1
Baseline	80.02 %
+Binarization	79.48 %
+Brown	80.99 %
+Prototype	79.82 %
+Collocation	80.23 %
+Clustering	80.24 %
+Clustering+Prototype	80.55 %
+Brown+Collocation	81.04 %
+Brown+Clustering	82.30 %
+Brown+Clustering+Prototype	81.19 %
+Brown+Clustering+Prototype+Collocation	80.96 %
+Brown+Clustering+Prototype+Collocation[a]	82.23 %
+Brown+Clustering+Prototype[a]	**82.44 %**
Carreras [5][b]	79.28 %
Carreras [5]	81.39 %
Finkel [9]	81.44 %
dos Santos [7]	82.21 %
Gillick [10]	82.95 %
Lample [12]	85.75 %
Yang [25]	85.77 %

[a] Brown clusters from English resource
[b] Did not take into in account gazetteers

many entities that share the same name and a similar context occur in texts from both languages, giving rise to features with higher predictive value.

4.6 Discussion

The first results for supervised Spanish NER using the CoNLL 2002 corpus considered a set of features with gazetteers and external knowledge [5] which turned out 81.39 % F1-score (see Table 6). However, without gazetteers and external knowledge results go down to 79.28 % (see Table 6).

It is worth noting that the knowledge injected to the previous learning model was *supervised*. We on the other hand have considered *unsupervised* external knowledge, while improving on those results. This is further substantiated by our exploring unsupervised features with the Stanford NER CRF model [9]. In this setting F-score of 81.44 % was obtained, again above Carreras [5].

More importantly, our work shows that an English resource (Brown clusters computed from English Wikipedia) can be used to improve Spanish NER with word representations as *(i)* entities in Spanish and English are often identical, and *(ii)* the resulting English Brown clusters for English entities correlate better with their entity types, giving rise to a better model.

Another point to note is that whilst binarization improves on English NER baselines Guo [11], the same does not work for Spanish. It seems that this approach adds instead noise to Spanish NER. Likewise, Collocations do not perform well for Spanish.

We also note that *word capitalization* has a distinct impact on our approach. With the following setting: English Brown clusters, Spanish cluster embeddings and *lower-cased* Spanish prototypes we got 0.78 % less F-score than with upper-cased prototypes. This is because the lower-cased prototypes will ignore the real context in which the entity appears (since a prototype is an instance of an entity class) and will be therefore mapped to the wrong word vector in the embedding (when computing cosine similarity). This seems to suggest that while prototypes are globally speaking useful, using Spanish data alone is not.

Finally, when comparing our approach to the current state-of-the-art using Deep Learning methods [7,10,12,25] (that extract features at the character, word and bytecode level to learn deep models), our work outperforms dos Santos [7] F-score and matches also Gillick [10].

5   Conclusions

This paper has explored unsupervised and minimally supervised features for Spanish NER, based on cross-lingual word representations within a CRF classification model. Our CRF model was trained over the Spanish CoNLL 2002 corpus, the Spanish Billion Word Corpus and English Wikipedia (2012 dump). This is a novel approach for Spanish. Our experiments show competitive results when compared to the current state-of-the-art in Spanish NER based on Deep Learning. In particular, we outmatched dos Santos [7].

Cross-lingual Word Representations have a positive impact on NER performance for Spanish. In the future, we would like to focus further on this aspect and consider more (large scale) cross-lingual datasets.

Acknowledgments. We are grateful to the Data and Web Science Group at University of Mannheim. Special thanks to Heiner Stuckenschmidt and Simone Ponzetto for their contributions and comments. This work was supported by the Master Program in Computer Science at Universidad Católica San Pablo and the Peruvian National Fund of Scientific and Technological Development through grant number 011-2013-FONDECYT.

References

1. Bhattarai, B.: Inducing cross-lingual word representations. Master's thesis, Multimodal Computing and Interaction, Machine Learning for Natural Language Processing. Universität des Saarlandes (2013)

2. Bouma, G.: Normalized (pointwise) mutual information in collocation extraction. In: Chiarcos, C., de Castilho, E., Stede, M. (eds.) Von der Form zur Bedeutung: Texte automatisch verarbeiten/From Form to Meaning: Processing Texts Automatically, Proceedings of the Biennial GSCL Conference 2009, pp. 31–40. Gunter Narr Verlag, Tübingen (2009)
3. Brown, P.F., deSouza, P.V., Mercer, R.L., Pietra, V.J.D., Lai, J.C.: Class-based n-gram models of natural language. Comput. Linguist. **18**(4), 467–479 (1992)
4. Cardellino, C.: Spanish Billion Words Corpus and Embeddings (2016). http:// crscardellino.me/SBWCE/
5. Carreras, X., Màrques, L., Padró, L.: Named entity extraction using adaboost. In: Proceedings of CoNLL-2002, Taipei, Taiwan, pp. 167–170 (2002)
6. Collobert, R., Weston, J.: Deep neural networks with multitask learning. In: Proceedings of the 25th International Conference on Machine Learning, ICML 2008, pp. 160–167. ACM, New York (2008)
7. dos Santos, C., Guimarães, V.: Boosting named entity recognition with neural character embeddings. In: Proceedings of the Fifth Named Entity Workshop, July, Beijing, China, pp. 25–33. Association for Computational Linguistics (2015)
8. Faruqui, M., Padó, S.: Training and evaluating a German named entity recognizer with semantic generalization. In: Proceedings of KONVENS 2010, Saarbrücken, Germany (2010)
9. Finkel, J.R., Grenager, T., Manning, C.: Incorporating non-local information into information extraction systems by gibbs sampling. In: Proceedings of the 43rd Annual Meeting on Association for Computational Linguistics, ACL 2005, Stroudsburg, PA, USA, pp. 363–370. Association for Computational Linguistics (2005)
10. Gillick, D., Brunk, C., Vinyals, O., Subramanya, A.: Multilingual Language Processing from Bytes. ArXiv e-prints (2015)
11. Guo, J., Che, W., Wang, H., Liu, T.: Revisiting embedding features for simple semi-supervised learning. In: Proceedings of the 2014 Conference on Empirical Methods in Natural Language Processing (EMNLP), October, Doha, Qatar, pp. 110–120. Association for Computational Linguistics (2014)
12. Lample, G., Ballesteros, M., Kawakami, K., Subramanian, S., Dyer, C: Neural architectures for named entity recognition. In: Proceedings of NAACL-HLT (NAACL 2016), San Diego, US (2016)
13. Liang, P.: Semi-supervised learning for natural language. Master's thesis, Department of Electrical Engineering and Computer Science. Massachusetts Institute of Technology (2005)
14. Mikolov, T., Chen, K., Corrado, G., Dean. J.: Efficient estimation of word representations in vector space. CoRR, abs/1301.3781 (2013a)
15. Mikolov, T., Sutskever, I., Chen, K., Corrado, G.S., Dean, J.: Distributed representations of words and phrases and their compositionality. In: Burges, C.J.C., Bottou, L., Welling, M., Ghahramani, Z., Weinberger, K.Q. (eds.) Advances in Neural Information Processing Systems 26, pp. 3111–3119 (2013b)
16. Okazaki, N.: CRFsuite: a fast implementation of conditional random fields (CRFs) (2007)
17. Passos, A., Kumar, V., McCallum, A.: Lexicon infused phrase embeddings for named entity resolution. In: Proceedings of the Eighteenth Conference on Computational Natural Language Learning, June, Ann Arbor, Michigan, pp. 78–86. Association for Computational Linguistics (2014)
18. Poulsen, S.: Collocations as a language resource. A functional and cognitive study in English phraseology. Ph.D. dissertation, Institute of Language and Communication. University of Southern Denmark (2005)

19. Ratinov, L., Roth, D.: Design challenges and misconceptions in named entity recognition. In: Proceedings of the Thirteenth Conference on Computational Natural Language Learning, CoNLL 2009, Stroudsburg, PA, USA, pp. 147–155. Association for Computational Linguistics (2009)
20. Sculley, D.: Combined regression and ranking. In: Proceedings of the 16th ACM SIGKDD International Conference on Knowledge Discovery and Data Mining, KDD 2010, pp. 979–988. ACM, New York (2010)
21. Sutton, C., McCallum, A.: An introduction to conditional random fields. Found. Trends Mach. Learn. **4**(4), 267–373 (2012)
22. Tjong Kim Sang, E.F.: Language-independent named entity recognition. In: Proceedings of the 6th Conference on Natural Language Learning - vol. 20, COLING-02, Stroudsburg, PA, USA, pp. 1–4. Association for Computational Linguistics (2002)
23. Turian, J., Ratinov, L., Bengio, Y.: A simple and general method for semi-supervised learning. In: Proceedings of the 48th Annual Meeting of the Association for Computational Linguistics, ACL 2010, Stroudsburg, PA, USA, pp. 384–394. Association for Computational Linguistics (2010)
24. Yang, Y., Pedersen, J.O.: A comparative study on feature selection in text categorization. In: Proceedings of the Fourteenth International Conference on Machine Learning, ICML 1997, San Francisco, CA, USA, pp. 412–420. Morgan Kaufmann Publishers Inc. (1997)
25. Yang, Z., Salakhutdinov, R., Cohen, W.: Multi-task cross-lingual sequence tagging from scratch. CoRR, abs/1603.06270 (2016)
26. Yu, M., Zhao, T., Dong, D., Tian, H., Yu, D.: Compound embedding features for semi-supervised learning. In: Human Language Technologies: Conference of the North American Chapter of the Association of Computational Linguistics, Proceedings, June 9–14, 2013, Westin Peachtree Plaza Hotel, Atlanta, Georgia, USA, pp. 563–568 (2013)
27. Yu, M., Zhao, T., Bai, Y., Tian, H., Yu. D.: Cross-lingual projections between languages from different families. In: Proceedings of the 51st Annual Meeting of the Association for Computational Linguistics (Volume 2: Short Papers), Sofia, Bulgaria, pp. 312–317. Association for Computational Linguistics (2013)

Machine Learning

Detection of Fraud Symptoms
in the Retail Industry

Rita P. Ribeiro[1,2(✉)], Ricardo Oliveira[3], and João Gama[2,3(✉)]

[1] Faculty of Sciences, University of Porto, Porto, Portugal
rjaoliveira@gmail.com
[2] LIAAD/INESC TEC, University of Porto, Porto, Portugal
rpribeiro@dcc.fc.up.pt
[3] Faculty of Economics, University of Porto, Porto, Portugal
jgama@fep.up.pt

Abstract. Data mining is one of the most effective methods for fraud detection. This is highlighted by 25 % of organizations that have suffered from economic crimes [1]. This paper presents a case study using real-world data from a large retail company. We identify symptoms of fraud by looking for outliers. To identify the outliers and the context where outliers appear, we learn a regression tree. For a given node, we identify the outliers using the set of examples covered at that node, and the context as the conjunction of the conditions in the path from the root to the node. Surprisingly, at different nodes of the tree, we observe that some outliers disappear and new ones appear. From the business point of view, the outliers that are detected near the leaves of the tree are the most suspicious ones. These are cases of difficult detection, being observed only in a given context, defined by a set of rules associated with the node.

Keywords: Outliers · Contextual outliers · Data mining

1 Introduction

Traditionally, the detection of fraud symptoms in retail is done manually, through ad-hoc research on corporate databases and focusing on only one variable of the data set. These techniques focus on quantitative and statistical aspects of the data, but they lack more efficient, effective and automatic multivariate analysis of data.

The aim of this work involves applying other techniques for obtaining fraud patterns, specifically techniques of data mining and machine learning to explore and "learn" about the data. These techniques allow the identification of different types of outliers referring to rare or unusual cases. Through statistical and graphical analysis applications, we can obtain a representation of the cases with a higher probability of being abnormal or strange, which may be or not a fraud. Still, no statistical analysis alone can ensure that a particular object under examination is fraudulent. It can only indicate that the object under study is more likely to be fraudulent than other objects [2].

M. Montes-y-Gómez et al. (Eds.): IBERAMIA 2016, LNAI 10022, pp. 189–200, 2016.
DOI: 10.1007/978-3-319-47955-2_16

2   Data Mining in the Detection of Fraud Symptoms

To do more and better with less has been one of the most spoken phrases in recent years, in the majority of departments within the organizations, and among fraud investigators. The current business environment in organizations requires increases in productivity at all organizational areas [3].

We are living in the information era, where organizations are overloaded with data. A growing amount of information is stored in databases, wherein transforming this data into knowledge creates a need to find powerful analysis tools [2]. As the use of information technologies increases, the use of these technologies in the detection of fraud becomes an urgent need. The goal is that inspectors can concentrate their efforts on areas of higher risk. The data analysis techniques used in the past were very focused on quantitative and statistical aspects regarding the characteristics of the used data. These techniques are useful in interpreting the data and can help gain further insight, building knowledge from there. Although these traditional data analysis techniques are able to extract knowledge from data, its process is created by analysts and consists in a manual creation process [2].

To overcome these limitations, a data analysis system should be developed using background knowledge and involving reasoning tasks from the provided data. To achieve this goal, researchers have focused on the machine learning area, which is a scientific discipline that exploits the construction and study of algorithms that "learn" from data. These algorithms operate from models based on inputs and use them to create predictions or decisions. Thus, emerged a new area often referred to as data mining and knowledge discovery. Data mining can be defined as the process of discovering patterns in data. This process can be automatic or semi-automatic (most common). Data mining is just a way to discover knowledge in large databases, which are usually owned by organizations [2]. Fraud detection is an important area for the practical application of data mining techniques. This is due to the economic and social consequences that are typically associated with these illegal activities [4]. Data mining brings the ability to consider and analyze thousands of transactions, reducing both time and cost and raising efficiency, compared to more traditional techniques forensic sampling [5]. The phenomenon of Big Data has brought a staggering volume, variety and velocity (3VS) business information, which changed the way leading enterprises manage their challenges of regulatory compliance and investigate strange behavior [6].

2.1   Supervised vs Unsupervised Learning

In terms of data mining strategies, there are two widely known learning techniques: supervised and unsupervised.

The supervised methods are implemented when there is a target variable in the learning process. Their intention is to carry out predictions from other input variables. In a fraud investigation, examples of records with fraud and non-fraud are needed, therefore all records available have to be classified as "fraudulent" or "non-fraudulent" [2]. After building the model using that data as the training set, the newly analyzed cases are then classified as "fraudulent" or "non-fraudulent". Of course, this implies that there is confidence in the classification of the training set data, which accounts for the sustainability of the model itself. This method is only applicable in fraud detection of a certain type: the type that has already occurred and there is data on the same event.

In contrast, unsupervised methods do not use previously classified records [7]. The unsupervised methods are applied over cases that do not have a target variable for the learning process, only input variables [8]. These methods can seek analysis of accounts, customers, suppliers, etc., that have behaviors considered "weird" for obtaining suspicious outputs or graphical anomalies.

For both methods, and as in any learning task, the obtained model should be validated in order to check whether the set of pre-established objectives have been achieved or not. According to that, a possible redesign or adjustment of the model may be required in order to guarantee more effective results. The use of either one of these methods gives only an indication of the likelihood of fraud. fraudulent than other objects [2]. Nevertheless, we should remark that in fraud detection it is impossible to be absolutely sure about the legitimacy of the intention of those who makes a transaction [9]. Thus, the evaluation step is generally not straightforward.

2.2   Outlier Detection

According to Hawkins [10], an outlier is an observation of a data set that is sufficiently different from the remaining observations to raise suspicions that has been caused by a different mechanism. The outlier detection refers to the problem of identifying patterns in the target data that are not compliant with an expected behavior. These anomalous patterns are referred to as anomalies, discordant observations, exceptions, faults, defects, aberrations, noise, errors and peculiarities in the various domains to which they apply. According to Chandola et al [11], outliers exist in practically all real datasets and can be classified into three categories, as follows.

Type 1: Global Outliers. The simplest type of outliers and the focus of most existing outlier detection schemes. The techniques that detect this type of outlier analyze the relationship of an individual observation in relation to the other observations. These outliers can be detected on any type of data.

Type 2: Contextual Outliers. This type of outlier is also about an individual instance, but it differs from those of Type 1 in the sense that it may not be considered an outlier if it is in a different context. So, the outliers of this type are defined depending on the context. The region of a store is an example of a context attribute.

Type 3: Collective Outliers. This type of outlier occurs when a subset of the instances in the observed data is peripheral or distant, relatively to the overall

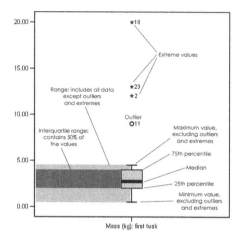

Fig. 1. Boxplot graph explanation [13]

data set. This type of outlier only has meaning when the data has a spatial or sequential nature. A characteristic of this type of outlier is its temporal/spatial continuity [12].

In this paper, our goal is to detect both global and contextual outliers that may indicate the symptom of a fraud.

One of the most common ways to detect global and univariate outliers is through the statistical analysis provided by a *boxplot*. This graph describes a variable under study, assuming a normal distribution. Figure 1 illustrates an example of a boxplot, which presents the values of a given continuous variable, assuming a normal distribution. The rectangle in the graph represents the range of 50 % of the most typical values of the distribution. The rectangle is divided in the amount corresponding to the lower quartile ($Q1$), the median ($Q2$) and the upper quartile ($Q3$). The value of IQR is the difference between the upper quartile ($Q3$) and the lower quartile ($Q1$). According to the boxplot, the outliers are values that exceed ($Q3 + 1.5 * IQR$) or are less than ($Q1 - 1.5 * IQR$). An outlier is considered *extreme* if its value is higher than ($Q3 + 3 * IQR$) or is less than ($Q1 - 3 * IQR$).

3   Outliers Detection on Returned Items in Stores

This study aims to implement data mining techniques to detect abnormal or unusual cases in real-world data of a large retail company. In particular, the goal is to identify outliers that might reveal the possibility of an occurrence of fraud on a data set regarding the process of returned items in stores.

After purchasing an item, the customer is entitled to a period of satisfaction during which the item can be returned. To do so, provided that the item is in perfect condition, it is only necessary to present proof of the purchase. The return

Table 1. Description of each observation in the returned items data set

Variable	Description	Type
Supervisor	Code of the supervisor from the total of 887 supervisors	Qualitative (categorical)
Store	Code of the store from the total of 40 stores	
Region	Region of the store North, Center, South	
WeekDay	Type of day: Business days, Weekend or Holidays	
Period	Period of day: Morning, Afternoon, Night	
ProdType	Product type: there are 18 different product types, such as Bakery, Fruits&Vegetables, Drink, Entertainment, etc.	
TotalNr	Total number of items returned by the supervisor	Quantitative (numerical)
AvgVal	Average value of the items returned by the supervisor	

registration is made by a store supervisor on a POS at store, reimbursing the customer. On the returned items process, there are two associated risks. One is that the return can be fictitious and the supervisor files the return to make a fraudulent refund to his advantage. For this case no item entries into physical stock, only on theoretical stock [14]. Another risk is the theft of the returned item, in which case the return was made by customer demand. However, after reimbursing the customer, the returned item gives input on the theoretical stock of the store, but physically it "disappears" [14].

On this study, the analyzed data concerns the transactions of returned items in 40 stores of a retailer. The extracted data period corresponds to three months of transactions, a total of 43286 transactions, which occurred between the months of December of 2014 and February of 2015. This period had two exceptional days (December, 25th and January, 1st), in which the stores were not open to the public. Table 1 presents a description of the data set used on this case study. Each observation represents a set of return transactions registered by a supervisor on a given store, aggregated by business day or weekend/holidays, day period and product type.

3.1 Univariate Analysis

For the identification of extreme outliers, we start by analysing the obtained boxplot on each of the target variables: TotalNr and AvgVal.

In Fig. 2a, we observe that the variable TotalNr has a high dispersion of values. The number of global extreme outliers is very high, more specifically

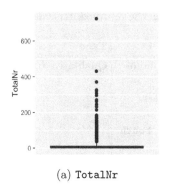   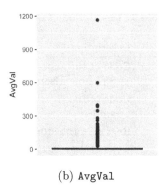

(a) `TotalNr`   (b) `AvgVal`

Fig. 2. Boxplot analysis for the univariate detection of outliers in the target variables.

1695 extreme outliers with values above 38 units. In Fig. 2b, we observe that the variable `AvgVal` shows a similar pattern with a high number of global extreme outliers. More specifically, 1467 extreme outliers with values above 30,5€.

Through the univariate analysis with boxplot many global extreme outliers were identified. However, this analysis can be considered very crude as does not reveal contextual outliers.

3.2 Bivariate Analysis

The goal at this point is to create an iterative model where the bivariate analysis is carried out for the combination of all the variables with each other. This analysis is intended to discover new outliers, different from those already identified by the univariate analysis. From all the possible 15 combinations of the six variables, we have selected two for illustration purposes.

Figure 3a shows the relation of total number of returned items with the day of week that the return was made. The distribution of registered returns for business days exhibits 944 extreme outliers. These correspond to transactions with more than 46 items returned. Still, all these outliers were also detected by the univariate analysis. Regarding weekends and holidays, there are 623 extreme outliers. These are transactions with more than 30 items returned. In these case, there are 307 new extreme outliers that were not identified in the univariate analysis. These new outliers correspond to transaction where the number of returned items is between 31 and 38.

A similar analysis was performed regarding the distribution of the average value of returned items (`AvgVal`) according to the period of the registration of the return. The obtained boxplot graphs are in Fig. 3b.

In the morning period, the registered transaction with values above 31,4€ are considered extreme outliers. Still, none of these 462 transactions is new, i.e., all of them were identified in the previous univariate analysis of `AvgVal` variable.

In the afternoon period, there are 566 transactions identified as extreme outliers. These transactions have an average return value above 29.99€. From these,

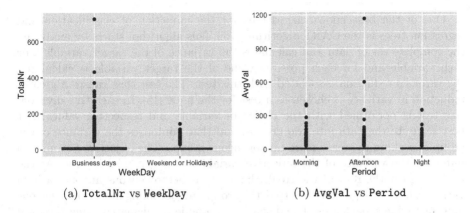

(a) TotalNr vs WeekDay   (b) AvgVal vs Period

Fig. 3. Boxplot analysis for two examples of bivariate detection of outliers

32 of them were not detected at univariate level. These new outliers correspond to transactions with an average value between 29.99€ and 30.5€.

Regarding the night period, there were 519 transactions with values above 29.86€ which are identified as extreme outliers. Again, in these set there are new outlier transactions that were not detected at univariate level. These are the 59 outlier transactions with average return value between 29.86€ and 30.5€.

In comparison to the univariate analysis, the bivariate analysis presents some advantages but also some disadvantages. As an advantage we can refer that the bivariate analysis allowed the detection of several new outliers, different from the ones observed in univariate analysis. Crossing the variables, even for pairs, can be better than analyzing just one variable each time. Still, the visual inspection of the bivariate analysis of all variables may require some extra effort. In data sets with many variables this can be quite tiresome, and not deliver the expected return. In this case the choice of the most important variables becomes a critical success factor.

3.3 Multivariate Analysis

At this point we intend to study the two quantitative variables, by combining all the qualitative variables simultaneously. Given the number of variables, the number of combinations is exponential, which makes the analysis of each possible combination humanly impossible. In this context, the use of a learning algorithm is of critical importance for the selection and analysis of the most relevant combinations. With the purpose of obtaining a partition of the dataset, we used a regression tree learning algorithm. Regression trees have the ability to seek possible solutions on complex problems by iteratively divide them into simpler problems [15]. Moreover, they represent a highly interpretable model. Our intuition is that the obtained data partition would help on the identification of outliers on specific partitions of the dataset.

One of the best known algorithms for the induction of classification and regression trees is the CART algorithm [16]. This algorithm starts by analyzing the complete set of data and calculates the variance of the target variable. For each variable and for each possible value of the target variable is calculated the variance reduction associated with this test. The test that causes a greater reduction in variance is chosen as a test for the node [15]. In each tree division, this test is performed with the choice of variable and respective values that best separate the data into two groups, and that minimizes the variance of the group in seeking to create more homogeneous groups. The objective becomes to minimize the variance of the objective variable, as it expands the tree. At the beginning (at the root of the tree), the group is heterogeneous and has a high variance, but as we go down the tree, the groups are becoming more homogeneous and the variance decreases. The deviation of a node is calculated using the total sum of squares: $SS_T = \sum (y_i - \bar{y})^2$, where y_i is target value in the observation i and $\bar{y}$ is the mean value of the target variable in the partition of data represented by the current node. Each division is made by maximizing between the impurity of the current node minus the sum of the impurity of the two child nodes, left (L) and right (R): $SS_T - (SS_L + SS_R)$. The process ends when it is no longer possible to divide more nodes in the tree as it is not possible to further decrease the variance of the respective group. If we divide this value by the number of observations in the node, we obtain the value of the variance.

We use the `rpart` library [17] from R [18] to learn the two regression trees: one for the target variable `TotalNr` and other for the target variable `AvgVal`. A critical parameter for `rpart` is the `cp` which controls the complexity of the tree. This value defines that any split in the tree should decrease the total lack of adjustment, before the tree be expanded. For the purpose of this study, we decided to set `cp=0`, meaning that we let the tree to be fully expanded. Still, for our outlier inspection, we limit the tree depth with `maxdepth=4`. After obtaining each of the trees (Figs. 4 and 5), we carried out for each node a boxplot graphical analysis. The goal is to inspect if the obtained boxplot reveals some new outliers or just the already known at the respective depth of the tree.

Let us take node 8 in the regression tree of Fig. 4 as an example. This node is reached by transactions in a subset of 33 stores and involving products from `Frozen`, `Bakery`, `Fishery`, `Pets&Plants` and `TakeAway`. At this node there are 127 new extreme outliers that were not detected in previous analysis (univariate and bivariate). Applying the rule described above, the total number of returned items is 4 units. However, the identified outlier cases represent returns from 17 to 30 units (cf. Figure 6a). Some of these values are up to 7 times higher than the average value of the number of returned items in that partition.

From the regression tree presented in Fig. 5, let us consider node 16 as an example. This node is reached by transactions involving products from `Fruits&Vegetables` and `Bakery`. At this node there are 111 new extreme outliers that were not detected in previous analysis (univariate and bivariate). For the associated data partition the average value of the returned items is 2.5. However, the identified outlier cases represent transactions where the average return

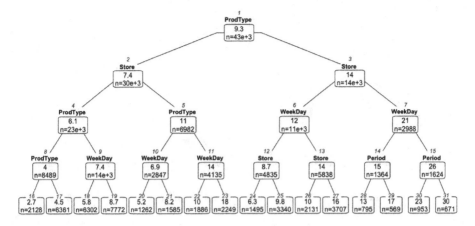

Fig. 4. Regression tree on `TotalNr` obtained with **rpart** [17] with `cp=0` and `maxdepth=4`

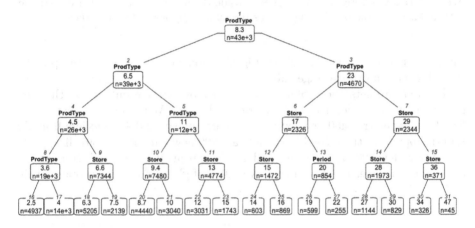

Fig. 5. Regression tree on `AvgVal` obtained with **rpart** [17] with `cp=0` and `maxdepth=4`

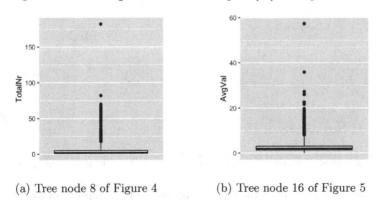

(a) Tree node 8 of Figure 4   (b) Tree node 16 of Figure 5

Fig. 6. Boxplot analysis of specific partitions in the generated regression trees for `TotalNr` and `AvgVal`

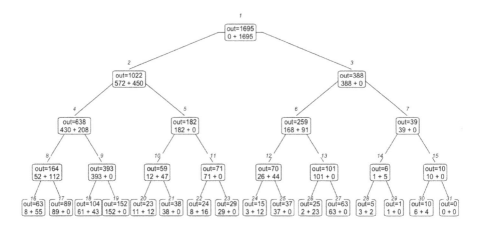

Fig. 7. Analysis of extreme outliers that "appear" and "disappear" in the regression tree on `TotalNr` (cf. Fig. 4). The numbers inside the nodes are the number of extreme outliers known in the previous node plus the number of new extreme outliers.

value is between 8.03 and 29.86 (cf. Fig. 6b). Some of these values are up to 13 times higher than the average value.

Figure 7 presents the evolution of number of extreme outliers as the tree grows for the regression tree on `TotalNr` (cf. Fig. 4). We can observe that at the first level of the tree 450 new extreme outliers appear, and 735 outliers from de root disappear; at the second level there are 299 new outliers and 591 disappear; at the third level 208 new outliers and 452 disappear. It is observed after the 1st division of the tree, it is found that over 40 % of global outliers disappear. This

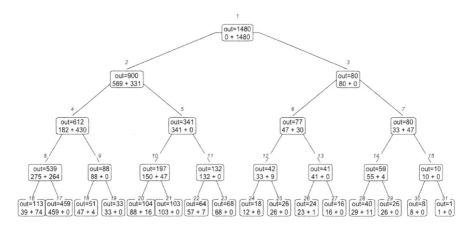

Fig. 8. Analysis of extreme outliers that "appear" and "disappear" in the regression tree on `AvgVal` (cf. Fig. 5). The numbers inside the nodes are the number of extreme outliers known in the previous node plus the number of new extreme outliers.

means that global outliers may disappear if been analysed in a certain context. Moreover, new outliers arise as the tree division creates groups (nodes) more homogeneous, revealing new contextual outliers unidentified so far.

Similar numbers for the other regression tree on Avg (cf. Figure 8). At the first level, 335 new extreme outliers emerge and 821 disappear, at second level 506 versus 377, and at the third level 324 versus 326.

4   Conclusions and Future Work

This work describes an exploratory study to detect fraud indicators by identifying outliers in transactional data. We collaborate with a retail company that provided the data and domain expertise.

We started our study by identifying extreme outliers in the available data using a well known technique: boxplots. Boxplot identifies outliers looking to the dispersion of data. The analysis, considering only the target variables, allowed a fast identification of several global outliers on a group of heterogeneous characteristics. Although it is a useful analysis, it lacks some more depth in the extraction of knowledge, since only looks for a variable at a time. Thus, the next step was to perform bivariate analysis, by combining the explanatory variables, two by two. The goal was to explore and identify new cases (outliers) not previously seen in univariate analysis. Finally, it was made multivariate analysis in order to study the two quantitative target variables, by combining the qualitative input variables, simultaneously. This analysis was performed by developing regression trees for the target variable. A regression tree grows by splitting the data into subgroups such that the variance of the target variable decreases. The splitting criteria are conditions based on the on the explanatory variables. The selected conditions are the ones that most reduce the variance of the target variable. Traversing the tree from the root to a leaf, the variance of the target variable in the corresponding subsets is always monotonically decreasing. We were expecting that the number of extreme outliers should decrease when the subgroups of data are more homogeneous. Surprisingly, we observe that, descending on the tree, some extreme outliers disappear but new extreme outliers emerge.

Of these three types of analysis, multivariate is undoubtedly the richest in terms of extraction of knowledge, as in addition to identifying global outliers, can descend to a very interesting detailed level, and reveal well *hidden* cases considered context of outliers. For example, it was only possible to identify several outliers in a context of returned items of product type like Fruits&Vegetables or Bakery, whose average returned value can be 13 times higher than the mean of average return value for transactions involving only those product types.

The expert confirmed the relevance of these outliers as symptoms of fraud. Nevertheless, it requires further investigation. We are now exploiting the proposed method as robust outlier detection using other datasets.

Acknowledgments. This work was supported by research project TEC4Growth - Pervasive Intelligence, Enhancers and Proofs of Concept with Industrial Impact/ NORTE-01-0145-FEDER-000020, financed by the North Portugal Regional Operational Programme (NORTE 2020), under the PORTUGAL 2020 Partnership Agreement, and through the European Regional Development Fund and by European Commission through the project MAESTRA (ICT-2013-612944).

References

1. Skalak, S.: Global economic crime survey. Technical report, PwC (2014)
2. Jans, M., Lybaert, N., Vanhoof, K: Data mining for fraud detection: toward an improvement on internal control systems? In: 30th Annual Congress European Accounting Association (EAA 2007)
3. Coderre, D.: Computer-Aided Fraud Prevention & Detection. Wiley, Hoboken (2009)
4. Torgo, L.: Data Mining with R: Learning with Case Studies, 1st edn. Chapman & Hall/CRC, Boca Raton (2010)
5. Bates, A.: Fraud risk management: developing a strategy for prevention,detection, and response, Technical report, KPMG Advisory Forensic (2006)
6. Stulb, D., Remnitz, D.: Big risks require big data thinking: global forensic data analytics survey 2014. Technical report, EY (2014)
7. Singh, K., Upadhyaya, S.: Outlier detection: applications and techniques. Int. J. Comput. Sci. Issues **9**(3), 307–323 (2012)
8. Kristin, R.N., Matkovsky, I.P.: Using data mining techniques for fraud detection. Technical report, SAS Institute Inc. and Federal Data Corporation (1999)
9. Phua, C., Lee, V.C.S., Smith-Miles, K., Gayler, R.W.: A comprehensive survey of data mining-based fraud detection research. CoRR abs/1009.6119 (2010)
10. Hawkins, D.: Identification of Outliers. Monographs on Applied Probability and Statistics. Chapman & Hall, New York (1980)
11. Chandola, V., Banerjee, A., Kumar, V.: Anomaly detection: a survey. ACM Comput. Surv. **41**(3), 1–58 (2009)
12. Gupta, M., Gao, J., Aggarwal, C.C., Han, J.: Outlier detection for temporal data: a survey. IEEE Trans. Knowl. Data Eng. **26**(9), 2250–2267 (2014)
13. Anglia Ruskin University: NuMBerS: numerical methods for biosciences students. http://web.anglia.ac.uk/numbers/. Accessed 02 May 2016
14. Wells, J.T.: Corporate Fraud Handbook: Prevention and Detection, 2nd edn. Wiley, Hoboken (2007)
15. Gama, J., Carvalho, A., Faceli, K., Lorena, C., Oliveira, M.: Extração de Conhecimento de Dados - Data Mining, 1st edn. Silabo (2012)
16. Breiman, L., Friedman, J.H., Olshen, R.A., Stone, C.J.: Classification and Regression Trees. Chapman & Hall, New York (1984)
17. Therneau, T., Atkinson, B., Ripley, B.: rpart: Recursive Partitioning and Regression Trees. R package version 4.1-10 (2015)
18. R Core Team: R: A Language and Environment for Statistical Computing. R Foundation for Statistical Computing, Vienna (2016)

A Machine Learning Model for Occupancy Rates and Demand Forecasting in the Hospitality Industry

William Caicedo-Torres[✉] and Fabián Payares

Department of Computer Science, Universidad Tecnológica de Bolívar,
Parque Industrial y Tecnológico Carlos Vélez Pombo,
Km 1 Vía Turbaco, Cartagena, Colombia
wcaicedo@unitecnologica.edu.co

Abstract. Occupancy rate forecasting is a very important step in the decision-making process of hotel planners and managers. Popular strategies as Revenue Management feature forecasting as a vital activity for dynamic pricing, and without accurate forecasting, errors in pricing will negatively impact hotel financial performance. However, having accurate enough forecasts is no simple task for a wealth of reasons, as the inherent variability of the market, lack of personnel with statistical skills, and the high cost of specialized software. In this paper, several machine learning techniques were surveyed in order to construct models to forecast daily occupancy rates for a hotel, given historical records of bookings and occupation. Several approaches related to dataset construction and model validation are discussed. The results obtained in terms of the Mean Absolute Percentage Error (MAPE) are promising, and support the use of machine learning models as a tool to help solve the problem of occupancy rates and demand forecasting.

Keywords: Machine learning · Forecasting · Hotel occupancy · Demand · Neural Networks · Ridge Regression · Kernel Ridge Regression

1 Introduction

In hospitality, occupancy rate forecasting has central importance in planning and decision making, since anticipating demand allows managers plan accordingly on issues as inventory, workforce, supplies, financial budgeting and pricing; all of this in order to maximize revenue and minimize costs.

Currently, the hospitality industry has adopted strategies to maximize income per room, and the set of these strategies fall into the discipline of Revenue Management, which allows selling each room at the highest price customers are willing to pay, so the highest income can be achieved [7].

In Revenue Management, forecasting is the essential element in pricing [11,16]. According to this, better pricing policies can be developed if better forecasting methods are developed. For this reason the main goal of this paper

© Springer International Publishing AG 2016
M. Montes-y-Gómez et al. (Eds.): IBERAMIA 2016, LNAI 10022, pp. 201–211, 2016.
DOI: 10.1007/978-3-319-47955-2_17

is to present the development and validation of a machine learning model to forecast room demand and occupation rate in hotels, taking into account that a sufficient low error is necessary to avoid a negative impact on revenue.

Basically, approaches to occupation forecasting can be grouped in two categories: historical booking models and advanced booking models [16]. Historical booking models try to solve the forecasting problem as a time series modelling problem, and advanced booking problems use reservations data, and the concept of "Pick-Up", which is the increment of bookings N from now to a day T (for which there are already K reservations) in the future, resulting in an occupation forecast of K + N for day T.

For instance, [3] used ARIMA and Holt-Winters exponential smoothing to forecast monthly hotel occupation, showing that both methods yield low Mean Squared Error. In [14], long-term forecast via Holt-Winters is combined with short-term forecast obtained from observations, to obtain ensemble predictions. [10] used Neural Networks for Time Series Prediction, where Japanese citizen travel to Hong Kong was forecasted with Neural Networks outperforming other techniques like Multiple Regression, Moving Average and Exponential Smoothing. Examples of the Pick-Up method can be found in [13], where sold rooms with Pick-Up are forecasted to 7, 14, 30, and 60 days of reserves, with day of week as additional input. [17] proposes a Monte-Carlo model for occupation forecasting at an Egyptian hotel.

As we have shown, statistical techniques can predict, sometimes with good accuracy, occupancy rates and demand. However, some of the aforementioned techniques require important statistical skills and lengthy procedures to be applied in order for then to function correctly (e.g. building and analyzing correlograms); or the use of expensive commercial software. We propose the use of Machine Learning algorithms to build predictive models for the occupancy rates and demand in the hospitality industry, that can be readily used by hotel personnel, without advanced training in statistics. Such models can be packaged into relatively inexpensive applications or executed on the cloud, to provide the hospitality sector with cost-effective forecasts.

The rest of the paper is organized as follows: First, a brief introduction to the algorithms used is given, then the general structure and characteristics of the data set are presented; next the experimental results are shown and discussed, followed by our conclusions.

2   Machine Learning Algorithms

In this work, four algorithms were considered and used. A brief description of each is given next.

2.1   Ridge Regression

Ridge Regression [9] is a regression algorithm that includes a complexity penalty (called a regularization term) in its cost function in order to better approximate the out-of-sample error. Ridge Regression penalizes the algorithm by a factor proportional to the Euclidean (L2) norm of its parameter vector,

$$J(\theta) = \sum_{i=1}^{N}(\theta^T x_i - y_i)^2 + \alpha||\theta||^2 \tag{1}$$

Where regularization factor α controls the relative importance of the complexity penalty. Solving for optimal parameters involve finding the derivative of the cost function and setting it to zero. According to this, the parameter vector that minimizes the error is

$$\theta = (X^T X + \lambda I)^{-1} X^T y \tag{2}$$

New predictions can be calculated as follows

$$h(x) = \theta^T x = \sum_{i=1}^{n} \theta_i x_i \tag{3}$$

2.2   Kernel Ridge Regression

Kernel Ridge Regression [12] is a Kernel Method that projects the original data to a feature space, without explicitly computing the transformations, via the Kernel Trick [6]. The transformation allows for a linear model to correctly adjust the data in a higher dimensional space, while remaining a well regularized regressor. As previously stated, Ridge Regression optimal weights are defined as

$$\theta = (X^T X + \lambda I)^{-1} X^T y \tag{4}$$

After some algebraic manipulation, we can state that

$$\theta = X^T (XX^T + \lambda I)^{-1} y \tag{5}$$

So, a prediction can be obtained by

$$h(x') = \theta^T x' = y^T (XX^T + \lambda I)^{-1} X x' \tag{6}$$

A Kernel function evaluates to a dot product of the form

$$f(x_1, x_2) = \phi(x_1).\phi(x_2) \tag{7}$$

without explicitly computing the transform ϕ. If we define $K_{i,j} = f(x_i, x_j)$ and $k_i = f(x_i, x')$, were x_i is the i-th row of matrix X, then

$$h(x') = \theta^T x' = y^T (K + \lambda I)^{-1} k \tag{8}$$

We fitted a model in a higher dimensional space induced by ϕ without incurring in additional computational complexity. Several Kernels can be used, being popular choices the polynomial Kernel

$$k(x_1, x_2) = (\gamma x_1^T x_2 + \beta)^\delta \tag{9}$$

and the Gaussian Kernel

$$k(x_1, x_2) = \exp(-\frac{||x_1 - x_2||^2}{2\delta^2}) \tag{10}$$

2.3   Multilayer Perceptron

The Multilayer Perceptron [15] is a feed forward neural network model capable of classification and regression. A multilayer perceptron is composed of fully connected layers of artificial neurons forming a connected graph (in the simplest case one hidden, one final layer), where neurons are modelled very loosely on biological neurons; and as a feed forward net, the output of each layer becomes the input of the next one. The activation of each neuron is given by

$$y = f(\sum_i^n w_i x_i + b) \tag{11}$$

where f is the activation function. Popular choices include the logistic sigmoid function $y = (1 + \exp(-x))^{-1}$, and the linear function $y = x$. The choice of activation function in the final layer dictates if the network will be able to adequately perform regression or classification.

Multilayer Perceptrons can be trained through the Backpropagation Algorithm [15]. The Backpropagation Algorithm is an application of gradient descent that takes into account the contribution to the error E by the neurons of the hidden layer. Through differential calculus the gradients of all neurons can be calculated and weights updated accordingly. Gradient descent update rule is

$$w_{t+1} = w_t - \eta \nabla E(w_t) \tag{12}$$

where w is the weight vector of any particular neuron. It remains then to specify the gradient. For neuron k in the final layer, the gradient components can be expressed as

$$\frac{\partial E(w)}{\partial w_{jk}} = \frac{\partial E}{\partial y_k} \frac{\partial y_k}{\partial z_k} \frac{\partial z_k}{\partial w_{jk}} \tag{13}$$

where w_{jk} are the weights connecting hidden layer neuron j to neuron k, z_k total input to neuron k, and y_k the output of neuron k. Partial $\frac{\partial E}{\partial y_k}$ depends on the cost function used to assess network performance (for regression and time series problems, the Mean Squared Error is often used, with Weight Decay as the complexity penalization strategy).

Assuming only one hidden layer, the gradient for neurons in that hidden layer can be expressed as

$$\frac{\partial E(w)}{\partial w_{ij}} = \frac{\partial E}{\partial y_j} \frac{\partial y_j}{\partial z_j} \frac{\partial z_j}{\partial w_{ij}} \tag{14}$$

where w_{ij} are the weights connecting input component i to neuron j, z_j total input to neuron j, and y_j the output of neuron j. The partial $\frac{\partial E}{\partial y_j}$ represents the variation of the error with respect to the output of neuron j, and can be expressed as

$$\frac{\partial E}{\partial y_j} = \sum_k \frac{\partial E}{\partial z_k} \frac{\partial z_k}{\partial y_j} \tag{15}$$

which captures the way neurons from the hidden layer affect the total error, by summing up the contributions from each connection to the neurons in the final layer. The algorithm gets its name from the fact that the partial

$$\frac{\partial E}{\partial z_k} = \frac{\partial E}{\partial y_k}\frac{\partial y_k}{\partial z_k} \tag{16}$$

was already calculated in the final layer and now re-used for the neurons in the hidden layer, i.e. it is being "back-propagated" from the final layer.

Now, with all the corresponding gradients known, the gradient descent update rule becomes

$$w_{j,k,t+1} = w_{j,k,t} - \eta\frac{\partial E}{\partial y_k}y_k(1 - y_k)y_j \tag{17}$$

for the weights of neurons in the final layer and

$$w_{i,j,t+1} = w_{i,j,t} - \eta\sum_k\frac{\partial E}{\partial z_k}w_{jk}y_j(1 - y_j)x_i \tag{18}$$

for the weights of neurons in the hidden layer.

2.4   Radial Basis Function Networks

Radial Basis Function (RBF) Networks [5] are Neural Networks with usually one hidden layer, that employ a set of functions called Radial Basis Functions as feature detectors. RBF nets can be used to perform regression and classification as well. Each neuron in the hidden layer uses a RBF as activation function ϕ, with the Gaussian function

$$\phi_j(x) = \exp(-\frac{||x - \mu_j||^2}{2\sigma_j^2}) \tag{19}$$

being a popular choice. The network output can be represented by

$$y(x) = \sum_{j=0}^{M}w_{kj}\phi_j(x) \tag{20}$$

with w_{jk} representing the weights connecting hidden layer neuron j to output layer neuron k.

Training consists in finding a set of values for all μ_j, σ_j, and some set of w_{kj} that maximize network performance. First, the Gaussian RBF parameters are selected according to some procedure (μ_j can be selected randomly or to coincide with any training point; and σ_j can be the maximum or average distance between data points), and then solving for w_{kj} to minimize an error function. For the Mean Squared Error as cost function, it can be shown that the gradient is

$$\frac{\partial E}{\partial w_{kj}} = \sum_n(\sum_{j=0}^{M}w_{kj}\phi_j(x_n) - y_{kn})\phi(x_n) \tag{21}$$

Setting the gradient to zero, we can solve analytically for the optimal weights. Expressed in matrix form,

$$w^T = \Phi^\dagger y \tag{22}$$

where

$$\Phi^\dagger = (\Phi^T \Phi)^{-1} \Phi^T \tag{23}$$

is the pseudo-inverse of Φ.

3   Model Development

3.1   Dataset Construction

Reservations and occupation data from July 1, 2008 to June 30, 2014 (2191 days), were gathered from a hotel in Cartagena, Colombia. Using these, three datasets were constructed using different schemes: The first, arranges as inputs the observations from earlier days $(x_{i+0}, x_{i+1}, ..., x_{i+n-1})$ and the output as the current day (x_{i+n}). Occupation in day i corresponds to x_i, and n represents the number of lag variables being used (a classic time series dataset). The second scheme organizes inputs as before but it also adds other variables as day of week, number of holidays present between the previous and following seven days, and whether or not it is tourism season. The third dataset is generated using the number of reservations made for the current day with several days in advance (7, 10, 15, 20, 30, 60 and 90 days), month of the year, day of the week, number of holidays present between the previous and following seven days, and whether or not it is tourism season. Datasets were then partitioned into three subsets, i.e. training, validation and test sets. The test set comprises the 8.26 % of the original data, corresponding to all the observations from year 2014 (181); 73.4 % of the data was used for the training set, and the remaining 18.34 % was used for the validation set.

3.2   Training and Validation

Training and validation were performed using Rolling Origin Update Cross Validation [8]. Forecasts are made using One-Step-Ahead (predict next day occupation) and Multi(h)-Step-Ahead procedures, with $h = 7$ (predict occupation seven days in advance), and model performance is measured through Mean Absolute Percentage Error (MAPE), defined as

$$MAPE = \frac{1}{n} \sum_{t=1}^{n} \left| \frac{y_t - h_t}{y_t} \right| \tag{24}$$

We used the Mlpy Python library [2] implementation of Ridge Regression and Kernel Ridge Regression, the ALGLIB library [4] implementation of Multilayer Perceptron and Thomas Rückstieß Radial Basis Function Networks implementation [1].

Table 1. Time series forecasting results.

Horizon	Algorithm	MAPE	Parameters
1-step-ahead	Kernel Ridge Regression (Gaussian)	11.2055	— 21 lag variables — $\delta = 2300$ — $\lambda = 0.015$
	Kernel Ridge Regression (Poly)	11.1799	— 20 lag variables — $\lambda = 0.1$ — $d = 1$ — $\beta = 0.1$ — $\gamma = 1$
	RBF Network	11.2301	— 20 lag variables — 150 hidden neurons — $\lambda = 0.0015$ — $\beta = 1.5 \times 10^{-7}$
7-step-ahead	Ridge Regression	19.5595	— 22 lag variables — $\lambda = 0.1$
	Kernel Ridge Regression (Gaussian)	19.6819	— 22 lag variables — $\delta = 3000$ — $\lambda = 0.15$
	Kernel Ridge Regression (Poly)	19.55972	— 22 lag variables — $\lambda = 0.1$ — $d = 1$ — $\beta = 0.1$ — $\gamma = 1$

Table 1 shows results for the first data set, where data is arranged as a time series. Table 2 shows results for the second data set, arranged as a time series plus additional variables.

Experiments show that usage of additional variables appear to slightly improve model performance and offer better results than using the time series

Table 2. Time series plus additional variables forecasting results.

Horizon	Algorithm	MAPE	Parameters
1-step-ahead	Kernel Ridge Regression (Poly)	10.2351	
			– 5 lag variables – $\lambda = 0.1$ – $d = 2$ – $\beta = 0.1$ – $\gamma = 1$
	RBF Network	11.23586	
			– 20 lag variables – 150 hidden neurons – $\lambda = 0.0015$ – $\beta = 1.5 \times 10^{-7}$
7-step-ahead	Ridge Regression	19.6819	
			– 23 lag variables – $\delta = 3000$ – $\lambda = 0.15$
	RBF Network	19.7113	
			– 22 lag variables – 150 hidden neurons – $\lambda = 0.0015$ – $\beta = 1.5 \times 10^{-7}$

Table 3. Ridge Regression and Kernel Ridge Regression results on the reservations dataset. It is not surprising that Kernel Ridge Regression and Ridge Regression obtained the same MAPE, given that the use of a Polynomial Kernel with $d = 1$ in Kernel Ridge Regression, equals to the standard Ridge Regression.

Algorithm	Kernel/Degree	λ	β	MAPE
Ridge Regression	1	0.15	n/a	8.6962
Ridge Regression	2	0.15	n/a	8.2012
Kernel Ridge Regression	Poly($d = 1$)	0.15	0.1	8.69622

data only in the one-step-ahead forecast problem, as Polynomial Kernel Ridge Regression using additional variables offered a MAPE of 10.2351 % versus 11.1799 % when using only the time series. However in the 7-step-ahead situation, best performance was exhibited by Ridge Regression using only time series data and no additional variables, with a MAPE of 19.5595 %.

In the case of the third dataset, inputs are comprised of reservations data at multiple dates before the actual date for which occupation is to be predicted. Tables 3 and 4 show best validation set results for Ridge Regression, Kernel Ridge Regression and Neural networks respectively.

Table 4. Neural Networks results on the reservations dataset.

Model	Hidden nodes	β	MAPE
Multilayer Perceptron	50	5	12.89174
RBF Network	150	1.5×10^{-9}	26.32086

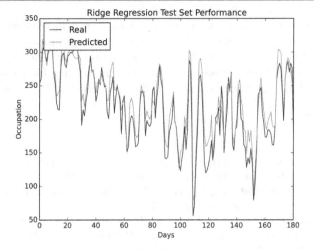

Fig. 1. Test set results for Ridge Regression with quadratic features.

As shown in Table 3, Ridge Regression with a second degree polynomial transform scored better than the other alternatives, with a MAPE of 8.2012 %.

3.3 Test

We decided to assess the performance of the best model on the reservations dataset, using the test set. Results show a MAPE of 8.65615 % on the test set, for the Ridge Regression model with quadratic polynomial features. Results are in line with the validation phase, and there is no evidence of overfitting as validation and test scores are quite close from each other. Figure 1 shows model behavior on the test set.

4 Conclusion

In this work, several machine learning techniques were compared. Different models were trained and validated using Ridge Regression, Kernel Ridge Regression, Multilayer Perceptron and Radial Basis Function Networks. Three data sets (each including training set, validation and test set) were constructed using occupation time series data, occupation times series data plus additional variables, and reservations data. Grid search was employed to find optimal parameters for the models. Results show a Ridge regression model with quadratic features trained on the reservations data set, outperforms the other models considered, with a validation set MAPE of 8.2012 % and a test set MAPE of 8.6561 %. Test set results show no evidence of overfitting. Also, it is worth noticing that models trained on time series plus additional variables data showed an modest increase in performance, compared to those trained on time series data only. The presence of additional inputs allowed the models to leverage contextual information and improve their predictions.

Finally, the use of bookings and reservations known in advance offered the best performance. The results obtained are promising and support the use of black-box Machine Learning based tools for estimating hotel occupation, which require little statistical expertise by the hotel staff; allowing for a more effective deployment of Revenue Management techniques in the hospitality sector.

References

1. Thomas rückstieß rbf network implementation. http://www.rueckstiess.net/research/snippets/show/72d2363e. Accessed 04 May 2016
2. Albanese, D., Visintainer, R., Merler, S., Riccadonna, S., Jurman, G., Furlanello, C.: mlpy: Machine learning python (2012)
3. Andrew, W.P., Cranage, D.A., Lee, C.K.: Forecasting hotel occupancy rates with time series models: an empirical analysis. J. Hospitality Tourism Res. **14**(2), 173–182 (1990). http://jht.sagepub.com/content/14/2/173.abstract
4. Bochkanov, S.: ALGLIB. http://alglib.net Accessed 26 Apr 2016
5. Broomhead, D., Lowe, D.: Multivariable functional interpolation and adaptive networks. Complex Syst. **2**, 321–355 (1988)
6. Cortes, C., Vapnik, V.: Support-vector networks. Mach. Learn. **20**(3), 273–297 (1995). http://dx.doi.org/10.1007/BF00994018
7. El-Gayar, N., Hendawi, A., Zakhary, A., El-Shishiny, H.: A proposed decision support model for hotel room revenue management. ICGST Int. J. Artif. Intell. Mach. Learn. **8**(1), 23–28 (2008)
8. Gilliland, M., Sglavo, U., Tashman, L.: Business Forecasting: Practical Problems and Solutions. Wiley, Hoboken (2016)
9. Hoerl, A.E., Kennard, R.W.: Ridge regression: biased estimation for nonorthogonal problems. Technometrics **42**(1), 80–86 (2000). http://dx.doi.org/10.2307/1271436
10. Law, R., Au, N.: A neural network model to forecast Japanese demand for travel to Hong Kong. Tourism Manag. **20**(1), 89–97 (1999)
11. Lee, A.O.: Airline reservations forecasting: probabilistic and statistical models of the booking process. Ph.D. thesis, Massachusetts Institute of Technology, Cambridge, MA (1990)
12. Murphy, K.P.: Machine Learning: A Probabilistic Perspective. The MIT Press, Cambridge (2012)
13. Phumchusri, N., Mongkolku, P.: Hotel room demand forecasting via observed reservation information. In: Kachitvichyanukul, V., Luong, H., Pitakaso, R. (eds.) Proceedings of the Asia Pacific Industrial Engineering & Management Systems Conference, pp. 1978–1985 (2012)
14. Rajopadhye, M., Ghalia, M.B., Wang, P.P., Baker, T., Eister, C.V.: Forecasting uncertain hotel room demand. Inf. Sci. **132**(1–4), 1–11 (2001). http://dx.doi.org/10.1016/S0020-0255(00)00082-7

15. Rumelhart, D.E., Hinton, G.E., Williams, R.J.: Learning internal representations by error propagation. In: Parallel Distributed Processing: Explorations in the Microstructure of Cognition, vol. 1, pp. 318–362. MIT Press, Cambridge (1986). http://dl.acm.org/citation.cfm?id=104279.104293
16. Weatherford, L.R., Kimes, S.E.: A comparison of forecasting methods for hotel revenue management. Int. J. Forecast. **19**(3), 401–415 (2003)
17. Zakhary, A., El Gayar, N., Atiya, A.F.: A comparative study of the pickup method and its variations using a simulated hotel reservation data. ICGST Int. J. Artif. Intell. Mach. Learn. **8**, 15–21 (2008)

A Machine Learning Model for Triage in Lean Pediatric Emergency Departments

William Caicedo-Torres[1]([✉]), Gisela García[1], and Hernando Pinzón[2]

[1] Department of Computer Science, Universidad Tecnológica de Bolívar,
Parque Industrial y Tecnológico Carlos Vélez Pombo,
Km 1 Vía Turbaco, Cartagena, Colombia
wcaicedo@unitecnologica.edu.co
[2] Hospital Infantil Napoleón Franco Pareja, Cartagena, Colombia

Abstract. High demand periods and under-staffing due to financial constraints cause Emergency Departments (EDs) to frequently exhibit over-crowding and slow response times to provide adequate patient care. In response, Lean Thinking has been applied to help alleviate some of these issues and improve patient handling, with success. Lean approaches in EDs include separate patient streams, with low-complexity patients treated in a so-called Fast Track, in order to reduce total waiting time and to free-up capacity to treat more complicated patients in a timely manner. In this work we propose the use of Machine Learning techniques in a Lean Pediatric ED to correctly predict which patients should be admitted to the Fast Track, given their signs and symptoms. Charts from 1205 patients of the emergency department of Hospital Napoleón Franco Pareja in Cartagena - Colombia, were used to construct a dataset and build several predictive models. Validation and test results are promising and support the validity of this approach and further research on the subject.

Keywords: Machine learning · Triage · Emergency department · Lean · Fast track · Neural networks · SVM · Logistic regression · PCA

1 Introduction

It is a well known fact that Emergency Departments (EDs) are under a lot of stress in current healthcare systems, and that there is a need for better strategies to use the available resources and effectively cope with demand [10]. In Colombia specifically, different issues as the distrust of patients about the quality of care they receive in primary care centers, complicate the situation even further and increase pressure on higher level healthcare facilities, as patients avoid the first level of care altogether and instead go directly to secondary or tertiary level care centers.

Triage is an essential tool to manage medical resources effectively and assign them to the patients that need it the most [8], and improvements on triage translate into lower waiting times and better performance overall. In triage,

M. Montes-y-Gómez et al. (Eds.): IBERAMIA 2016, LNAI 10022, pp. 212–221, 2016.
DOI: 10.1007/978-3-319-47955-2_18

patients are classified according to the urgency of their condition, and the waiting time depends on the classification given by triage personnel. It is worth noticing that in pediatrics, triage presents greater difficulty, as patients cannot always refer symptoms accurately, or even refer them at all to physicians or nurses.

On the other hand, Lean is a set of practices and principles developed and applied by Toyota Motor Corporation in its production system [23]. Principles include elimination of waste, Just in Time delivery, solving problems at their source, and the automation of defect inspection. Given its success, there have been several efforts to apply these principles to ED management [10]. One of the process changes implemented by the application of Lean in the ED, is the separation of patients in two streams on the basis of their complexity, rather than acuity or urgency [12,14,18]. The first stream, called a Fast Track or Express Care Track, receives low complexity patients, while the Regular Track receives patients that need more specialized care. EDs with a Fast Track also perform Triage at a different stage, with a different goal: Separate patients into those going to the Fast Track and those to be treated at the Regular Track. This separation must be performed taking into account possible outcomes, and the likelihood of complications that require advanced care.

In the spirit of Lean thinking, the automation of Triage and separation of patient streams could optimize patient flow in the ED even further. Machine Learning could help to construct a solution that automatically assign patients to the fast Track reliably, using datasets built upon medical charts and records. This work present the results of the training, validation and testing of several Machine Learning classifiers developed to classify patients according to their complexity, in order to assign those of lower complexity into the Fast Track inside a Pediatric ED. The rest of the paper is organized as follows: First, a brief introduction to the algorithms used is given, then the general structure and characteristics of the dataset are presented; next the experimental results are shown and discussed, followed by our conclusions.

2   Machine Learning Algorithms

In this study, several Machine Learning techniques were considered. We used Logistic Regression, Support Vector Machines with Polynomial and Gaussian kernels and the Multilayer Perceptron Neural Network. All models were built using the Orange Data Mining Suite [6].

2.1   Logistic Regression

Logistic Regression is a linear classification algorithm [16] widely used in medicine, where a logistic sigmoid function is coupled to a linear regression model. Results can be interpreted as the probability for the input to belong to the positive class, $p(Y = 1|x; \theta)$. In this study, Logistic Regression was trained with Cross-Entropy loss. A L2 regularization penalty was introduced to account for model complexity and avoid overfitting:

$$min \ -\frac{1}{N} \sum_{i=1}^{N} [y_i log h_\theta(x_i) + (1 - y_i) log(1 - h_\theta(x_i))] + \lambda ||\theta||^2 \tag{1}$$

where

$$h_\theta(x) = \frac{1}{1 + e^{-\theta^t x}} \tag{2}$$

It is possible to increase its capabilities by applying a polynomial transform to the input, in which case the decision boundary can be non-linear and handle more difficult problems.

2.2 Support Vector Machine

The Support Vector Machine (SVM) is a classification model invented in its original form by Vapnik et al. [5]. SVMs try to find the maximum margin hyperplane to provide a robust separator for the problem classes while being tolerant of misclassification as well, in order to handle non-linearly separable problems (soft-margin SVM). Its primal formulation minimizes the structural risk to avoid overfitting:

$$min \ \frac{1}{N} \sum_{i=1}^{N} \xi_i + \lambda ||w||^2 \tag{3}$$

Subject to:

$$y_i(\boldsymbol{x_i} \cdot w + b) \geq 1 - \xi_i \quad \text{and} \quad \xi_i \geq 0, \forall i \tag{4}$$

where

$$\xi_i = max(0, 1 - y_i(w \cdot \boldsymbol{x_i} + b)), \text{ the hinge loss.} \tag{5}$$

The Lagrangian dual of the optimization problem reveals an interesting structure:

$$max_\alpha \ \sum_{i=1}^{N} \alpha_i - \frac{1}{2} \sum_{i=1}^{N} \sum_{j=1}^{N} y_i \alpha_i (\boldsymbol{x_i} \cdot \boldsymbol{x_j}) y_j \alpha_j \tag{6}$$

Subject to:

$$\sum_{i=1}^{N} \alpha_i y_i = 0 \quad \text{and} \quad 0 \leq \alpha_i \leq \frac{1}{2n\lambda}, \forall i \tag{7}$$

Support vectors are the inputs with non-zero associated Lagrangian Multipliers (α), and all lie on the classification margin, hence the name.

The solution to the dual (optimal multipliers $\{\alpha_{i,o}\}_{i=1}^{l}$) allows us to formulate the maximum margin hyperplane for classification. The optimal weight vector is given by

$$\boldsymbol{w_o} = \sum_{i=1}^{l} \alpha_{i,o} y_i \boldsymbol{x_i} \tag{8}$$

so the hyperplane is then

$$\boldsymbol{w_o}^t \boldsymbol{x} + b_o = \sum_{i=1}^{l} \alpha_{i,o} y_i \boldsymbol{x_i} \cdot \boldsymbol{x} + b_o = 0 \tag{9}$$

To handle non-linear decision boundaries, the solution structure can be exploited: the dot product $x_i \cdot x_j$ can be replaced by a product of the transformed versions of the original inputs, $\phi(x_i) \cdot \phi(x)$. The Kernel trick [5] provides a way to compute the dot product without performing the transformations explicitly using Kernel functions, which by definition represent dot products in high dimensional spaces [5]. In this way, the SVM remains a well regularized classifier and can handle high dimensional feature transformations with computational efficiency. In this study, a Gaussian kernel,

$$K(x_i, x_j) = exp(-\frac{||x_i - x_j||^2}{2\sigma^2}) \tag{10}$$

and a Polynomial kernel $(\gamma x_1^T x_2 + \beta)^\delta$ were used.

2.3 Multilayer Perceptron

The Multilayer Perceptron [20] is a feed forward neural network model capable of classification and regression. A multilayer perceptron is composed of fully connected layers of artificial neurons forming a connected graph (in the simplest case one hidden, one final layer), where neurons are modelled very loosely on biological neurons; and as a feed forward net, the output of each layer becomes the input of the next one. The activation of each neuron is given by

$$y = f(\sum_i^n w_i x_i + b) \tag{11}$$

where f is the activation function. Popular choices include the logistic sigmoid function $y = (1 + \exp(-x))^{-1}$, and the linear function $y = x$. The choice of activation function in the final layer dictates if the network will be able to adequately perform regression or classification.

Multilayer Perceptrons can be trained through the Backpropagation Algorithm [20]. The Backpropagation Algorithm is an application of gradient descent that takes into account the contribution to the error E by the neurons of the hidden layer. Through differential calculus the gradients of all neurons can be calculated and weights updated accordingly. Gradient descent update rule is

$$w_{t+1} = w_t - \eta \nabla E(w_t) \tag{12}$$

where w is the weight vector of any particular neuron. It remains then to specify the gradient. For neuron k in the final layer, the gradient components can be expressed as

$$\frac{\partial E(w)}{\partial w_{jk}} = \frac{\partial E}{\partial y_k} \frac{\partial y_k}{\partial z_k} \frac{\partial z_k}{\partial w_{jk}} \tag{13}$$

where w_{jk} are the weights connecting hidden layer neuron j to neuron k, z_k

total input to neuron k, and y_k the output of neuron k. Partial $\frac{\partial E}{\partial y_k}$ depends on the cost function used to assess network performance (for regression and time series problems, the Mean Squared Error is often used, with Weight Decay as the complexity penalization strategy).

Assuming only one hidden layer, the gradient for neurons in that hidden layer can be expressed as

$$\frac{\partial E(w)}{\partial w_{ij}} = \frac{\partial E}{\partial y_j} \frac{\partial y_j}{\partial z_j} \frac{\partial z_j}{\partial w_{ij}} \tag{14}$$

where w_{ij} are the weights connecting input component i to neuron j, z_j total input to neuron j, and y_j the output of neuron j. The partial $\frac{\partial E}{\partial y_j}$ represents the variation of the error respect to the output of neuron j, and can be expressed as

$$\frac{\partial E}{\partial y_j} = \sum_k \frac{\partial E}{\partial z_k} \frac{\partial z_k}{\partial y_j} \tag{15}$$

which captures the way neurons from the hidden layer affect the total error, by summing up the contributions from each connection to the neurons in the final layer. The algorithm gets its name from the fact that the partial

$$\frac{\partial E}{\partial z_k} = \frac{\partial E}{\partial y_k} \frac{\partial y_k}{\partial z_k} \tag{16}$$

was already calculated in the final layer and now re-used for the neurons in the hidden layer, i.e. it is being "back-propagated" from the final layer.

Now, with all the corresponding gradients known, the gradient descent update rule becomes

$$w_{j,k,t+1} = w_{j,k,t} - \eta \frac{\partial E}{\partial y_k} y_k (1 - y_k) y_j \tag{17}$$

for the weights of neurons in the final layer and

$$w_{i,j,t+1} = w_{i,j,t} - \eta \sum_k \frac{\partial E}{\partial z_k} w_{jk} y_j (1 - y_j) x_i \tag{18}$$

for the weights of neurons in the hidden layer.

3   Model Development

3.1   Dataset Construction

Medical records from 2013 were analyzed, looking for general trends about the kind of conditions patients exhibited in the ED of Hospital Infantil Napoleón Franco Pareja, leading to the identification of 17132 records. Table 1 shows ED records grouped by type of condition reported. According to these, we identified infectious disease (including prominently respiratory infectious disease), other respiratory disease, and fever with no obvious source as the main cause of consultation. Then, a review of the relevant medical literature was conducted to identify predictors of complexity as possible inputs to a classifier. As a result of this review, a number of Pediatric Value Scales were identified, which included the Rochester Criteria for Febrile Infants [11], the Young Infant Observational Scale (age > 90 days) [3], the Yale Observational Scale [15], Wood-Downes-Ferres [7], Westley-Mintegui [17], Scarfone Pulmonary Index [21], Glasgow scale (age > 3 years) [9], Raimondi - Glasgow (age < 3 years) [9], the Santolaya scale [19], Modified Glasgow for meningococcal sepsis [4], the ASA Pyshical Status Grading System [1], the Oucher pain scale [2], and behavioral and facial pain scales [22].

Then a prospective cohort was constructed, with 1205 patients enrolled from March 12, 2014, to August 7, 2014. For each patient, information for all variables considered in the aforementioned scales were collected, yielding a dataset with 1205 instances and 197 variables. 51 variables from different scales were dropped due to duplication between scales or complete lack of variability, for a total of 147 variables to be used in the dataset. Patients were followed until they would whether be admitted to hospitalization/advanced care, or discharged from the ED. We decided to use admission to hospitalization/advanced care as a proxy for complexity (our response variable), with approximately 70 % of instances (846) belonging to the negative class - i.e. discharged without admission to hospitalization/advanced care, and the remaining 30 % (359) to the positive class.

3.2   Training and Validation

The dataset was partitioned into training set (80 %, 964 instances) and test set (20 %, 241 instances) using stratified random sampling to preserve the original ratio between classes. Models were trained to assign patients to the Fast Track (negative class) and to the Regular Track (positive class). 10-Fold Cross validation was employed to validate classifier performance on the training set. Given the high dimensionality of the training set, it was decided to perform dimensionality reduction using Principal Component Analysis, to find a lower dimensionality transformation to ease the computation involved in learning the classifiers and improve the chances of generalization. Principal Component Analysis [13] identifies a basis of orthogonal vectors that capture most of the original variability, corresponding to the Eigenvectors of the Covariance Matrix. Figure 1 shows the variability captured by the principal components of the dataset.

PCA shows that the 74 components with largest eigenvalues retain 95 % of all the variability of the original dataset, while the 85.023 % and the 81.098 % of variability is captured by the most important 54 and 48 components, respectively.

Cross Validation results show the use of 74 components offered the best performance (the dimensionality of the dataset was reduced nearly in half). Performance was evaluated via standard measures as Precision (Positive Predictive Value), Sensitivity (Recall, True Positive Rate), and the F1 score. This measures allow us to adequately assess performance, given the fact our dataset is unbalanced.

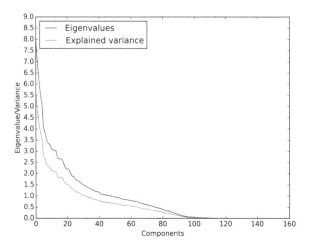

Fig. 1. Variance percentage explained by principal components

Table 1. Conditions reported in the ED in 2013

Diagnosis	Number of records	Percentage (%)
Infectious	6350	37.07
Trauma	2478	14.46
Respiratory infectious	2478	14.46
Surgical	1660	9.69
Respiratory	1290	7.53
Other	958	5.59
Not specified	824	4.81
Neurological	556	3.25
Metabolic	367	2.14

Each classifier was trained using the lower dimensional transformation induced by the 48, 54 and 74 first principal components, ranked by Eigenvalue.

10-Fold Cross Validation results are presented in Tables 2, 3 and 4. We observe the model with highest sensitivity was the Multilayer Perceptron, with a score of 0.846 using the 74 component dataset. Table 5 shows parameters used to learn the models with best performance.

3.3   Test

Consequently we chose the Multilayer Perceptron to be tested using the test set, which was transformed using the transformation matrix with 74 components obtained by PCA on the training set. We held out the test set when computing the principal components in order to avoid data leakage. Test results for the

Table 2. 10-Fold cross validation results with 48 principal components

Model	Sensitivity	Precision	F1
Perceptron	0.835	0.835	0.835
Gaussian SVM	0.805	0.832	0.818
Logistic regression	0.79	0.8	0.8

Table 3. 10-Fold cross validation results with 54 principal components

Model	Sensitivity	Precision	F1
Perceptron	0.838	0.856	0.847
Gaussian SVM	0.805	0.828	0.816
Logistic regression	0.795	0.805	0.800

Table 4. 10-Fold cross validation results with 74 principal components

Model	Sensitivity	Precision	F1
Perceptron	0.846	0.851	0.848
Gaussian SVM	0.808	0.831	0.819
Logistic regression	0.810	0.823	0.817

Table 5. Parameters used in each model (74 components). Note that Orange uses LibSVM to learn an SVM and defines the Gaussian kernel as $exp(-\gamma||x_i - x_j||^2)$

Model	Hidden neurons	γ	λ
Perceptron	250	n/a	0.3
Gaussian SVM	n/a	0.3	n/a
Logistic regression	n/a	n/a	10

Table 6. Test set results

Model	Sensitivity	Precision	F1
Perceptron	0.81	0.89	0.84

Multilayer Perceptron are shown in Table 6. Results indicate a 81 % sensitivity for the model, which signals good discriminative properties. Additionally, test set scores are close to the validation ones, which suggest there is no overfitting occurring.

4 Conclusion

The ED serves as the entry point to the healthcare system for a sizeable portion of the total of patients, putting considerable pressure on resources and staff. Triage then constitutes itself as a very important process to adequately direct resources in a timely manner to those in most need. Any improvement made to the triage process will translate into better ED performance overall and increased patient satisfaction. Lean thinking offers new ideas and directions that when applied to ED management have shown to be effective to improve resource allocation and patient flow. One of the key changes brought about by applying Lean to the ED is the introduction of differential patient streams including a so-called Fast Track which receives low complexity patients according to a modified Triage process. In this paper we stated a new approach to Triage in a Lean Pediatric Emergency Department, based on Machine Learning models capable of classifying patients according to their likelihood of admission as a proxy for complexity. Several classifiers were trained and tested using a dataset comprised of features extracted through the application of different scales from the relevant literature.

Results showed a Multilayer Perceptron Neural Network outperforming the other techniques considered with a 10-Fold cross validated sensitivity of 84.6 %, and a test set sensitivity of 81 %. Results are promising and suggest that a Machine Learning based triage tool can function as a decision support system capable of enhancing patient flow and resource allocation, placing ED patients in the correct track. We believe Machine Learning can impact in a very beneficial way the quality of care in the healthcare system.

References

1. Aplin, S., Baines, D., DE Lima, J.: Use of the ASA physical status grading system in pediatric practice. Paediatr. Anaesth. **17**(3), 216–222 (2007)
2. Beyer, J.E., Turner, S.B., Jones, L., Young, L., Onikul, R., Bohaty, B.: The alternate forms reliability of the Oucher pain scale. Pain Manag. Nurs. **6**(1), 10–17 (2005)
3. Bonadio, W.A., Hennes, H., Smith, D., Ruffing, R., Melzer-Lange, M., Lye, P., Isaacman, D.: Reliability of observation variables in distinguishing infectious outcome of febrile young infants. Pediatr. Infect. Dis. J. **12**(2), 111–114 (1993)
4. Carrol, E., Riordan, F., Thomson, A., Sills, J., Hart, C.: The role of the Glasgow meningococcal septicaemia prognostic score in the emergency management of meningococcal disease. Arch. Dis. Child. **81**(3), 278 (1999). http://www.ncbi.nlm.nih.gov/pmc/articles/PMC1718049/
5. Cortes, C., Vapnik, V.: Support-vector networks. Mach. Learn. **20**(3), 273–297 (1995). http://dx.doi.org/10.1007/BF00994018
6. Demšar, J., Curk, T., Erjavec, A., Gorup, Č., Hočevar, T., Milutinovič, M., Možina, M., Polajnar, M., Toplak, M., Starič, A., Štajdohar, M., Umek, L., Žagar, L., Žbontar, J., Žitnik, M., Zupan, B.: Orange: data mining toolbox in python. J. Mach. Learn. Res. **14**, 2349–2353 (2013). http://jmlr.org/papers/v14/demsar13a.html
7. Ferres, J.: Comparison of two nebulized treatments in wheezing infants. Eur. Respir. J. **1**, 306 (1988)
8. FitzGerald, G.: Triage revisited. Emerg. Med. **10**(4), 291–293 (1998). http://dx.doi.org/10.1111/j.1442-2026.1998.tb00694.x

9. Herndon, R.: Handbook of Neurologic Rating Scales, 2nd edn. Demos Medical Publishing, New York (2006). http://books.google.com.co/books?id=w1yPmehSZ2cC

10. Holden, R.J.: Lean thinking in emergency departments: a critical review. Ann. Emerg. Med. **57**(3), 265–278 (2010). http://dx.doi.org/10.1016/j.annemergmed.2010.08.001

11. Huppler, A.R., Eickhoff, J.C., Wald, E.R.: Performance of low-risk criteria in the evaluation of young infants with fever: review of the literature. Pediatrics **125**(2), 228–233 (2010). http://pediatrics.aappublications.org/content/125/2/228

12. Ieraci, S., Digiusto, E., Sonntag, P., Dann, L., Fox, D.: Streaming by case complexity: evaluation of a model for emergency department fast track. Emerg. Med. Australas. **20**(3), 241–249 (2008). http://dx.doi.org/10.1111/j.1742-6723.2008.01087.x

13. Jolliffe, I.: Principal Component Analysis. Springer Series in Statistics. Springer, Berlin (2002). http://books.google.com.co/books?id=TtVF-ao4fI8C

14. Kelly, A.M., Bryant, M., Cox, L., Jolley, D.: Improving emergency department efficiency by patient streaming to outcomes-based teams. Aust. Health Rev. **31**(1), 16–21 (2007). http://www.publish.csiro.au/paper/AH070016

15. McCarthy, P.L., Sharpe, M.R., Spiesel, S.Z., Dolan, T.F., Forsyth, B.W., DeWitt, T.G., Fink, H.D., Baron, M.A., Cicchetti, D.V.: Observation scales to identify serious illness in febrile children. Pediatrics **70**(5), 802–809 (1982). http://pediatrics.aappublications.org/content/70/5/802

16. McCullagh, P., Nelder, J.: Generalized Linear Models, 2nd edn. Chapman & Hall/CRC Monographs on Statistics & Applied Probability. Taylor & Francis, Abingdon-on-Thames (1989). http://books.google.co.uk/books?id=h9kFH2_FfBkC

17. Mintegui, R.S., Sanchez, E.J., Benito, F.J., Angulo, B.P., Gastiasoro, C.L., Ortiz, A.A.: Usefulness of oxygen saturation in the assessment of children with moderated laryngitis. An. Esp. Pediatr. **45**(3), 261–263 (1996)

18. Ng, D., Vail, G., Thomas, S., Schmidt, N.: Applying the lean principles of the Toyota production system to reduce wait times in the emergency department. CJEM **12**(1), 50–57 (2010)

19. Paganini, H., de Santolaya, P., Álvarez, M., Araña Rosaínz, M.D.J., Arteaga Bonilla, R., Bonilla, A., Caniza, M., Carlesse, F., López, P., Dueñaas de Chicas, L., de León, T., del Pont, J.M., Melgar, M., Naranjo, L., Odio, C., Rodríguez, M., Scopinaro, M.: Diagnóstico y tratamiento de la neutropenia febril en niños con cáncer. consenso de la sociedad latinoamericana de infectología pediátrica. Revista chilena de infectología **28**, 10–38 (2011). http://www.scielo.cl/scielo.php?script=sci_arttext&pid=S0716-10182011000400003&nrm=iso

20. Rumelhart, D.E., Hinton, G.E., Williams, R.J.: Parallel distributed processing: explorations in the microstructure of cognition. In: Learning Internal Representations by Error Propagation, vol. 1., pp. 318–362. MIT Press, Cambridge (1986). http://dl.acm.org/citation.cfm?id=104279.104293

21. Scarfone, R.J., Fuchs, S.M., Nager, A.L., Shane, S.A.: Controlled trial of oral prednisone in the emergency department treatment of children with acute asthma. Pediatrics **92**(4), 513–518 (1993)

22. Velasco-Pérez, G.: Escalera analgésica en pediatría. Acta pediátrica de México **35**, 249–255 (2014). http://www.scielo.org.mx/scielo.php?script=sci_arttext&pid=S0186-23912014000300011&nrm=iso

23. Womack, J.P., Jones, D.T., Roos, D.: The Machine That Changed the World: The Story of Lean Production. The MIT International Motor Vehicle Program. HarperCollins, New York (1991). https://books.google.de/books?id=Jz4zog27W7gC

An Empirical Validation of Learning Schemes Using an Automated Genetic Defect Prediction Framework

Juan Murillo-Morera[1]([✉]), Carlos Castro-Herrera[1],
Javier Arroyo[2], and Rubén Fuentes-Fernández[2]

[1] Doctoral Program in Computer Science,
University of Costa Rica, San José, Costa Rica
juan.murillomorera@ucr.ac.cr, carlos.castro.herrera@intel.com
[2] Department of Software Engineering and Artificial Intelligence,
University Complutense of Madrid, Madrid, Spain
{javier.arroyo,ruben}@fdi.ucm.es

Abstract. Today, it is common for software projects to collect measurement data through development processes. With these data, defect prediction software can try to estimate the defect proneness of a software module, with the objective of assisting and guiding software practitioners. With timely and accurate defect predictions, practitioners can focus their limited testing resources on higher risk areas. This paper reports a benchmarking study that uses a genetic algorithm that automatically generates and compares different learning schemes (preprocessing + attribute selection + learning algorithms). Performance of the software development defect prediction models (using AUC, Area Under the Curve) was validated using NASA-MDP and PROMISE data sets. Twelve data sets from NASA-MDP (8) and PROMISE (4) projects were analyzed running a $M \times N$-fold cross-validation. We used a genetic algorithm to select the components of the learning schemes automatically, and to evaluate and report those with the best performance. In all, 864 learning schemes were studied. The most common learning schemes were: data preprocessors: Log and CoxBox + attribute selectors: Backward Elimination, BestFirst and LinearForwardSelection + learning algorithms: NaiveBayes, NaiveBayesSimple, SimpleLogistic, MultilayerPerceptron, Logistic, LogitBoost, BayesNet, and OneR. The genetic algorithm reported steady performance and runtime among data sets, according to statistical analysis.

Keywords: Software quality · Fault prediction models · Genetic algorithms · Learning schemes · Learning algorithms · Machine learning

1 Introduction

Software fault prediction has been an important research topic within the software engineering field for more than 30 years [1]. Software measurement data

© Springer International Publishing AG 2016
M. Montes-y-Gómez et al. (Eds.): IBERAMIA 2016, LNAI 10022, pp. 222–234, 2016.
DOI: 10.1007/978-3-319-47955-2_19

collected during the development process include valuable information about the project's status, progress, quality, performance, and evolution. These data are commonly used as input to fault prediction models. The static code attributes as McCabe [2], Halstead [3] and Line of Code can be used to predict defects. These static code attributes are relatively simple to calculate, and can be easily automated. The metrics cited before are **module-based**, where a module is defined as the smallest unit of functionality in a program, such as a **function** or **method**. Fault prediction models try to identify defect prone software modules [4]. The main goal of generating these predictions is to allow software engineers to focus development and testing activities on the most fault-prone parts of their code, thereby improving software quality and making a better use of the limited time and resources [5,6]. The study and construction of these techniques have been the core of the *fault prediction modeling* research area, and the subject of many previous works, where the right selection of machine learning algorithms has an important role in the prediction process [7,8].

This paper analyzes 8 Data Preprocessing techniques (DP), 6 Attribute Selectors (AS), and 18 machine Learning Algorithms (LA) representing different kinds of models. Its main objective is to evaluate the frequency of selection and performance of several learning schemes, with the main focus on machine learning algorithms. The learning schemes are selected by a genetic algorithm. We have selected the genetic algorithms for three mainly reasons. According to the literature, [9] identified that "there are very few studies that examine the effectiveness of evolutionary algorithms", as an open area for future work. She pointed out that "future studies may focus on the predictive accuracy of evolutionary algorithms for software fault prediction". Further, the selection of the learning schemes is a combinatorial problem. Finally, this research is a maximization problem, both typical sceneries of the genetic algorithms.

The categories according to [10] of the machine learning algorithms analyzed in this paper include **Bayes** (NaiveBayes, BayesNet, BayesianLogistic Regression and NaiveBayesSimple), **Functions** (Logistic, SimpleLogistic and MultilayerPerceptron), **Meta** (LogitBoost, MIBoost, Bagging and Dagging), **Rules** (OneR and ZeroR) and **Trees** (J48, RandomForest, REPTree, NBTree and RandomTree). Each technique is applied to twelve data sets within the domain of software defect prediction. The structure of the rest of the article is as follows. Section 2 presents research questions. Section 3 discusses the related work and state of the art. The proposed framework is explained in Sect. 4. The genetic configuration is described in Sect. 5 and the experimental design in Sect. 6. Section 7 reports the frequency and performance of machine learning algorithms. Section 8 presents the hypothesis and statistical test. Section 9 addresses the threats to validity. Finally, Sect. 10 discusses conclusions and presents future work.

2   Research Questions

This section lists the main research questions that we set out to answer.

- RQ-1 Which data preprocessors are selected for each data set?
- RQ-2 Which attribute selectors are selected for each data set?

- RQ-3 Which machine learning algorithms are selected for each data set?
- RQ-4 Which machine learning algorithms are not selected per data set?
- RQ-5 Which machine learning algorithms report the best Area Under the Curve (AUC) per data set?

3   Related Work

In 2007, Menzies et al. [11], published a study in which they compared the performance of two machine learning techniques (Rule Induction and Naive Bayes) to predict software components containing defects. They claimed that how attributes are used to build predictors is much more important than which particular attributes are used and the choice of the learning method is far more important than which subset of available data is used for learning.

In 2011, Song et al. [1] published a study in which they proposed a fault prediction framework based on Menziess research. They analyzed 12 learning schemes. They argued that, although how is more important than which", the choice of a learning scheme should depend on the combination of data preprocessing techniques, attribute selection methods, and learning algorithms.

In 2015, following this lead, we published a study [12] that proposed a genetic fault prediction framework based on Song's architecture. In that work, we selected the learning schemes automatically. This automation allows exploring many more possibilities in order to find better learning schemes for **each data set**. The genetic framework had a 3.2 % improvement over Song's framework.

Our main contribution here is how to select the best learning schemes automatically for a specific data set, with a main focus on machine learning algorithms according their performance (AUC) using a genetic framework.

4   Proposed Framework

4.1   Learning Schemes Generator

To build the prediction models, we mainly followed the framework proposed by Song et al. [1]. The main difference is that our work uses a genetic algorithm to select parts of the learning scheme. This mean (preprocessing + attribute selection + learning algorithms), instead of using a group of pre-established combinations, as Song does. In our framework, each part of the learning scheme is a part of the chromosome used (for details see section). Our framework consists of two components: (1) Learning Scheme Generator-Evaluator and (2) Defect Prediction. The first one builds the chromosomes and selects the best genetically. The second component generates the final predictor that contains the learning scheme selected by the genetic framework. The final predictor uses the learning scheme previously selected and classifies the fault-prone modules (true/false).

The Learning Scheme Generator-Evaluator is responsible for generating, evaluating, and selecting the different learning schemes. Selection is done through the elitism technique of the genetic algorithm. Figure 1 shows its components.

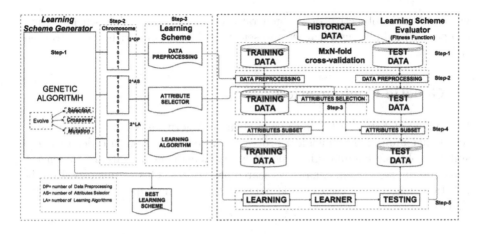

Fig. 1. Learning Schemes Generator Adapted from [12]

The main steps of this process are the following:

1. The population of individuals (chromosomes) is generated and transformed (operators: selection, crossover and mutation) for the generator component (Fig. 1, **Generator, Step-1**).
2. Chromosomes are genetically created (Fig. 1, **Generator, Step-2**) and represented by learning schemes (Fig. 1, **Generator, Step-3**).
3. Historical data were randomized and divided into a training set and a test set. This is done using a $M \times N$-fold cross-validation (Fig. 1, Evaluator **Step-1** and also under Algorithm 2, lines 6–16 - Specifically lines 8 and 9).
4. The selected data pre-processing technique is applied to both the training and the test set (Fig. 1, Evaluator **Step-2**), thus resulting in modified training and test data. This step is represented with the so-called learning method (see below Algorithm 2, line 10).
5. The chosen attribute selection technique is applied only to the training set (Fig. 1, Evaluator **Step-3**) and the best subset of attributes is chosen.
6. The selected attributes are then extracted for both the training and test set (Fig. 1, Evaluator **Step-4** and below Algorithm 2, line 10).
7. The Learning algorithm is built using the training set, and it is evaluated with the test set. This is performed at (Fig. 1, Evaluator **Step-5** and Algorithm 2, line 10).

Finally, the best learning scheme, in our case the chromosome (with its data pre-processing, attribute selector and learning algorithm), is selected by the genetic algorithm described in Algorithm 1.

4.2 Defect Prediction

The second component of the proposed framework is the defect prediction stage. The main objective of this component is building a predictor for new data, while the main objective of the generator-evaluator of learning schemes (first component) is building a learner and selecting the best one with regard to their AUC.

5   Genetic Configuration

In the field of Artificial Intelligence, a Genetic Algorithm (GA) is a search approach that mimics the biological process of natural selection in order to find a suitable solution in a multidimensional space. Genetic algorithms are, in general, substantially faster than exhaustive search procedures. This section describes the genetic configuration used in this paper: Chromosome, Operators and Fitness Function.

5.1   Chromosome

The chromosome of a GA represents the set of any possible combinations in the search space. It is commonly represented as a binary chain of 0 s and 1s. In this paper the chromosome consists of three parts: Data Pre-Processing (DP), Attribute Selector (AS), and Learning Algorithms (LA); effectively constructing a triplet of <DP, AS, LA>. We represented DP, AS and LA by binary chains of bits: For DP, we considered 8 possibilities with a chain of 3 bits ($2^3 = 8$); for AS, 6 possible techniques represented by 3 bits ($2^3 = 8$); and for LA, 18 different possibilities, which require 5 bits ($2^5 = 32$). For further details on the considered techniques and their coding, see Sect. 6.2. With this chromosome representation, the goal of the GA is to find the chromosome that maximizes the fitness function.

5.2   Fitness Function

Algorithms 1 and 2 describe the details for the implementation of the fitness function. The final value of the AUC is calculated in the prediction phase. (Algorithm 3).

Algorithm 1 shows the $N - PASS$ stage, where the fitness score of a chromosome is calculated as the average of 10 runs. Before this algorithm, the whole data set is split randomly into histData (90 % used for training and testing) and newData (10 %), and we repeat this for each $PASS$.

The coding process generates a chain of bits (genotype). On the other hand, the decoding process (phenotype) is calculated with the fitness function into the evaluation function (Algorithm 1, line 5). We selected the DP, AS, and LA conforming to the phenotype of each part of the chromosome, represented by the learning scheme used into (Algorithm 2, line 10). The final phenotype is represented by the AUC (performance) of a specific learning scheme (chromosome), calculated into Algorithm 3.

Algorithm 2 shows the evaluation phase. This algorithm performs a $M \times N$-fold cross-validation, where multiple rounds are executed with the different partitions in order to reduce variability. Validation results are averaged over the rounds.

Algorithm 1. Fitness Function

Require: *individual* : *Chromosome*
Ensure: *AUC* ← *Real*
1: *LS* ← *individual.getlS*()
2: *i, AUC* ← 0
3: *NPASS* ← 10
4: **while** *i* < *NPASS* **do**
5:     *AUC* ← *AUC* + *Evaluation*(*histData*[*i*], *LS*)
6:     *i* ← *i* + 1
7: **end while**
8: **return** *AUC/NPASS*

Algorithm 2. Evaluation

Require: *histData* : *Instances*, *lS* : *learningScheme*
Ensure: *EvalAUC* ← *Real*
1: *trainDS, testDS* : *instances*
2: *eval* : *evaluation*
3: *i, j, EvalAUC* ← 0
4: *M, N, folds* ← 10
5: *histData* ← *histData.random*()
6: **while** *i* < *M* **do**
7:     **while** *j* < *N* **do**
8:         *trainDS* ← *histData.getFold*(*j*)
9:         *testDS* ← *histData.getAllFoldsExcept*(*j*)
10:         *eval* ← *learning*(*trainDS, lS*)
11:         *EvalAUC* ← *EvalAUC* + *evaluateModel*(*eval, testDS*)
12:         *j* ← *j* + 1
13:     **end while**
14:     *i* ← *i* + 1
15: **end while**
16: **return** *EvalAUC/*(*M* ∗ *N*)

Finally, Algorithm 3 illustrates the prediction phase. This algorithm generates the final value of AUC using the unseen newData and the best learning scheme (lS) calculated in the evaluation phase.

5.3   Operators

The operators of selection, reproduction, crossover, and mutation used here were configured using the default values provided by WEKA in its genetic search. [10]: Population size = 20, Crossover probability = 0.6, Mutation probability = 0.033 − 0.01. and Elitism = True.

Algorithm 3. Prediction

Require: $historicalData[], newData[] : Instances, lS : learningScheme$
Ensure: $PredAUC$
 1: $i \leftarrow 0$
 2: $NPASS \leftarrow 10$
 3: **while** $i < NPASS$ **do**
 4:     **if** $TypePrep <> NONE$ **then**
 5:         $newData[i] \leftarrow ApplyPreProcessing(newData[i])$
 6:     **end if**
 7:     $PredAUC \leftarrow PredUAC + prediction(historical[i], newData[i], lS)$
 8: **end while**
 9: **return** $PredAUC/NPASS$

---

6   Empirical Study

6.1   Data Sets

To conduct the experiments, we used the following data sets: AR1, AR3, AR4, and AR6 from the PROMISE repository, and KC1, KC3, KC4, MW1, PC1, PC2, PC3, PC4 from the NASA-MDP. For further details, see [1].

6.2   Learning Schemes

As stated in Sect. 6.2, the learning schemes consist of three parts: Data Pre-processing, Attribute Selector, and Machine Learning Algorithm (mentioned as learning algorithm). This section presents a detail of the different techniques used for each part. In all, we tested 864 different learning schemes [10].

– *Data Pre-processing (DP)*: None, Log, BoxCox $\lambda = -2$ $(BC - 2)$, $\lambda = -1$ $(BC - 1)$, $\lambda = -0.5$ $(BC - 0.5)$, $\lambda=0.5$ (BC0.5), $\lambda=1$ (BC1), and $\lambda=2$ (BC2).
– *Attribute selector (AS)*: Backward Elimination(BE), Forward Selection(FS), BestFirst(BF), LinearForwardSelection(LFS), RankSearch(RS) and Genetic-Search(GS).
– *Learning Algorithm (LA)*: NaiveBayes (NB), BayesNet (BN), Bayesian-LogisticRegression (BLR), NaiveBayesSimple (NBS), Logistic (LOG), SimpleLogistic (SL), MultilayerPerceptron (MP), Bagging (BAG), Dagging (DAG), LogitBoost (LGB), MIBoost (MIB), OneR (OneR), ZeroR (ZeroR), J48 (J48), RandomForest (RF), REPTree(REPT), NBTree (NBT) and RandomTree (RT).

6.3   Experimental Design

The experimental process is described as follows:

1. In our experiment, we used $n = 12$ data sets (see Sect. 6). The experimental process had the following characteristics:

(a) The search space presented a total of 864 combinations (see Sect. 6.2).
(b) In the generation-evaluation phase we used the following configuration: Mutation $(0.033 - 0.01)$ and Crossover (0.60). We used Roulette, as operator of selection. We set elitism to true.
(c) In the generation-evaluation phase, we applied a strategy for the selection of attributes called Wrapper [10]. It was used with the objective of selecting the attributes for each subset using an internal cross-validation. Wrappers generally provide better results than filters, but they are more computationally intensive [1].
(d) We executed a total of 20 generations with 20 genetic individuals using the standard configuration of WEKA's geneticSearch [10]. Each genetic individual was represented by a learning scheme.
(e) We set $N - PASS = 10$, and calculated the AUC average after $N - PASS$ runs. For each $PASS$, we selected 90 % of the data as historical at random.
(f) An $N = 10 \times M = 10$-fold cross-validation was used to evaluate each learning scheme. Modules were selected at random. Furthermore, the evaluation metrics AUC, Recall and Precision, and their average were calculated after $N \times M$-fold cross-validation.
(g) The fitness function of each genetic individual was executed in 1000 hold-out experiments, $(N - PASS = 10)$ and $N = 10 \times M = 10$-fold cross-validation. The mean of the 1000 AUC measures was reported as the evaluation performance per genetic individual. The historicalData (90 %) was preprocessed considering preprocessing techniques. Then, the predictor was used to predict defects with the newData (10 %), which was preprocessed the same way as the historical data.
(h) Finally, steps (a) to (g) were executed 10 times in order to study the steadiness of our framework.

7   Results and Analysis

For analysis purposes, we will use the research questions from Sect. 2. Our inquiry focuses on each data set and between groups (NASA-MDP and PROMISE). Our results represent 10 executions of the genetic framework (for further details, see Sect. 6).

- **RQ-1 Which data preprocessors are selected for each data set?**
  Table 1 shows the results of the different data preprocessors selected. We show the frequency of selection of each data preprocessor. According to these results, there is no predominant data preprocessor per data set. However, NASA-MDP reported a Log as predominant, and BoxCox was predominant between the groups (NASA-MDP and PROMISE).
- **RQ-2 Which attribute selectors are selected for each data set?**
  Table 2 shows the results of the different attribute selectors selected. The attribute selector predominant for NASA-MDP is BE, while those for PROMISE are BF and LFS.

Table 1. Data preprocessors

Dataset	Data preprocessors	Frequency
PC1	Log-None-BC-2-Others	(30 %)(20 %)(20 %)(30 %)
PC2	BC0.5-BC1-Log-Others	(30 %)(30 %)(20 %)(20 %)
PC3	Log-BC-0.5-BC0.5-Others	(30 %)(30 %)(20 %)(20 %)
PC4	None-BC0.5-BC1-Others	(20 %)(20 %)(20 %)(40 %)
AR1	None-Log-BC-1-Others	(30 %)(20 %)(20 %)(30 %)
AR3	None-BC-2-BC-0.5-Others	(30 %)(20 %)(20 %)(30 %)
AR4	BC1-BC-1-BC0.5-Others	(40 %)(30 %)(20 %)(10 %)
AR6	BC0.5-BC-1-BC1-Others	(30 %)(20 %)(20 %)(30 %)
KC1	Log-BC-2-BC1-Others	(40 %)(20 %)(20 %)(20 %)
KC3	BC1-BC-2-BC-0.5-Others	(40 %)(20 %)(20 %)(20 %)
MC2	BC-2-BC1-BC-1-Others	(40 %)(30 %)(10 %)(20 %)
MW1	Log-BC-2-BC0.5-Others	(30 %)(20 %)(10 %)(40 %)

- **RQ-3 Which machine learning algorithms are selected for each data set?** Table 3 shows the results of the different machine learning algorithms selected. The machine learning algorithms predominant for NASA-MDP were: LOG, LB, and BN, while those for PROMISE were LOG, OneR, and NB.
- **RQ-4 Which machine learning algorithms are not selected for the data sets?** The learning algorithms that were never selected by NASA-MDP dataset were: BLR, DAG, MIB, OneR, ZeroR, J48, REPT, NBT and RT. For the PROMISE dataset, they were: BN, BLR, SL, MIB, REPT, NBT and RT. The learning algorithms never selected for both groups were: BLR, MIB, REPT, NBT, and RT.
- **RQ-5 Which machine learning algorithms report the best AUC per data set?** Table 4 shows the learning algorithms with the best AUC (according to n=10 executions). BAG and LB were the learning algorithms with the best performance. In addition to the LG algorithm, they presented the best result for both data set groups (NASA-MDP and PROMISE).

8   Statistical Analysis

We applied a non-parametric test called Wilcoxon rank sum test (Wilcoxon Mann Whitney). We evaluated three hypotheses. (1) To evaluate the performance of our genetic framework against an exhaustive framework (baseline). (2) To analyze the evaluation and prediction phases according to their performance, and (3) To evaluate the runtime of the genetic framework against the runtime exhaustive framework (baseline). Table 5 presents a summary of the performance and runtime of our genetic framework. For the first hypothesis, $p_{value} = 0.5057 > \alpha = 0.05$. This means that we did not find a statistically significant difference between the exhaustive framework (baseline)

Table 2. Attribute selectors

Dataset	Attribute selectors	Frequency
PC1	BE-LFS-GS	(40 %)(40 %)(20 %)(-%)
PC2	RS-FS-BE-Others	(40 %)(40 %)(10 %)(10 %)
PC3	BE-FS-RS-Others	(60 %)(10 %)(10 %)(20 %)
PC4	BE-FS-LFS-Others	(30 %)(20 %)(20 %)(30 %)
AR1	RS-FS-BF-Others	(30 %)(20 %)(20 %)(30 %)
AR3	BF-GS-LFS-Others	(30 %)(30 %)(20 %)(20 %)
AR4	FS-LFS-RS-Others	(30 %)(30 %)(10 %)(30 %)
AR6	BF-LFS-BE-Others	(30 %)(30 %)(20 %)(20 %)
KC1	FS-LFS-RS-Others	(30 %)(20 %)(20 %)(30 %)
KC3	GS-LFS-FS-Others	(30 %)(20 %)(20 %)(30 %)
MC2	BE-BF-RS-Others	(30 %)(30 %)(20 %)(20 %)
MW1	LFS-BF-GS-Others	(50 %)(20 %)(20 %)(10 %)

and our framework, with regard to performance. For the second hypothesis, $p_{value} = 0.47 > \alpha = 0.05$. This means that we did not find a statistically significant difference between the evaluation phase and prediction phase, with regard to performance. This means that our approach is very reliable. For the third hypothesis, $p_{value} = 0.01529 < \alpha = 0.05$. This means that we found a statistically significant difference between the exhaustive runtime (baseline) and the genetic runtime, reporting a better runtime for the genetic framework.

Table 3. Learning algorithms

Dataset	Learning algorithms	Frequency
PC1	LOG-NB-BN-Others	(30 %)(20 %)(20 %)(30 %)
PC2	LOG-NBS-NB-Others	(50 %)(20 %)(10 %)(20 %)
PC3	LB-SL-LOG-Others	(60 %)(20 %)(10 %)(10 %)
PC4	LOG-BN-NB-Others	(30 %)(20 %)(10 %)(40 %)
AR1	LOG-MP-NBS-Others	(40 %)(30 %)(10 %)(20 %)
AR3	OneR-NB-LOG-Others	(40 %)(20 %)(20 %)(20 %)
AR4	BN-LOG-MP-Others	(20 %)(20 %)(10 %)(50 %)
AR6	NB-DAG-LB-Others	(30 %)(20 %)(20 %)(30 %)
KC1	BN-NB-BAG-Others	(30 %)(20 %)(20 %)(30 %)
KC3	LOG-LOG-SL-Others	(50 %)(30 %)(10 %)(10 %)
MC2	LOG-NB-MP-Others	(30 %)(20 %)(20 %)(30 %)
MW1	NBS-SL-MP-Others	(20 %)(20 %)(20 %)(40 %)

Table 4. Learning algorithms with the best AUC

Dataset	Learning algorithm	Best prediction-AUC
PC1	BAG	0.8183
PC2	LOG	0.6863
PC3	LGB	0.8012
PC4	LGB	0.8430
AR1	LOG	0.4546
AR3	LOG	0.4678
AR4	NB	0.7356
AR6	LGB	0.5058
KC1	BAG	0.8064
KC3	SL	0.6456
MC2	NBS	0.6899
MW1	NBS	0.6979

9   Threats to Validity

Internal validity: Statistical results show that our framework is steady according to the results reported. However, more experiments are required to validate other configurations.

Table 5. Genetic framework performance

	AUC (Performance)		RunTime (milliseconds)		Genetic-Framework	
	GEN	EXH	GEN	EXH	EVAL	PRED
PC1	0.81	0.83	7.5e+6	3.57e+7	0.80	0.81
PC2	0.68	0.70	5.1e+6	2.778e+7	0.67	0.68
PC3	0.80	0.82	4.5e+6	1.0032e+8	0.78	0.80
PC4	0.84	0.84	1.098e+7	1.29e+8	0.83	0.84
AR1	0.45	0.49	2.4e+6	3.96e+6	0.42	0.45
AR3	0.46	0.45	1.26e+6	3.6e+6	0.44	0.46
AR4	0.73	0.76	1.86e+6	4.2e+6	0.72	0.73
AR6	0.50	0.55	1.32e+6	3.72e+6	0.48	0.50
KC1	0.80	0.80	4.08e+6	5.556e+7	0.79	0.80
KC3	0.64	0.71	2.22e+6	8.16e+6	0.62	0.64
MC2	0.68	0.70	2.88e+6	5.1e+6	0.65	0.68
MW1	0.69	0.70	1.878e+7	1.278e+7	0.63	0.69

External validity: The results of this study only consider public data sets (NASA-MDP and PROMISE). More experimentation is needed with real life projects presenting more missing values, imbalance data, outliers, among others.

Construction validity: Wrappers generally provide better results than filters, but they are more computationally intensive. In our proposed framework, we use the wrapper evaluation method.

10   Conclusions and Future Work

In all, 864 learning schemes were analyzed. The most common learning schemes were: **data preprocessors:** Log and CB + **attribute selectors:** BE, BF and LFS + **machine learning algorithms:** NB, NBS, SL, MP, LG, LB, BN and OneR. The first group of data sets, where the framework reported better performance were: PC3, KC3, PC1 and PC4 with an AUC between (0.8012–0.8430). On the other hand, the second group of data sets with better performance were: AR1, AR3, and AR6 with an AUC between (0.4546–0.5058). Finally, the third group with better performance were: KC3, MC2, MW1 and AR4 with an AUC between (0.6456–0.7356). The machine learning algorithm with the best performance was: LGB and with the worse performance was: LOG. The function category (LOG, LGB, SL) represented the (58.33 %) of the learning algorithms selected, while Bayes category represented the (25 %). The Meta category represented the rest (16.67 %). According to the statistical analysis, our framework presented steadiness and reliability in its results. We did not find a statistically significant difference between the evaluation and the prediction phase. Further, we did not find a statistically significant difference between the exhaustive framework (baseline) and our framework.

In the future, we plan to work with other kinds of data sets. For example private data sets or projects. Furthermore, we will like to work with more learning schemes providing more attribute selector methods, machine learning algorithms, and data pre-processing. Additionally, we plan to conduct more experimentation with different genetic configurations, changing the number of populations and generations used, increasing the $N - PASS$ parameter, experimenting with other performance metrics, such as precision, recall, balance, among others. Finally, we will execute an analysis of the treatments (864) in order to find interactions between the factors (learning schemes).

Acknowledgments. This research was supported by University of Costa Rica, National University of Costa Rica and Ministry of Science, Technology and Telecommunications (MICITT).

References

1. Song, Q., Jia, Z., Shepperd, M., Ying, S., Liu, J.: A general software defect-proneness prediction framework. IEEE Trans. Softw. Eng. **37**, 356–370 (2011)
2. McCabe, T.J.: A complexity measure. IEEE Trans. Softw. Eng. 308–320 (1976)

3. Halstead, M.H.: Elements of software science. IEEE Trans. Softw. Eng. (1977)
4. Wang, H., Khoshgoftaar, T.M., Napolitano, A.: Software measurement data reduction using ensemble techniques. Neurocomputing **92**, 124–132 (2012)
5. Hall, T., Beecham, S., Bowes, D., Gray, D., Counsell, S.: A systematic literature review on fault prediction performance in software engineering. IEEE Trans. Softw. Eng. **38**, 1276–1304 (2012)
6. Arisholm, E., Briand, L.C., Johannessen, E.B.: A systematic and comprehensive investigation of methods to build and evaluate fault prediction models. J. Syst. Softw. **83**, 2–17 (2010)
7. Malhotra, R.: Comparative analysis of statistical and machine learning methods for predicting faulty modules. Appl. Soft Comput. **21**, 286–297 (2014)
8. Shepperd, M., Song, Q., Sun, Z., Mair, C.: Data quality: some comments on the nasa software defect datasets. IEEE Trans. Softw. Eng. **39**, 1208–1215 (2013)
9. Malhotra, R.: A systematic review of machine learning techniques for software fault prediction. Appl. Soft Comput. **27**, 504–518 (2015)
10. Witten, I.H., Frank, E.: Data Mining: Practical Machine Learning Tools and Techniques. Morgan Kaufmann, Burlington (2005)
11. Menzies, T., Greenwald, J., Frank, A.: Data mining static code attributes to learn defect predictors. IEEE Trans. Softw. Eng. **33**, 2–13 (2007)
12. Murillo-Morera, J., Jenkins, M.: A software defect-proneness prediction framework: a new approach using genetic algorithms to generate learning schemes. In: The 27th International Conference on Software Engineering and Knowledge Engineering, SEKE 2015, Wyndham Pittsburgh University Center, Pittsburgh, PA, USA, 6–8 July 2015, pp. 445–450 (2015)

Machine Learning Approaches to Estimate Simulated Cardiac Ejection Fraction from Electrical Impedance Tomography

Tales L. Fonseca$^{(\boxtimes)}$, Leonardo Goliatt, Luciana C.D. Campos,
Flávia S. Bastos, Luis Paulo S. Barra, and Rodrigo W. dos Santos

Federal University of Juiz de Fora, Juiz de Fora, Brazil
tales.lima@ice.ufjf.br, {leonardo.goliatt,luciana.campos,flavia.bastos,
luis.barra,rodrigo.weber}@ufjf.edu.br

Abstract. The ejection fraction (EF) is a parameter that represents the amount of blood pumped out of each ventricle in each cardiac cycle and can be used for analyzing the heart failure. There are several diagnostic tests to determine whether the person has heart failure but some are expensive tests and they do not allow obtaining continuous estimations of EF. However, use the Electrical Impedance Tomography (EIT) with Regression Models is an alternative to obtain continuous estimations of EF. The quality of EIT is that it allows a quick diagnosis about the heart's health, combining low cost and high portability. This paper it proposed four regression models, using the electrical measures from EIT, to estimate the EF : Gaussian Processes (GP), Support Vector Regression (SVR), Elastic Net Regression (ENR) and Multivariate Adaptive Regression Splines (MARS). The overall evaluation of results show that all models achieved competitive results and the method SVR has produced better results than the others tested.

Keywords: Ejection fraction · Machine learning · Electric impedance tomography

1 Introduction

The ejection fraction (EF) is a measurement in determining how well the heart is pumping out blood. An EF of 60 % means that 60 % of the total amount of blood in the left ventricle is pushed out with each heartbeat. If the heart is contracting abnormally, this parameter presents values outside the range between 50 and 70 %. The cardiac EF have a high correlation with the functional state of the heart, specially to determining of the severity of systolic heart failure, which is associated to some heart diseases, such as aortic regurgitation, coronary artery disease among others [2,14,21] and identifies high-risk patients [20]. The volume of blood within a ventricle immediately before a contraction is known as the end-diastolic volume (EDV). Likewise, the volume of blood left in a ventricle at the end of contraction is known as the end-systolic volume (ESV). The difference

© Springer International Publishing AG 2016
M. Montes-y-Gómez et al. (Eds.): IBERAMIA 2016, LNAI 10022, pp. 235–246, 2016.
DOI: 10.1007/978-3-319-47955-2_20

between EDV and ESV represents a ratio between the ventricles full and emptied, which allows many variables such as stroke volume (SV). The SV describes the volume of blood ejected from the right and left ventricles with each heartbeat and the EF is the fraction of the end-diastolic volume that is ejected with each beat [8], that is, it is stroke volume (SV) divided by end-diastolic volume (EDV):

$$EF = \frac{SV}{EDV} = \frac{EDV - ESV}{EDV} \tag{1}$$

The volume of the ventricle in diastole and systole cycles can be replaced by the areas of the ventricle to simplify the calculation of EF, without loss of quality. The reason is that the areas that represent the ventricles are based on a cross section of the heart cavities and it is assumed to be proportional to their volumes. Thus, the EF can be calculated as follows:

$$EF = \frac{EDA - ESA}{EDA} \tag{2}$$

where EDA is the area of the end of diastole, while ESA means the area of the end of systole. Several non-invasive techniques can be applied to determine the EF like echocardiogram, cardiac catheterization, cardiac magnetic resonance, computerized tomography, and others. Such techniques are capable of produce high definition images for well-accurate diagnostics, but they are techniques with high costs, one important reason that it not allow they can be used for continuous monitoring of the heart. In this work, the Electrical Impedance Tomography (EIT) technique was chosen because it does not use ionizing radiation, has low costs and better portability. In the context related to biomedical engineering, the work [13] has discussed the feasibility of EIT for continuous monitoring of cardiac ejection fraction, and other related works [18,19] have shown preliminary results on the same subject.

Computational models associated with data analysis techniques provide a powerful tool to understand the processes involved in generating the EF and the relationship with other myocardial mechanisms. In the literature, some efforts have been made to build simpler models to improve the understanding and interpretation of the EF [10]. In addition, reliable estimations of cardiac EF has been obtained using computational intelligence techniques to predict EF, such as hybrid heuristics [15] and Neural Networks [5,17].

In this paper we propose the use of four different machine learning models in the determination of Ejection Fraction: Gaussian Processes (GP), Support Vector Regression (SVR), Elastic Net Regression (ENR) and Multivariate Adaptive Regression Splines (MARS). Their performance was compared with the computational intelligence methodologies [5,17] and simulation strategies [12,13] that deal with the same problem. All the tests were carried out based on a synthetic dataset. The results presented in this paper are competitive with those found on the literature and suggest that Support Vector Regression can be used with the EIT as a useful alternative of diagnostic tools that offer continuous and non-invasive estimations of cardiac ejection fraction. This paper is organized as follows. Section 2 describes the dataset and the prediction methods and parameter tuning strategies. Section 3 presents the results and the discussions and the conclusions are given in Sect. 4.

2   Materials and Methods

2.1   Simulated Dataset

The adjustments of the machine learning models were made with a simulated data set due to the lack of a real medical data base. The data set generation is divided by three steps [13]:

1. it was carried out a parametrization of a magnetic resonance image;
2. it was carried out the calculation of the electrical potentials;
3. they were carried out generations of new ventricles control points.

An extended x-spline curve is used to represent the boundary of a ventricular cavity [3]. Using this spline, were defined 15 control points - 7 points for the left ventricle and 8 points for the right ventricle (see Fig. 1). As done in [13] the electrical potentials were calculated based on an equation that takes in account: the lungs shape, the torso shape, the amount of blood and their conductivities.

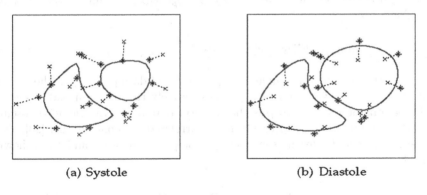

(a) Systole   (b) Diastole

Fig. 1. Control points of both ventricles [17]

A database of the electrical potentials synthetically calculated was constructed in function of the known resistivity and conductivity distribution of the body, based on two magnetic resonance images: one at the systole cycle and other at the diastole cycle [5]. The images were segmented manually in five regions: both lungs, both heart ventricles and a torso boundary as shown in Fig. 2. New control points were created to generate a set of different heart configurations with pathological and healthy conditions.

It was used a parameter t where $t = 0$ represents the systole cycle and $t = 1$ represents the diastole cycle. In the synthetical database they were used 11 parameters t_i in the interval $[0, 0.1, 0.2, \ldots, 1]$. For each t_i, 19 different configurations were created:

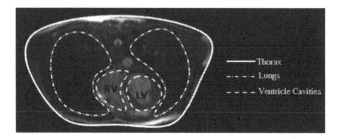

Fig. 2. The ressonance magnetic image segmented [17]

- one with the exactly t_i, without changing any control points.
- 6 configurations changing the control points of the left ventricles.
- 6 configurations changing the control points of the right ventricles.
- 6 configurations changing the control points of both ventricles.

2.2   Regression Models

Elastic Net Regression – ENR. Generalized Linear regression [7] fits a linear model with coefficients $\mathbf{w} = [w_0, w_1, w_2, \ldots, w_p]$ which minimizes the residual sum of squares

$$\min_{\mathbf{w}} \frac{1}{2N} \|\mathbf{Xw} - \mathbf{y}\|_2^2 + \alpha\rho\|\mathbf{w}\|_1 + \frac{\alpha(1-\rho)}{2}\|\mathbf{w}\|_2^2$$

where N in the number of samples, $\mathbf{X}$ is the array of data points, $\mathbf{y}$ is the vector of outputs, $\alpha \geq 0$, $\|\mathbf{w}\|_2$ and $\|\mathbf{w}\|_1$ are respectively the L_2-norm and L_1-norm of the parameter vector, and ρ is the L_1-ratio parameter. Elastic-net is useful when there are multiple features which are correlated with one another [23], also consisting in a tool for features extracting. The parameters of ρ and α are shown in Table 1.

Gaussian Processes – GP. Gaussian Process (GP) or Kriging models [16] are composed of a regression trend model and stochastic process

$$\hat{y}(\mathbf{x}) = \sum_{k=1}^{P} \beta_k f_k(\mathbf{x}) + Z(\mathbf{x}) = \mathbf{f}(\mathbf{x})^T \boldsymbol{\beta} + Z(\mathbf{x}) \tag{3}$$

where $\mathbf{f}(\mathbf{x}) = [f_1(\mathbf{x}), \ldots, f_P(\mathbf{x})]^T$ is a set of P regression basis functions, $\boldsymbol{\beta} = [\beta_1, \ldots, \beta_P]$ is a vector of unknown coefficients, $Z(\mathbf{x})$ is a is assumed to be a Gaussian process with zero mean and covariance $\text{cov}\,[Z(\mathbf{x}^i), Z(\mathbf{x}^j)] = \sigma^2 \mathbf{R}(\boldsymbol{\theta}, \mathbf{x}^i, \mathbf{x}^j)$ where σ^2 is the process variance, $\mathbf{R}(\boldsymbol{\theta}, \mathbf{x}^i, \mathbf{x}^j)$ is the correlation function, $\boldsymbol{\theta}$ is the correlation function parameter and x_k and x_l are components of the sample $\mathbf{x}$. In this paper we have used the linear correlation $R(\theta_j, x_k, x_l) = \max(0, 1 - \theta_j |x_k - x_l|)$ and the cubic correlation function $R(\theta_j, x_k, x_l) = 1 - 3\xi_j^2 + 2\xi_j^3$, $\xi_j = \min(1, \theta_j |x_k - x_l|)$.

The GP prediction is expressed as $\hat{y}(\mathbf{x}) = \mathbf{f}(\mathbf{x})^T\hat{\boldsymbol{\beta}} + \mathbf{r}(\mathbf{x})^T\mathbf{R}^{-1}(\mathbf{y} - \mathbf{F}\hat{\boldsymbol{\beta}})$ where $\hat{\boldsymbol{\beta}}$ is the least-squares estimate of $\boldsymbol{\beta}$ given by $\hat{\boldsymbol{\beta}} = (\mathbf{F}^T\mathbf{R}^{-1}\mathbf{F})^{-1}\mathbf{F}^T\mathbf{R}^{-1}\mathbf{y}$. The parameter $\mathbf{y} = [y_1, \ldots, y_N]^T$ is the set of true responses at known samples, $\mathbf{F} = [\mathbf{f}(\mathbf{x}_1), \ldots, \mathbf{f}(\mathbf{x}_N)]^T$, and $\mathbf{r}(\mathbf{x}) = [\mathbf{R}(\boldsymbol{\theta}, \mathbf{x}^1, \mathbf{x}), \ldots, \mathbf{R}(\boldsymbol{\theta}, \mathbf{x}^N, \mathbf{x})]^T$. The correlation parameters $\boldsymbol{\theta} = [\theta_1, \ldots, \theta_N]$ can be determined by Maximizing Maximum Likelihood Estimator (MLE)

$$\max_{\theta_1, \ldots, \theta_N} \left[-\frac{N\ln\hat{\sigma}^2 + \ln\|\mathbf{R}\|}{N} \right], \quad \theta_L \leq \theta_k \leq \theta_U, \quad k = 1, \ldots, N$$

where $\hat{\sigma}^2 = (\mathbf{y} - \mathbf{F}\hat{\boldsymbol{\beta}})^T\mathbf{R}^{-1}(\mathbf{y} - \mathbf{F}\hat{\boldsymbol{\beta}})/N$ is the estimated process variance. The goal of the MLE method is to find optimal $\boldsymbol{\theta}$ that maximizes the likelihood function based on all observations. The parameters of GP process are shown in Table 1.

Multivariate Adaptive Regression Splines – MARS. Multivariate Adaptive Regression Splines (MARS) [6] uses two-sided truncated power functions as spline basis functions, described by the following equations

$$[-(x - t)]_+^q = \begin{cases} (t - x)^q & \text{if } x < t \\ 0 & \text{otherwise} \end{cases}, \qquad [+(x - t)]_+^q = \begin{cases} (t - x)^q & \text{if } x \geq t \\ 0 & \text{otherwise} \end{cases}$$
$$(4)$$

where $q \geq 0$ is the power to which the splines are raised and which determines the degree of smoothness of the resultant function estimate.

The output in MARS is given by $\hat{y}(\mathbf{x}) = c_0 + \sum_{m=1}^{M} c_m\mathbf{B}_m(\mathbf{x})$, where c_0 is constant, $\mathbf{B}_m(\mathbf{x})$ is m-th basis function, c_m is the coefficient of m-th basis function, $\mathbf{x}$ is a sample from database and M is number of basis functions. To determine which basis functions should be used in the model, MARS implements a stepwise procedure which employs the generalized cross-validation (GCV) criterion given by

$$\text{GCV}(M) = \frac{\frac{1}{N}\sum_{i=1}^{N}(y_i - \hat{y}(\mathbf{x}^i))^2}{(1 - C(M)/N)^2}$$

where $C(M)$ is a complexity penalty $C(M) = \gamma M + M + 1$, where M is the number of basis functions, and the parameter γ is a penalty for each basis function. Table 1 shows the parameters of q and γ used in the exahustive search used in this paper.

Support Vector Regression – SVR. Support Vector Regression (SVR) [22] maps the input vectors $\mathbf{x} = [x_1, \ldots, x_N]$ into a high dimensional space where a linear machine build an optimal function $f(\mathbf{x})$ that minimizes the functional $J = \frac{1}{N}\sum_{i=1}^{N} L_\varepsilon$, where

$$L_\varepsilon(y - f(\mathbf{x})) = \begin{cases} 0 & \text{if } |y - f(\mathbf{x}| \leq \varepsilon \\ |y - f(\mathbf{x}| & \text{otherwise} \end{cases}$$
$$(5)$$

and ε is a SVR parameter. The minimization problem can be transformed into a dual problem

$$\begin{aligned}
\max_{\alpha,\alpha*} \quad & -\tfrac{1}{2}\sum_{i=1}^{N}\sum_{j=1}^{N}(\alpha_i - \alpha_i^*)(\alpha_j - \alpha_j^*)K(\mathbf{x}_i,\mathbf{x}_j) \\
& -\varepsilon\sum_{i=1}^{N}(\alpha_i + \alpha_i^*) - \varepsilon\sum_{i=1}^{N} y_i(\alpha_i + \alpha_i^*) \\
\text{subject to} \quad & \begin{cases} \sum_{i=1}^{N} y_i(\alpha_i + \alpha_i^*) = 0 \\ 0 \geq \alpha_i \geq C,\ 0 \geq \alpha_i^* \geq C,\ i = 1,\dots,N \end{cases}
\end{aligned} \tag{6}$$

where α_i and α_i^* are the weights which determine the influence of each data point on the model, $K(x_i, x_j)$ is the kernel function, and C is the regularization parameter.

The SVR prediction is given by $f(\mathbf{x}) = \sum_{i=1}^{N}(\alpha_i + \alpha_i^*)K(x_i, x_j) + b$ where b is a constant threshold, $K(x_i, x)$ is the radial basis kernel function of the form $K(x_i, x) = \sum_{i=1}^{m}\exp(-\gamma\|x_i - x_j\|^2))$ and γ is the bandwidth parameter. The parameters C, ε and γ was carried out using the grid search method described in Table 1.

2.3 Parameter Settings Using Cross-Validation

Cross-validation is a sampling statistical technique used to evaluate the ability of generalization of a model [9]. In this paper we have used the K-Fold cross-validation which divides the training set into K subsets of equal size. From the K subsets, K−1 are used for training and the remaining set is used for testing. This procedure is repeated K times, using a different test set in each iteration. Figure 3 shows an example of the 6-fold cross-validation.

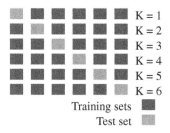

Fig. 3. Division of the dataset in K = 6 subsets (folds).

The model's generalization performance and estimation accuracy depends on a good setting of hyper-parameters. We have used an exhaustive search procedure [11] in the set of hyper-parameters of each model to find those that lead to lower mean absolute error. Table 1 shows the sets of hyper-parameters used in grid search with cross-validation.

Table 1. Hyper-parameters used in grid search with cross-validation.

Model	Parameters	Hyper-parameter sets
GP	Regression type	Constant, linear
	Correlation type	Linear, cubic
MARS	Polynomial degree (q)	[1,2,3,4,5]
	Penalty parameter (γ)	[1, 10, 100, 1000]
ENR	α	1, 10^{-1}, 10^{-2}, 10^{-3}, 10^{-4}, 10^{-5}, 10^{-6}
	ρ	0.70, 0.80, 0.85, 0.90, 0.95, 0.98, 1.00
SVR	C	$10^{-2}, 10^{-1}, 10^{0}, 10^{1}, 10^{2}, 10^{3}, 10^{4}, 10^{5}, 10^{6}, 10^{7}, 10^{8}$
	γ	$10^{-5}, 10^{-4}, 10^{-3}, 10^{-2}, 10^{-1}, 10^{0}, 10^{1}, 10^{2}, 10^{3}, 10^{4}$
	ε	$10^{-6}, 10^{-5}, 10^{-4}, 10^{-3}, 10^{-2}, 10^{-1}, 10^{1}$

3   Results and Discussion

In this section we present the results obtained for the regressions models described in Sect. 2. The four regression models, GP, SVR, ENR and MARS return the End Diastolic Area and the End Systolic Area as output, which are used to calculating the EF according to Eq. (2). In order to obtain consistent and reliable results we ran each computational experiment 30 times using 6-fold cross-validation with shuffled data generated by different random seeds. The numerical experiments described here were conducted based in scikit-learn framework [11] and implementations adapted from [1,6].

In this paper we have used the following metrics: R^2 score, Root Mean Squared Error ($RMSE$), Model Efficiency (MEF) and Mean Absolute Percentage Error ($MAPE$). These criteria can be written as

$$R^2 = 1 - \frac{\sum_{i=0}^{N-1}(y_i - \hat{y}_i)^2}{\sum_{i=0}^{N-1}(y_i - \bar{y})^2}$$
$$RMSE = \sqrt{\frac{1}{N}\sum_{i=0}^{N-1}(y_i - \hat{y}_i)^2}$$
$$MEF = 1 - \frac{(N-1)\sum_{i=0}^{N-1}(y_i - \hat{y}_i)^2}{(N-p)\sum_{i=0}^{N-1}(y_i - \bar{y})^2}$$
$$MAPE = 100 \times \frac{1}{N}\sum_{i=0}^{N-1}\frac{|y_i - \hat{y}_i|}{|y_i|}$$

where $\hat{y}_i$ is the estimated target output, y_i is the corresponding target output, N is the number of samples, p is the number of model parameters, and $\bar{y}$ is the mean of the vector $[y_1, \ldots, y_N]$.

Figures 4 and 5 show the values of $MAPE$ for End Diastolic Area (EDA) and End Systolic Area (ESA). Among all prediction models, we observe that GP obtained the worst mean performance for EDA and ESA. MARS reached a slightly better mean performance when compared to ENR, however SVR has produced clearly better results.

Table 2 shows the performance metrics for all models. When comparing the results according to R^2 values, one can observe the best and the worst averaged

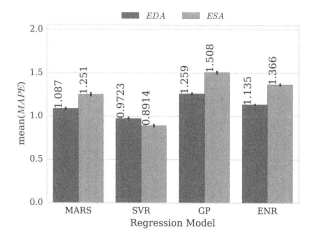

Fig. 4. Averaged $MAPE$ results for tEnd Diastolic Area (EDA) and End Systolic Area (ESA).

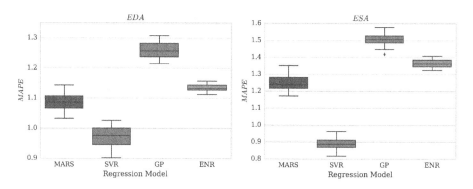

Fig. 5. Boxplots of $MAPE$ for End Diastolic Area (EDA) and End Systolic Area (ESA).

results were obtained by SVR (0.9578) and GP (0.8920), respectively. When comparing the $RMSE$ values, the lowest averaged $RMSE$ was produced by SVR, with mean 0.0170 and standard deviation $5.59(10^{-4})$. Again, the GP method achieved the worst performance, with average 0.0272. A similar behavior can be observed for averaged values for MEF: SVR ans GP obtained the best and the worst results, respectively. The good predictive performance of SVR can be observed on the values of metric MEF. Although SVR has 3 parameters to be adjusted, while ENR, GP and MARS have 2 parameters, SVR has achieved higher values of MEF.

Boxplots comparing the $RMSE$ and R^2 metrics for predicted EF values are shown in Fig. 6. Among all models, SVR presented better predictive performance, leading to the smallest values of $RMSE$ (left) and higher R^2 (right) while MARS

Table 2. Performance metrics for predicted Ejection Fraction. The first column shows the machine learning techniques and column R^2 shows the coefficient of determination, while the third column presents the values of $RMSE$. The Model Efficiency metric appears in the last column.

Regression model	R^2	$RMSE$	MEF
ENR	0.9244 (0.001899)	0.0227 (0.0002846)	0.9856 (0.0004185)
GP	0.8920 (0.00461)	0.0272 (0.0005812)	0.9815 (0.0008942)
MARS	0.9359 (0.003861)	0.0209 (0.0006237)	0.9866 (0.0007804)
SVR	0.9578 (0.002749)	0.0170 (0.000559)	0.9894 (0.001294)

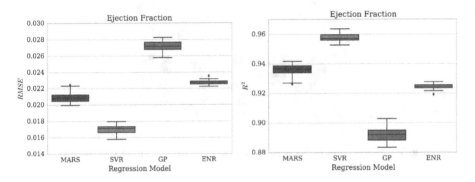

Fig. 6. Ejection fraction: comparison among all models according to $RMSE$ and R^2 metrics.

and ENR has produced similar results for $RMSE$ and R^2. We remark the comparison between ENR and GP. ENR is a regularized linear regression model, which is simpler than GP in mathematical formulation as well as in implementation aspects. However, ENR produced better results for EDA, ESA and EF as shown in Figs. 4 and 6. The poorer performance of GP can be related to the dimensionality of the problem ($d = 104$), since its theoretical efficiency depends exponentially on dimension of the domain over which the function is defined [4].

The overall results show the better predictive performance of SVR when compared to ENR, MARS and GP. Figure 7 displays the distribution of SVR parameters along the 30 runs. Analyzing the results for parameter C, associated to the model complexity, we can see that it was tuned in the set $[10^3, 10^4, 10^5]$ for all runs. Observing the parameter settings for SVR in Table 1 and considering the grid search procedure, we can infer that smaller values of parameter C did not produced accurate models for EDA and ESA and higher values of C resulted in too complex models, which makes the optimization problem (6) harder. The parameter γ controls the width of the radial basis function. For output EDA, γ falls in the range $[1, 10]$ and for ESA, $\gamma = 10$ in 27 out of 30 runs, which means ESA requires radial basis functions with larger widths. The parameter ε

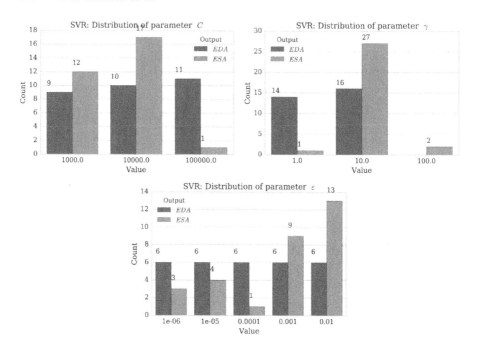

Fig. 7. Distribution of SVR parameters (C, γ and ε) along the 30 runs

controls the width of the ε-insensitive zone according to Eq. (5). It is used to fit the training data and it affects the complexity and the generalization capability of the prediction model. For all runs, the grid search with cross-validation tuned the parameter ε in the set $[10^{-6}, 10^{-5}, 10^{-4}, 10^{-3}, 10^{-2}]$ for both outputs EDA and ESA. The value of ε determines the level of accuracy of the prediction model and we can observe that higher values of ε were set for ESA along the 30 runs. Higher values of ε reflects the better performance of SVR predictions of ESA (when compared to EDA) as can be seen in Figs. 4 and 5.

Table 3 shows a comparison of results obtained in this paper and results collected from the literature. We observe that all models presented in this paper lead to smaller averaged $MAPE$ for EDA when compared to [12] and [5]. Considering ESA, SVR produced better results than [5], which implements an ensemble of neural networks. We highlight the outstanding result for ESA presented in [12]. However, the predictions in that paper was obtained by solving an inverse problem using a numerical method which requires higher computational burden when compared with SVR predictions. Although accurate for ESA, the method proposed in [12] can lead to drawbacks in a scenario of continuous monitoring of the heart due to the requirement of solving an inverse problem. Moreover, the calculation of EF demands accurate estimations of both systolic and diastolic areas and higher levels of error in EDA or ESA could compromise the EF predictions.

Table 3. Mean absolute percentage error (*MAPE*) for *EDA* and *ESA*. Results obtained in [12] were produced using a Boundary Element Method simulation model and the results in [5] were produced by a ensemble of neural networks

Variable	ENR	GP	MARS	SVR
EDA	1.135 (0.0109)	1.259 (0.0279)	1.087 (0.0287)	0.9723 (0.0331)
ESA	1.366 (0.0233)	1.508 (0.0370)	1.251 (0.0474)	0.8914 (0.0323)
Variable	Ref. [12]	Ref. [5]		
EDA	2.41	4,49		
ESA	0.09	1.06		

4   Conclusions

In this paper we compare four machine learning techniques to estimate ejection fraction using known electrical potentials from Electrical Impedance Tomography (EIT). The results obtained from all models developed herein show that SVR consistently outperforms MARS, ENR and GP techniques, leading to accurate results for End of Diastole and End of Systole Areas, as well as for Ejection Fraction, and suggests that SVR can be used with the EIT to quickly estimate the EF in a scenario of continuous monitoring.

Acknowledgments. The authors would like to thank the anonymous referees for their comments, which helped us to improve a previous version of this paper. The authors would like to thank the Brazilian agencies FAPEMIG (grant 01606/15), CNPq and CAPES for financial support.

References

1. Py-earth documentation. http://contrib.scikit-learn.org/py-earth/. Accessed 01 Aug 2016
2. Bekeredjian, R., Grayburn, P.A.: Valvular heart disease: aortic regurgitation. Circulation **112**(9), 125–134 (2005)
3. Blanc, C., Schlick, C.: X-splines: a spline model designed for the end-user. In: Proceedings of the 22nd Annual Conference on Computer Graphics and Interactive Techniques. SIGGRAPH 1995, pp. 377–386. ACM, New York (1995)
4. Djolonga, J., Krause, A., Cevher, V.: High-dimensional gaussian process bandits. In: Advances in Neural Information Processing Systems, pp. 1025–1033 (2013)
5. Filho, R.G.N.S., Campos, L.C.D., dos Santos, R.W., Barra, L.P.S.: Determination of cardiac ejection fraction by electrical impedance tomography using an artificial neural network. In: Castro, F., Gelbukh, A., González, M. (eds.) MICAI 2013, Part II. LNCS, vol. 8266, pp. 130–138. Springer, Heidelberg (2013)
6. Friedman, J.H.: Multivariate adaptive regression splines. Ann. Stat. **19**, 1–67 (1991)
7. Hastie, T., Tibshirani, R., Friedman, J.: The Elements of Statistical Learning - Data Mining, Inference, and Prediction, 2nd edn. Springer, New York (2009)

8. Hozumi, T., Shakudo, M., Shah, P.: Quantitation of left ventricular volumes and ejection fraction by biplane transesophageal echocardiography. Am. J. Cardiol. **72**, 356–359 (1993). Elsevier Inc

9. Kohavi, R., et al.: A study of cross-validation and bootstrap for accuracy estimation and model selection. IJCAI **14**, 1137–1145 (1995)

10. MacIver, D.H., Adeniran, I., Zhang, H.: Left ventricular ejection fraction is determined by both global myocardial strain and wall thickness. IJC Heart Vasculature **7**, 113–118 (2015)

11. Pedregosa, F., Varoquaux, G., Gramfort, A., Michel, V., Thirion, B., Grisel, O., Blondel, M., Prettenhofer, P., Weiss, R., Dubourg, V., et al.: Scikit-learn: machine learning in python. J Mach. Learn. Res. **12**, 2825–2830 (2011)

12. Peters, F.C., Barra, L.P.d.S., dos Santos, R.W.: Determination of cardiac ejection fraction by electrical impedance tomography. INTECH Open Access Publisher (2011)

13. Peters, F.C., Barra, L.P.S., dos Santos, R.W.: Determination of cardiac ejection fraction by electrical impedance tomography - numerical experiments and viability analysis. In: Allen, G., Nabrzyski, J., Seidel, E., van Albada, G.D., Dongarra, J., Sloot, P.M.A. (eds.) ICCS 2009, Part I. LNCS, vol. 5544, pp. 819–828. Springer, Heidelberg (2009)

14. Regeer, M.V., Versteegh, M.I., Klautz, R.J., Schalij, M.J., Bax, J.J., Marsan, N.A., Delgado, V.: Comparison of left ventricular volume and ejection fraction and frequency and extent of aortic regurgitation after operative repair of type a aortic dissection among three different surgical techniques. Am. J. Cardiol. **117**(7), 1167–1172 (2016)

15. Ribeiro, M.H., dos Santos, R.W., Barra, L.P.S., Peters, F.C.: Simulation study on the determination of cardiac ejection fraction by electrical impedance tomography using a hybrid heuristic approach. J. Med. Imag. Health Inf. **4**(1), 113–121 (2014)

16. Sacks, J., Welch, W.J., Mitchell, T.J., Wynn, H.P.: Design and analysis of computer experiments. Stat. Sci. **4**, 409–423 (1989)

17. Filho, R.G.N.S., Campos, L.C.D., dos Santos, R.W., Barra, L.P.S.: Artificial neural networks ensemble applied to the electrical impedance tomography problem to determine the cardiac ejection fraction. In: Bazzan, A.L.C., Pichara, K. (eds.) IBERAMIA 2014. LNCS, vol. 8864, pp. 734–741. Springer, Heidelberg (2014)

18. da Silva Barra, L.P., Peters, F.C., de Paula Martins, C., Barbosa, H.J.C.: Computational experiments in electrical impedance tomography. In: XXVII Iberian Latin American Congress on Computational Methods in Engineering (2006)

19. da Silva Barra, L.P., dos Santos, R.W., Peters, F.C., Santos, E.P., Barbosa, H.J.C.: Parallel computational experiments in electrical impedance tomography. 18th Symp. Comput. Archit. High Perform. Comput. **1**, 7–13 (2006)

20. Squeri, A., Gaibazzi, N., Reverberi, C., Caracciolo, M.M., Ardissino, D., Gherli, T.: Ejection fraction change and coronary artery disease severity: a vasodilator contrast stress-echocardiography study. J. Am. Soc. Echocardiogr. **25**(4), 454–459 (2012)

21. Upadhya, B., Taffet, G.E., Cheng, C.P., Kitzman, D.W.: Heart failure with preserved ejection fraction in the elderly: scope of the problem. J. Mol. Cell. Cardiol. **83**, 73–87 (2015). Perspectives on Cardiovascular Aging: Top to Bottom

22. Were, K., Bui, D.T., Øystein, B.D., Singh, B.R.: A comparative assessment of support vector regression, artificial neural networks, and random forests for predicting and mapping soil organic carbon stocks across an afromontane landscape. Ecol. Ind. **52**, 394–403 (2015)

23. Zou, H., Hastie, T.: Regularization and variable selection via the elastic net. J. R. Stat. Soc.: Ser. B (Stat. Methodol.) **67**(2), 301–320 (2005)

Machine Learning Models for Early Dengue Severity Prediction

William Caicedo-Torres[1][(✉)], Ángel Paternina[2], and Hernando Pinzón[2]

[1] Department of Computer Science, Universidad Tecnológica de Bolívar Parque Industrial y Tecnológico Carlos Vélez Pombo, Km 1 Vía Turbaco, Cartagena, Colombia wcaicedo@unitecnologica.edu.co

[2] Hospital Infantil Napoleón Franco Pareja Cartagena, Cartagena, Colombia

Abstract. Infection by dengue-virus is prevalent and a public health issue in tropical countries worldwide. Also, in developing nations, child populations remain at risk of adverse events following an infection by dengue virus, as the necessary care is not always accessible, or health professionals are without means to cheaply and reliably predict how likely is for a patient to experience severe Dengue. Here, we propose a classification model based on Machine Learning techniques, which predicts whether or not a pediatric patient will be admitted into the pediatric Intensive Care Unit, as a proxy for Dengue severity. Different Machine Learning techniques were trained and validated using Stratified 5-Fold Cross-Validation, and the best model was evaluated on a disjoint test set. Cross-Validation results showed an SVM with Gaussian Kernel outperformed the other models considered, with an 0.81 Receiver Operating Characteristic Area Under the Curve (ROC AUC) score. Subsequent results over the test set showed a 0.75 ROC AUC score. Validation and test results are promising and support further research and development.

Keywords: Machine learning · Dengue · Severity · Children · PICU · Naive Bayes · SVM · Logistic regression

1 Introduction

Infection by dengue-virus is a public health issue in tropical countries wolrdwide [4]. Moreover, in developing nations, child populations remain at risk of adverse events following infection as the necessary care is not always accessible, or because health professionals are without means to predict cheaply and reliably how likely is for a patient to experience severe dengue. Studies conducted to identify risk factors for severe dengue have singled-out several epidemiological features that tend to be associated with severity, as secondary infection [10], gender [1,23], pregnancy [17] and under-nutrition [12]. Symptoms, signs, as well as hematological and physiological parameters have been used to triage patients based on their propensity to experience severe forms of the disease, with varied

© Springer International Publishing AG 2016
M. Montes-y-Gómez et al. (Eds.): IBERAMIA 2016, LNAI 10022, pp. 247–258, 2016.
DOI: 10.1007/978-3-319-47955-2_21

performance [5,21,22,25]. In recent years, several approaches involving blood levels of certain viral proteins as NS1 and immunological markers as the terminal complement complex SC5b-9 have shown promise to be used as predictors of dengue severity and adverse outcomes [2,15,16,26]. However, their deployment and usage are sometimes hindered by socio-economical conditions in poor countries. Statistical techniques widely used in medicine as well as Machine Learning models have been applied to this problem with some success [6,22,24], although some of these prediction models included certain laboratory test results that are not readily available in remote or poor areas. Computational approaches like these are interesting, since they can be deployed using personal devices like laptops, smartphones or tablets in an inexpensive and practical way; and once trained, statistical and Machine Learning models can offer a prediction in seconds. In this paper we compare some Machine Learning algorithms trained using simple and readily available clinical data for early prediction of dengue severity in children. We examined popular models as Logistic Regression [11,19], Support Vector Machines [9,14], and Naive Bayes Classifiers [11] to asses their performance predicting whether or not a pediatric patient will experience severe dengue. In order to simplify our prediction target, we used the admission into the pediatric Intensive Care Unit (PICU) as proxy for dengue severity as patients with severe dengue are often admitted to intensive care to manage their condition properly; so admission becomes a representative benchmark for severity in the scope of this study. A wrapper Feature Selection procedure was conducted using a Recursive Feature Elimination strategy, with ROC AUC score as the target to be optimized.

2   Methods and Materials

2.1   Participants

A retrospective cohort study was carried out, which included children arriving to Hospital Infantil Napoleón Franco Pareja emergency department from January 1st of 2013 to August 14th of 2014, between 2 and 18 years of age, which had not yet received intra-hospital treatment for the current symptoms. Children with suspicious clinical manifestations for dengue were preselected according to the senior pediatrics and infectology staff assessment, and diagnoses were confirmed via an IgM or NS1 rapid test positive result. A total of 742 patients were initially included, according to the guidelines of the study.

2.2   Dataset

Initial predictor selection was based on the current literature, and simple, ready to obtain ones were preferred, in order to learn models based on features that could be easily available to medical practitioners in poor or remote areas. According to this, the predictor variables included were temperature (in C°), hematocrit (percentage), time with symptoms before emergency admission (in days), age (in

Table 1. Patient data.

	Negative class (Non-severe)	Positive class (Severe)	Total
Median temperature (Q1–Q3), C°	37.0 (36.6–38.0)	38.0 (37.0–38.5)	37.0 (36.6–38.0)
Median hematocrit (Q1–Q3), %	36.5 (34.0–39.3)	37.0 (33.1–42.0)	36.5 (34.0–39.4)
Median time with symptoms before emergency admission (Q1–Q3), days	5.0 (4.0–6.0)	5.0 (4.0–6.5)	5.0 (4.0–6.0)
Median age (Q1–Q3), years	8.0 (5.0–11.0)	7.0 (6.0–9.0)	8.0 (5.0–11.0)
Median platelets (Q1–Q3), per 1,000 cells/mm3	113.0 (69.0–165.0)	57.5 (30.0–98.5)	106.0 (64.0–163.0)
Median white blood cells (Q1–Q3), per 1,000 cells/mm3	4.6 (3.3–6.7)	5.3 (3.4–8.3)	4.7 (3.3–6.8)
Mucosal bleeding, %	35 (6.3)	4 (10.0)	39 (6.6)
Lethargy, %	3 (0.5)	7 (17.5)	10 (1.7)
Pleural effusion, %	16 (2.9)	7 (17.5)	23 (3.9)
Ascites, %	8 (1.4)	4 (10.0)	12 (2.0)
Presence of vomit, %	288 (51.9)	29 (72.5)	317 (53.3)
Edema, %	13 (2.3)	2 (5.0)	15 (2.5)
Liver enlargement, %	15 (2.7)	8 (20.0)	23 (3.9)
Splenomegaly, %	1 (0.2)	1 (2.5)	2 (0.3)
Time to discharge (Q1–Q3), days	1.5 (0.7–3.0)	n/a	n/a
Time to critical care (Q1–Q3), days	n/a	2.5 (1.5–3.9)	n/a

years), platelet count (cells per mm^3), white blood cell count (cells per mm^3), mucosal bleeding collected at admission (gingivorrhagia, hematemesis or epistaxis), lethargy (Glasgow comma scale ≤ 13), pleural effusion, ascites, presence of vomit, edema (excluding pleural effusion or ascites), liver enlargement and splenomegaly. All predictor information was collected at admission to the emergency department. The target variable selected was admission to the PICU. Median time to admission to the PICU for patients with severe dengue was 2.5 days (IQR = 2.4), and median time to discharge for patients not admitted to the PICU was 1.5 days (IQR = 2.3). Overall median time inside the institution was 1.61 days (IQR = 2.39) (Table 1).

After collection and preliminary analysis, 595 patients had complete information for the 14 predictor variables considered. Missing data was assumed to be missing-at-random, and we decided against imputation at this point, so the study used the complete case database. 40 children were admitted to the PICU, representing 6.7 % of all cases. Distribution thus shows heavily imbalanced classes. Data was partitioned into disjoint training and test sets, and the training set was then standardized ($\mu = 0, \sigma = 1$). The relevant transformation parameters were

Table 2. Dataset partition

Class	Training	Test	Total
Non severe	444	111	555
Severe	32	8	40
Total			**595**

stored to be applied later for test set standardization. Partition was performed before standardization to avoid data leakage (Table 2).

2.3   Learning Models

In this study, several Machine Learning techniques were considered. We used L2 Regularized Logistic Regression with and without polynomial features, Support Vector Machines with Linear and Gaussian kernel, and Naive Bayes Classifier with Gaussian priors. All models were implemented using the Scikit-Learn Python Machine Learning library [20]. Models were learned using the whole predictor set and the reduced one, to compare feature selection results.

Logistic Regression. Logistic Regression is a linear classification algorithm [18] widely used in medicine where a logistic sigmoid function is coupled to a linear regression model. Results can be interpreted as the probability for the input to belong to the positive class, $p(Y = 1|x; \theta)$. In this study, Logistic Regression was trained with Cross-Entropy loss. A L2 regularization penalty was introduced to account for model complexity and avoid overfitting:

$$min -\frac{1}{N} \sum_{i=1}^{N} [y_i log h_\theta(x_i) + (1 - y_i) log(1 - h_\theta(x_i))] + \lambda ||\theta||^2 \qquad (1)$$

where

$$h_\theta(x) = \frac{1}{1 + e^{-\theta^t x}} \qquad (2)$$

It is possible to increase its capabilities by applying a polynomial transform to the input, in which case the decision boundary can be non-linear and handle more difficult problems.

Support Vector Machine. The Support Vector Machine (SVM) [7] is a classification model invented in its original form by Vapnik et al. SVMs try to find the maximum margin hyperplane to provide a robust separator for the problem classes while being tolerant of misclassification as well, in order to handle non-linearly separable problems (soft-margin SVM). Its primal formulation minimizes the structural risk to avoid overfitting:

$$min \ \frac{1}{N} \sum_{i=1}^{N} \xi_i + \lambda ||w||^2 \tag{3}$$

Subject to:

$$y_i(\boldsymbol{x_i} \cdot w + b) \geq 1 - \xi_i \quad \text{and} \quad \xi_i \geq 0, \forall i \tag{4}$$

where

$$\xi_i = max(0, 1 - y_i(w \cdot \boldsymbol{x_i} + b)), \text{ the hinge loss.} \tag{5}$$

The Lagrangian dual of the optimization problem reveals an interesting structure:

$$max_\alpha \ \sum_{i=1}^{N} \alpha_i - \frac{1}{2} \sum_{i=1}^{N} \sum_{j=1}^{N} y_i \alpha_i (\boldsymbol{x_i} \cdot \boldsymbol{x_j}) y_j \alpha_j \tag{6}$$

Subject to:

$$\sum_{i=1}^{N} \alpha_i y_i = 0 \quad \text{and} \quad 0 \leq \alpha_i \leq \frac{1}{2n\lambda}, \forall i \tag{7}$$

Support vectors are the inputs with non-zero associated Lagrangian Multipliers (α), and all lie on the classification margin, hence the name.

The solution to the dual (optimal multipliers $\{\alpha_{i,o}\}_{i=1}^{l}$) allows us to formulate the maximum margin hyperplane for classification. The optimal weight vector is given by

$$\boldsymbol{w_o} = \sum_{i=1}^{l} \alpha_{i,o} y_i \boldsymbol{x_i} \tag{8}$$

so the hyperplane is then

$$\boldsymbol{w_o^t} \boldsymbol{x} + b_o = \sum_{i=1}^{l} \alpha_{i,o} y_i \boldsymbol{x_i} \cdot \boldsymbol{x} + b_o = 0 \tag{9}$$

To handle non-linear decision boundaries, the solution structure can be exploited: the dot product $\boldsymbol{x_i} \cdot \boldsymbol{x_j}$ can be replaced by a product of the transformed versions of the original inputs, $\phi(\boldsymbol{x_i}) \cdot \phi(\boldsymbol{x})$. The Kernel trick [7] provides a way to compute the dot product without performing the transformations explicitly using Kernel functions, which by definition represent dot products in high dimensional spaces [7]. In this way, the SVM remains a well regularized classifier and can handle high dimensional feature transformations with computational efficiency. In this study, a Linear kernel,

$$K(\boldsymbol{x_i}, \boldsymbol{x_j}) = \boldsymbol{x_i} \cdot \boldsymbol{x_j} \tag{10}$$

and a Gaussian kernel,

$$K(\boldsymbol{x_i}, \boldsymbol{x_j}) = exp(-\frac{||\boldsymbol{x_i} - \boldsymbol{x_j}||^2}{2\sigma^2}) \tag{11}$$

were used.

Naive Bayes. Naive Bayes [8] is a learning model based on the Bayes Theorem [3], and operates under strong assumptions of features being independent given the instance class, which rarely hold in reality; hence the name "naive"(although Naive Bayes tend to work well in practice [8]). Naive Bayes is a model of conditional probability which assigns a posterior class probability to each instance in the dataset, $p(C_k|x_1,...,x_n)$. This is accomplished using Bayes Theorem:

$$p(C_k|\boldsymbol{x}) = \frac{p(\boldsymbol{x}|C_k)p(C_k)}{p(\boldsymbol{x})} \tag{12}$$

Exploiting the conditional independence assumption, we can use the chain rule of conditional probability to compute the posterior class probability. By definition:

$$p(x_1, x_2, ..., x_n|C_k) = p(x_1|C_k)p(x_2, ..., x_n|C_k) \tag{13}$$

by repeated application of the chain rule:

$$p(x_1, x_2, ..., x_n|C_k) = p(x_1|C_k)p(x_2|C_k, x_1)...p(x_n|C_k x_1, x_2, ..., x_{n-1}) \tag{14}$$

and because of the conditional independence assumption,

$$p(x_1, x_2, ..., x_n|C_k) = p(x_1|C_k)p(x_2|C_k)...p(x_n|C_k) = \prod_{i=1}^{n} p(x_i|C_k) \tag{15}$$

then, according to the Naive Bayes Classifier, the posterior class probabilities are given by

$$p(C_k|x_1, x_2, ..., x_n) = \frac{1}{Z}p(C_k) \prod_{i=1}^{n} p(x_i|C_k) \tag{16}$$

where $Z = p(x_1, x_2, ..., x_n)$ is a normalizing factor depending only on features x_i. The Naive Bayes Classifier selects the class with maximum posterior probability. In this study, Gaussian priors were used, with each feature mean and variance estimated through Maximum Likelihood.

3   Results

3.1   Training and Validation

All learning models were trained using 5-Fold Stratified Cross-Validation. ROC AUC results are reported using both the whole predictor set and the set obtained by Recursive Feature Elimination.

L2 Regularized Logistic Regression. For logistic regression models, we considered the original dataset and additional polynomial features, up to the 5th degree using Grid Search. Best performance was exhibited by a model with no polynomial features, a regularization factor (λ) of 980, and a predictor subset found by Recursive Feature Elimination. Features identified by this procedure were temperature, hematocrit, time with symptoms before emergency admission, age, platelet count, white blood cell count, lethargy, pleural effusion and liver enlargement. Figures 1 and 2 show results for a Logistic Regression classifier with all predictors and with a selected subset of them, respectively.

Support Vector Machine. Figure 3 shows results for an SVM with Gaussian Kernel, and the results for a Linear Kernel are shown in Fig. 4. The Gaussian SVM with all predictors yielded better results compared to a Linear SVM using all predictors as well. Gamma and cost parameters were selected via Grid Search.

Naive Bayes. Naive Bayes was trained using the whole predictor set and the subset used with Logistic Regression. Naive Bayes with selected predictors performed significantly better. Figures 5 and 6 show validation results.

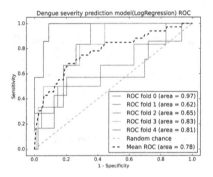

Fig. 1. L2 Regularized Logistic Regression, all predictors ($\lambda = 980$).

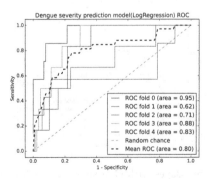

Fig. 2. L2 Regularized Logistic Regression, selected predictors ($\lambda = 980$).

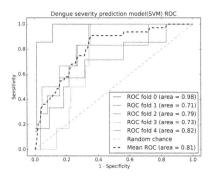

Fig. 3. SVM with Gaussian Kernel ($\gamma = 0.0404, c = 0.2828$).

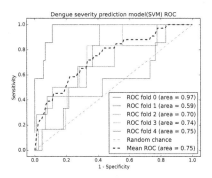

Fig. 4. SVM with Linear Kernel ($c = 0.01011$).

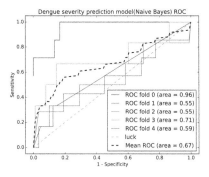

Fig. 5. Naive Bayes with Gaussian priors, all predictors.

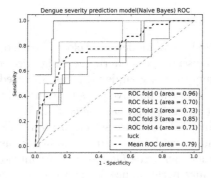

Fig. 6. Naive Bayes with Gaussian priors, selected predictors.

3.2 Test

Table 3 shows the best Cross Validation results for each classifier. According to these, the SVM with Gaussian Kernel outperformed the other models (ROC AUC = 0.81), with Logistic Regression close behind (ROC AUC = 0.80). We decided then to assess the SVM performance on the test set. Results are shown in Fig. 7.

Table 3. Best Cross Validation results for each model. Gaussian SVM comes on top with a 0.81 ROC AUC

Model	Parameters	ROC AUC
Linear SVM	$c = 0.01011$	0.75
Gaussian SVM	$\gamma = 0.0404, c = 0.2828$	**0.81**
Logistic Regression	$\lambda = 980$	0.80
Naive Bayes with Gaussian priors	n/a	0.79

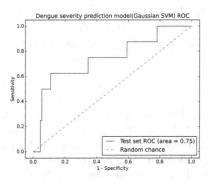

Fig. 7. Test results for SVM with Gaussian Kernel.

4   Discussion

Cross Validation results show all models very close to each other in terms of ROC AUC score, with the Gaussian SVM offering a slight edge over the competitors. The use of linear techniques (low complexity) was preferred given the relatively low count of examples from the positive class (patients admitted to the PICU), being the Gaussian Kernel a special case since it converges to linear behavior if necessary (as $\sigma \to \inf$) [13]. Feature selection improved the results for both Logistic Regression and Naive Bayes, and in the latter case, improvement was substantial. On the other hand, the Gaussian SVM introduced a non-linearity that possibly eclipsed any performance boost we could have obtained by feature selection, and instead used all predictors to get maximum performance. Additional research is needed to ascertain exactly the cause of this behavior.

Additionally, ROC curves show that for a fixed specificity of 0.7, all models exhibit sensitivities superior to 70 %, which is very interesting from a clinical standpoint. Test results for the SVM show the expected performance dip, with a 0.75 ROC AUC, and almost 80 % sensitivity at 65 % specificity. It is worth noticing these results while interesting and suggestive of the good diagnostic capabilities of the model, were obtained using a highly imbalanced dataset, so with a larger data set they could be most likely improved on.

5   Conclusions

Cheap, fast and reliable prediction of dengue severity in children can be an important tool in countries with prevalence of infection. In this paper, a Machine Learning based approach to dengue severity prediction in children was proposed. Data from patients of Hospital Infantil Napoleón Franco Pareja was collected and used to train and validate several classification models. Regularization and class weighting were used in order to overcome poor classification results and avoid overfitting. Additionally, a Recursive Feature Elimination feature selection procedure was performed to improve predictive accuracy. 5-Fold Stratified Cross Validation results show good prediction capabilities and a model based on a Gaussian SVM offered the best results measured by Receiver Operating Characteristic Area Under the Curve score (validation-AUC = 0.8, test-AUC = 0.75). Results obtained are promising and support further research and development.

References

1. Anders, K.L., Nguyet, N.M., Van Vinh Chau, N., Hung, N.T., Thuy, T.T., Lien, L.B., Farrar, J., Wills, B., Hien, T.T., Simmons, C.P.: Epidemiological factors associated with dengue shock syndrome and mortality in hospitalized dengue patients in Ho Chi Minh City, Vietnam. Am. J. Trop. Med. Hyg. **84**(1), 127–134 (2011). http://www.ncbi.nlm.nih.gov/pmc/articles/PMC3005500/

2. Avirutnan, P., Punyadee, N., Noisakran, S., Komoltri, C., Thiemmeca, S., Auetha-vornanan, K., Jairungsri, A., Kanlaya, R., Tangthawornchaikul, N., Puttikhunt, C., Pattanakitsakul, S.N., Yenchitsomanus, P.T., Mongkolsapaya, J., Kasinrerk, W., Sittisombut, N., Husmann, M., Blettner, M., Vasanawathana, S., Bhakdi, S., Malasit, P.: Vascular leakage in severe dengue virus infections: a potential role for the nonstructural viral protein ns1 and complement. J. Infect. Dis. **193**(8), 1078–1088 (2006)
3. Bayes, M., Price, M.: An essay towards solving a problem in the doctrine of chances. By the Late Rev. Mr. Bayes, F.R.S. communicated by Mr. Price, in a letter to John Canton, A.M.F.R.S. Philos. Trans. **53**, 370–418 (1763). http://rstl. royalsocietypublishing.org/content/53/370.short
4. Bhatt, S., Gething, P.W., Brady, O.J., Messina, J.P., Farlow, A.W., Moyes, C.L., Drake, J.M., Brownstein, J.S., Hoen, A.G., Sankoh, O., Myers, M.F., George, D.B., Jaenisch, T., Wint, G.W., Simmons, C.P., Scott, T.W., Farrar, J.J., Hay, S.I.: The global distribution and burden of dengue. Nature **496**(7446), 504–507 (2013). http://www.ncbi.nlm.nih.gov/pmc/articles/PMC3651993/
5. Cao, X.T.P., Ngo, T.N., Wills, B., Kneen, R., Nguyen, T.T.H., Ta, T.T.M., Tran, T.T.H., Doan, T.K.L., Solomon, T., Simpson, J.A., White, N.J., Farrar, J.J.: Evaluation of the world health organization standard tourniquet test and a modified tourniquet test in the diagnosis of dengue infection in Vietnam. Trop. Med. Int. Health **7**(2), 125–132 (2002)
6. Carrasco, L.R., Leo, Y.S., Cook, A.R., Lee, V.J., Thein, T.L., Go, C.J., Lye, D.C.: Predictive tools for severe dengue conforming to world health organization 2009 criteria. PLoS Negl. Trop. Dis. **8**(7), e2972 (2014). http://dx.doi.org/10.1371
7. Cortes, C., Vapnik, V.: Support-vector networks. Mach. Learn. **20**(3), 273–297 (1995). http://dx.doi.org/10.1007/BF00994018
8. Domingos, P., Pazzani, M.: On the optimality of the simple bayesian classifier under zero-one loss. Mach. Learn. **29**(2–3), 103–130 (1997). http://dx.doi.org/10.1023/A:1007413511361
9. Gomes, A.L.V., Wee, L.J.K., Khan, A.M., Gil, L.H.V.G., Marques, E.T.A.J., Calzavara-Silva, C.E., Tan, T.W.: Classification of dengue fever patients based on gene expression data using support vector machines. PLoS One **5**(6), e11267 (2010)
10. Guzman, M., Alvarez, M., Halstead, S.: Secondary infection as a risk factor for dengue hemorrhagic fever/dengue shock syndrome: an historical perspective and role of antibody-dependent enhancement of infection. Arch. Virol. **158**(7), 1445–1459 (2013). http://dx.doi.org/10.1007/s00705-013-1645-3
11. Huy, N.T., Thao, N.T.H., Ha, T.T.N., Lan, N.T.P., Nga, P.T.T., Thuy, T.T., Tuan, H.M., Nga, C.T.P., Van Tuong, V., Van Dat, T., Huong, V.T.Q., Karbwang, J., Hirayama, K.: Development of clinical decision rules to predict recurrent shock in dengue. Crit. Care **17**(6), R280–R280 (2013). http://www.ncbi.nlm.nih.gov/pmc/articles/PMC4057383/
12. Kalayanarooj, S., Nimmannitya, S.: Is dengue severity related to nutritional status? SE Asian J. Trop. Med. Public Health **36**(2), 378–384 (2005)
13. Keerthi, S.S., Lin, C.J.: Asymptotic behaviors of support vector machines with gaussian kernel. Neural Comput. **15**(7), 1667–1689 (2003). http://dx.doi.org/10.1162/089976603321891855
14. Kesorn, K., Ongruk, P., Chompoosri, J., Phumee, A., Thavara, U., Tawatsin, A., Siriyasatien, P.: Morbidity rate prediction of dengue hemorrhagic fever (DHF) using the support vector machine and the aedes aegypti infection rate in similar climates and geographical areas. PLoS ONE **10**(5), e0125049 (2015). http://dx.doi.org/10.1371/journal.pone.0125049

15. Libraty, D.H., Endy, T.P., Houng, H.S.H., Green, S., Kalayanarooj, S., Suntayakorn, S., Chansiriwongs, W., Vaughn, D.W., Nisalak, A., Ennis, F.A., Rothman, A.L.: Differing influences of virus burden and immune activation on disease severity in secondary dengue-3 virus infections. J. Infect. Dis. **185**(9), 1213–1221 (2002)
16. Libraty, D.H., Young, P.R., Pickering, D., Endy, T.P., Kalayanarooj, S., Green, S., Vaughn, D.W., Nisalak, A., Ennis, F.A., Rothman, A.L.: High circulating levels of the dengue virus nonstructural protein NS1 early in dengue illness correlate with the development of dengue hemorrhagic fever. J. Infect. Dis. **186**(8), 1165–1168 (2002)
17. Machado, C.R., Machado, E.S., Rohloff, R.D., Azevedo, M., Campos, D.P., de Oliveira, R.B., Brasil, P.: Is pregnancy associated with severe dengue? A review of data from the Rio de Janeiro surveillance information system. PLoS Negl. Trop. Dis. **7**(5), e2217 (2013)
18. McCullagh, P., Nelder, J.: Generalized Linear Models. Chapman & Hall/CRC Monographs on Statistics & Applied Probability, 2nd edn. Taylor & Francis, Abingdon (1989). https://books.google.co.uk/books?id=h9kFH2_FfBkC
19. Moraes, G.H., de Fatima Duarte, E., Duarte, E.C.: Determinants of mortality from severe dengue in Brazil: a population-based case-control study. Am. J. Trop. Med. Hyg. **88**(4), 670–676 (2013)
20. Pedregosa, F., Varoquaux, G., Gramfort, A., Michel, V., Thirion, B., Grisel, O., Blondel, M., Prettenhofer, P., Weiss, R., Dubourg, V., Vanderplas, J., Passos, A., Cournapeau, D., Brucher, M., Perrot, M., Duchesnay, E.: Scikit-learn: machine learning in python. J. Mach. Learn. Res. **12**, 2825–2830 (2011)
21. Phuong, C.X.T., Nhan, N.T., Kneen, R., Thuy, P.T.T., van Thien, C., Nga, N.T.T., Thuy, T.T., Solomon, T., Stepniewska, K., Wills, B.: Clinical diagnosis and assessment of severity of confirmed dengue infections in vietnamese children: is the world health organization classification system helpful? Am. J. Trop. Med. Hyg. **70**(2), 172–179 (2004)
22. Potts, J.A., Gibbons, R.V., Rothman, A.L., Srikiatkhachorn, A., Thomas, S.J., Supradish, P.O., Lemon, S.C., Libraty, D.H., Green, S., Kalayanarooj, S.: Prediction of dengue disease severity among pediatric thai patients using early clinical laboratory indicators. PLoS Negl. Trop. Dis. **4**(8), e769 (2010)
23. Shekhar, K.C., Huat, O.L.: Epidemiology of dengue/dengue hemorrhagic fever in Malaysia-a retrospective epidemiological study 1973–1987. Part II: dengue fever (DF). Asia Pac. J. Public Health **6**(3), 126–133 (1992)
24. Tanner, L., Schreiber, M., Low, J.G.H., Ong, A., Tolfvenstam, T., Lai, Y.L., Ng, L.C., Leo, Y.S., Thi Puong, L., Vasudevan, S.G., Simmons, C.P., Hibberd, M.L., Ooi, E.E.: Decision tree algorithms predict the diagnosis and outcome of dengue fever in the early phase of illness. PLoS Negl. Trop. Dis. **2**(3), e196 (2008). http://dx.plos.org/10.1371
25. Trung, D.T., Thao, L.T.T., Dung, N.M., Ngoc, T.V., Hien, T.T., Chau, N.V.V., Wolbers, M., Tam, D.T.H., Farrar, J., Simmons, C., Wills, B.: Clinical features of dengue in a large vietnamese cohort: intrinsically lower platelet counts and greater risk for bleeding in adults than children. PLoS Negl. Trop. Dis. **6**(6), e1679 (2012)
26. Vaughn, D.W., Green, S., Kalayanarooj, S., Innis, B.L., Nimmannitya, S., Suntayakorn, S., Endy, T.P., Raengsakulrach, B., Rothman, A.L., Ennis, F.A., Nisalak, A.: Dengue viremia titer, antibody response pattern, and virus serotype correlate with disease severity. J. Infect. Dis. **181**(1), 2–9 (2000)

Early Prediction of Severe Maternal Morbidity Using Machine Learning Techniques

Eugenia Arrieta Rodríguez[1], Francisco Edna Estrada[1],
William Caicedo Torres[2], and Juan Carlos Martínez Santos[2(⊠)]

[1] E.S.E Clínica de Maternidad Rafael Calvo, Cartagena, Colombia
sios@maternidadrafaelcalvo.gov.co
[2] Universidad Tecnológica de Bolívar, Cartagena, Colombia
jcmartinezs@unitecnologica.edu.co
http://www.maternidadrafaelcalvo.gov.co/,
http://www.unitecnologica.edu.co/

Abstract. Severe Maternal Morbidity is a public health issue. It may occur during pregnancy, delivery, or puerperium due to conditions (hypertensive disorders, hemorrhages, infections and others) that put in risk the women's or baby's life. These conditions are really difficult to detect at an early stage. In response to the above, this work proposes using several machine learning techniques, which are considered most relevant in a bio-medical setting, in order to predict the risk level for Severe Maternal Morbidity in patients during pregnancy. The population studied correspond to pregnant women receiving prenatal care and final attention at E.S.E Clínica de Maternidad Rafael Calvo in Cartagena, Colombia. This paper presents the preliminary results of an ongoing project, as well as methods and materials considered for the construction of the learning models.

Keywords: Severe maternal morbidity · Machine learning · Logistic regression

1 Introduction

The term Severe Maternal Morbidity (SMM) includes a set of complications that can have a severe adverse effect on women and baby health, and happen during pregnancy, delivery, or puerperium. When any of these appear, it is necessary to provide the patient with immediate attention, in order to avoid death [10]. Although maternal health outcomes have shown positive variation, complications of pregnancy still are an important public health issue. Each year around 585.000 women die during pregnancy, delivery or puerperium worldwide [5], and annually close to 50 million complications in maternal health are registered, and approximately 300 million women suffer from short and long-term illnesses and injuries related to pregnancy, childbirth and postpartum [10]. Currently, there is an epidemiological surveillance strategy which consists in identifying SMM cases, reporting them to the public surveillance system (SIVIGILA) [16], and following them

M. Montes-y-Gómez et al. (Eds.): IBERAMIA 2016, LNAI 10022, pp. 259–270, 2016.
DOI: 10.1007/978-3-319-47955-2_22

up. This allows to characterize SMM and have a better understanding about the main factors of risk in the population and devise policies to help lower incidence. However, the number of SMM cases continue to be very high.

Studies conducted to identify causes of SMM show that this condition is related with hypertensive disorder, hemorrhage, and infections. The main risk factors associated with occurrence of SMM are black race, obesity, multi parity and backgrounds of previous cesarean sections and presence of co-morbidities [3,6,7,14].

The development of adequately sensitive and specific predictive tests for these outcomes has received significant focus in perinatal research. According to the literature, machine learning approaches are used frequently to identify patterns and make predictions. Specially in medicine, logistic regression [11,13], support vector machines [1], neural networks [9] have been used successfully.

The World Health Organization and PanAmerican Health Organization during the last decades have tried to reduce mortality and Severe Maternal Morbidity. For this, the action plan of 2012–2017 was proposed [3]. It consists in strengthening information systems and monitoring of maternal health in the countries of the region. The reduction of maternal mortality is a millennium goal and a national purpose. Actions, as epidemiological surveillance, the availability of statistical data, and the identification of risk factors related to these events, have contributed to its decrease.

Institutions and doctors strive to avoid SMM, because it is not easy to detect and prevent such situations. Especially, when the volume of pregnant women is quite high in a day, or when novice doctors do not have enough experience. Even with the implementation of the above actions, failure to meet the stated goal persist. Because of that it is necessary to implement new mechanisms for early warning and monitoring of SMM cases.

This paper proposes the use of machine learning techniques to build a risk classifier for SMM. With this, we expect to have early detection of morbidity cases, providing support for medical staff in decision-making, to enable a timely intervention of patients. This would help reduce the risk the mother and baby may have during this stage, and in turn to reduce social and economic repercussions.

The paper is organized as follows. Section 2 presents related work. Section 3 shows the methods and materials used in our approach. Section 4 shows the preliminary results of our on-going research. Finally, Sect. 5 states the conclusions of this paper.

2   Related Work

Every time, it is more frequent to find the use of machine learning in medicine, especially for classification problems. Some studies of prediction of at least one of the major diseases associated with SMM are mentioned below.

In Poon et al. [13], authors show work about early prediction of hypertensive disorders during first-trimester pregnancy, in the population of London, UK.

Logistic regression was used and a detection rate around 90 % for early preeclampsia was obtained, and a false-positive rate of 5 %.

In Park et al. [11], authors present an algorithm based on multiple logistic regression to predict the risk of preeclampsia in an Australian population. The algorithm correctly predicted preeclampsia in 95 % of women with a 10 % false positive rate.

In Nanda et al. [8], authors present a model for prediction of gestational diabetes mellitus in the first-trimester of pregnancy, based on bio markers and some maternal features. The use of logistic regression gave them a 74.1 % of correct predictions, with 20 % false positive rate.

In Farran et al. [4], authors implemented logistic regression, k-nearest neighbours (k-NN), multifactor dimensionality reduction and support vector machines for predicting diabetes, hypertension and comorbidity. The techniques were satisfactory implemented and similar results were obtained.

According to the reviewed literature there is evidence that the results obtained from the implementation of machine learning for classification problems in medicine are quite satisfactory. The authors of this paper have not been able to find similar studies or proposals, using machine learning as a tool to help to avoid or mitigate the risk of SSM.

3   Methods and Materials

3.1   Participants

A retrospective cohort study was done through clinical histories of prenatal controls obtained between 2014 and 2015. The population selected for this study include patients with ages between 12 and 45 years who had at least one control at E.S.E Clínica de Maternidad Rafael Calvo and whose labor was cared for in this institution.

Cohort patients were classified according SMM outcome, in two groups: patients who did not present SMM, and patients which presented SMM (which also were reported to the public healthcare surveillance system). For the first group, we used random sampling, and for the second group, we used convenience sampling. This method is known as mixed sampling [2].

3.2   Data Set

The construction of the machine learning model was based on features or risk factors. These factors were selected according to the risk factor characterization described by Latin American Center of Perinatology (Centro Latino Americano de Perinatología, CLAP), compared to the 2015 SMM protocol from the Colombia Ministry of Health and Social Protection (Ministerio de Salud y de Protección Social) [3]. The selected predictor set was then supplemented with socio demographic data and the gynecological and obstetrical history for each patient, Tables 1 and 2.

Table 1. Socio demographic characteristics

Feature	Options
Age in years	Younger than 14, between 14 and 19, between 19 and 34, older than 35
Ethnicity	Native Colombians, Gypsy, Raizal, Palenquero, Black, Mulatto, Afro-descendants, Other
Scholarship	Basic Primary, Secondary Basic, Technical, University, None
Socio economic strata	Strata 1, Strata 2, Strata 4, Strata 5, Strata 6, Unknown
Health care regulation	Contributory Regime (CR), Subsidized Regime (SR)
Origin	Capital city, Village, Rural zone
Marital status	Single, Married, Domestic partnership, Separated, Widowed

The selected population included treated patients in gynecological antenatal consultations at the ESE clinic during pregnancy, and whose labor was cared for, between years 2014 and 2015.

For the data set construction we used Google Forms. Two forms were designed, the first one to record Obstetrics Gynecology (OBGYN) and socio-demographic background data, and the second one to record diagnoses for each gestational week, using the International Classification of Diseases (ICD-10) codification [15].

The data set is being built with help of sixth year medicine students. They were trained on SMM, review of medical records and filling out Google forms. The manual review of prenatal medical records of each patient is necessary, because the information is sometimes scattered and not totally centralized on the hospital information system. To perform preliminar training and validation, two patient groups were generated according to outcomes (SMM and not SMM). First group was sub-sampled in order to reduce class imbalance, given that the number of patients that exhibited SMM is lower than the non SMM group. The collection process is still underway, and subsequent validation will be carried to monitor progress and performance of the predictive models.

3.3 Statistical Data Analysis

The filtering features allow to select the set of variables that represent variability in the occurrence of SMM. Once the variables are defined, and the database constructed, we proceeds to do an analysis statistical to obtain a database only with the information of the variables that we considered predictors for the model. We use descriptive statistics to identify the frequency with which diagnoses are presented, and multi-factor analysis of variance (ANOVA) to determine which variables are more likely to be considered predictive for the model that we want to implement. It was tested with different levels of confidence for the group of variables most likely had to influence the behavior of the response variable.

Table 2. Gynecological and obstetrical history

Feature	Options
Maternal parity	Nulliparous, Multiparous
Pregnancy spacing less than two years	Yes, No
Multiple birth	Simple, Twins, Triplets or more
Prenatal care	0, More than 0 less than 4, More than 4
Gestational age in first-prenatal care	First-trimester, Second-trimester, Third-trimester
Micronutrient intake	Yes, No
Personal history of preeclampsy	Yes, No
Pregnancy induced hypertension PIH	Yes, No
Chronic hypertension	Yes, No
Superimposed preeclampsy	Yes, No
Diabetes	Yes, No
Autoimmune disorders	Yes, No
Human immunodeficiency virus HIV	Yes, No
Congenital syphilis	Yes, No
Hepatitis B	Yes, No
Previous perinatal mortality	Yes, No
Incompatible with life VIP	Yes, No
Maternal causes VIP	Yes, No
Sexual abuse VIP	Yes, No
Urinary tract infection, UTI	Yes, No
Drinking/Smoking	Yes, No
Illicit and nonillicit drug use	Yes, No
Anemia in pregnancy	Yes, No
TORCH infections	Yes, No
Obesity in pregnancy	Yes, No
Under-nutrition during pregnancy	Yes, No

3.4   Learning Models

This section lists some of the machine learning techniques most commonly used for classification or pattern recognition.

Logistic Regression. The logistic regression has been historical an important tool for data analysis in medical investigation and epidemiology. This allows differentiating between some classes, in terms of a set of numerical variables, as a predictor. The basic goals of a logistic regression model are:

– Get an unbiased or adjusted estimate of the relationship between variable dependent (or result) and an independent variable.
– Simultaneously evaluate several factors that are allegedly related somehow (or not) with the dependent variable.
– Build a model and get a hypothesis prediction purpose or calculating risk.

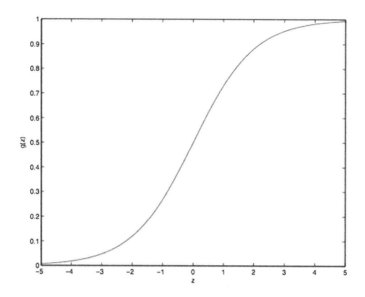

Fig. 1. Sigmoid function

Logistic Regression is a linear classification algorithm widely used in medicine where a logistic sigmoid function is coupled to a linear regression model. The Sigmoid function used to make the prediction algorithm is shown in Fig. 1.

Results can be interpreted as the probability for the input to belong to the positive class, $p(Y = 1|x; \theta)$. In this study, Logistic Regression was trained with Cross-Entropy loss. A L2 regularization penalty was introduced to account for model complexity and avoid over-fitting:

$$min \; - \frac{1}{N} \sum_{i=1}^{N} [y_i log h_\theta(x_i) + (1 - y_i) log(1 - h_\theta(x_i))] + \lambda ||\theta||^2 \qquad (1)$$

where

$$h_\theta(x) = \frac{1}{1 + e^{-\theta^t x}} \qquad (2)$$

It is possible to increase its capabilities by applying a polynomial transform to the input, in which case the decision boundary can be non-linear and handle more difficult problems.

In this work, we used L2 Regularized Logistic Regression with polynomial transform of 2nd degree. The model was implemented using the Scikit-Learn Python Machine Learning Library, version 2.7.11. [12]

The model was trained with Stratified K-Folds cross validation iterator. The algorithm was probe in a server in a High Performance Computing Laboratory through a batch system to run tasks (HTCondor).

4   Results

4.1   Statistical Analysis

This paper presents preliminary results of the project that is still ongoing. Achieving determine the variables set related to the occurrence of extreme maternal morbidity. The entire population is 1838 patients. 72 belong to the first group, and the remain 1766 belong to the second group. For a 95 % confidence level with a confidence interval of 4.5, we obtained a total sample of 377 patients. Once the system is trained, we will validate the obtained data in 145 patients. Initially, we performed a descriptive data frequency analysis of the patients who developed SMM.

We perform an analysis of frequency of the diagnosis. As we can see in Fig. 2, the amount of diagnostics is very large and the graph of frequency showed a data distribution dispersed and difficult to interpret. It was decided to group them according to the ICD-10, leaving a moderate amount of variables. The new results are shown in the Table 3.

Followed by this, we did the frequency analysis by trimester. In the first trimester only the diagnostic group Z30-Z39 had high frequency. For the second trimester, the results show that the most frequent group are Z30-Z39, O30-O48, O20-O29, E65-E68, and N70-N77. In the third trimester is obtained that the diagnostic groups most frequently are Z30-Z39, O30-O48, O10-O16, N70-N77, O95-O99, D50-D64, and O60-O75.

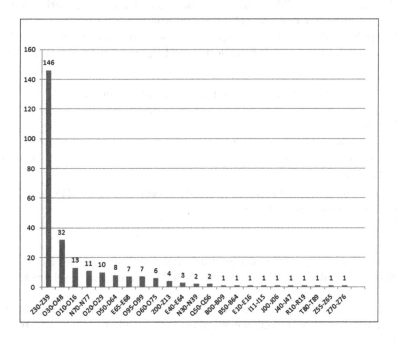

Fig. 2. Diagnostics group frequency

Table 3. Diagnostics frequency analysis

CODE (ICD-10)	Diagnostic group
Z30-Z39	Persons encountering health Services
O30-O48	Maternal care related to the fetus and amniotic cavity and possible delivery problems
O10-O16	Oedema, proteinuria and hypertensive disorders in pregnancy, childbirth and the puerperium
N70-N77	Inflammatory diseases of female pelvic organs
O20-O29	Other maternal disorders predom1inantly related to pregnancy
D50-D64	Nutritional anaemias
E65-E68	Obesity and other hyperalimentation
O95-O99	Other obstetric conditions, not elsewhere classified
O60-O75	Complications of labour and delivery

After having a notion of the data trend, we decided to carry out an analysis of variance ANOVA to the data of the patients with SMM and not SMM. It was organized per trimester. Similarly, the diagnostics were grouped according to SMM. The analysis of the first trimester was done initially with levels of confidence of 95 %, 90 % and finally 85 %. The results indicate that none of the variables have a high probability of being connected with the SMM. It notes that the diagnosis Z30-Z39 shows more likely to have any connection with the response variable, but it is the default diagnostic.

The analysis of the data obtained for the second trimester was tested with the 90 %, 80 % and 75 % confidence levels, and the results are shown in Table 4.

Table 4. Anova second trimester 75 % Confidence

Description	Feature	P-value
Obesity and other hyperalimentation	E65-E68	0.0661
Persons encountering health services for examination and investigation	Z00-Z13	0.2411
Persons with potential health hazards related to socioeconomic and psychosocial circumstances	Z55-Z65	0.2196

The analysis of variance of the third trimester was tested with the 95 %, 85 % and 80 % confidence levels. Results with 80 % confidence level are shown in Table 5.

Finally to make the analysis of ANOVA without taking into account the trimester was tested with confidence levels of 95 % and 80 %. Results are shown in Table 6.

Table 5. ANOVA third trimester with 80 % confidence

Description	Feature	P-value
Oedema, proteinuria and hypertensive disorders in pregnancy, childbirth and the puerperium	O10-O16	0.0383
Complications of labour and delivery	O60-O75	0.0408
Obesity and other hyperalimentation	E65-E68	0.109
Acute upper respiratory infections	J00-J06	0.1232
Symptoms and signs involving the digestive system and abdomen	R10-R19	0.1769

Table 6. ANOVA with the whole pregnancy period with 80 % confidence

Description	Feature	P-value
Oedema, proteinuria and hypertensive disorders in pregnancy, childbirth and the puerperium	O10-O16	0.0277
Complications of labour and delivery	O60-O75	0.0408
Symptoms and signs involving the digestive system and abdomen	R10-R19	0.0725
Acute upper respiratory infections	J00-J06	0.1391
Disorders of other endocrine glands	E20-E35	0.1505

Table 7. Anova personal history with 90 % confidence level

Feature	P-value
VIP Maternal causes	0.0009
Preeclampsy	0.0013
Scholarship	0.0227
Diabetes	0.0318
Origin	0.0422
Socio economic strata	0.0697
Perinatal mortality	0.0821
Htacronica	0.0937

In the Table 7 is observed the analysis of ANOVA for the personal history of the patients performed with a confidence level of 90 %. It identifies the factors that have the most significant statistical effect on the occurrence of the SMM.

4.2 Training

The data set was divided into training set and test set, the first corresponding to 80 % (178 instances) and the other 20 % (44 instances). The learning model was trained using 5-fold Stratified Cross-Validation. We Used L2 Regularized

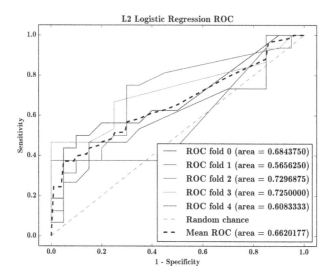

Fig. 3. SMM prediction model ROC

Logistic Regression to control the model complexity. To attack the problem of the disparity of the classes is it penalizes classifier 10 times for each error in the positive kind. The process of training takes 20 min in a server using one processor core. Figure 3 show the result for a Logistic Regression classifier with the selected predictors. The subsets are represented by a Receiver Operation Characteristic Score Area Under the Curve (ROC AUC). It is a graphical representation of the sensitivity vs. (1 - specificity).

To interpret the curve ROC is recommended to analyze the following intervals: $[0.5 - 0.6]$ Bad test, $[0.6 - 0.75]$ Regular test $[0.75 - 0.9]$ Good test, $[0.9 - 0.97]$ Very Good test, $[0.97 - 1]$ Excellent test.

In the Fig. 3, the ROC shows that the average area under the curve is 0.66. Then, the results are considerate Regular. To improve it, we suggest to try the following options:

1. Take the original data set and to use Recursive Feature Elimination for logistic regression, in this way obtain the best predictors to the model.
2. To Prove with polynomial transform higher than 2nd degree.
3. If the above options fail, we need to try a vector support machine.
4. After, we need to compare the results of ROC graphics for polynomial transform and support vector machine and select the model that shows the best especificity and sensibility.

5 Conclusions

Severe maternal morbidity remains a public health-care problem that affects much the pregnant women population, and in many cases, it is possible to avoid this. The problem lies in the early identification of risk patients who have finished in SMM. In response to the above mentioned, this paper presents the usage of the logistic regression for SMM detection. It is a pattern recognition technique commonly used in the medical field to solve problems of classification, prediction and identification of patterns. By the using of this technique, it is expected to build a tool for risk identification or risk classification of a patient having SMM. The goal is to provide a timely and adequate attention to each patient depending on the risk level to be determined.

With the implemented logistic regression model, we obtained regular results, for this, we to continue to proving others techniques of machine learning from to obtain a model with best results.

Acknowledgements. Special thanks for their cooperation to the High-Performance Computing Laboratory (HPCLab) at Universidad Tecnológica de Bolívar and to research group on maternal safety of Center of research for maternal health, Perinatal and women at E.S.E Clínica de Maternidad Rafael Calvo.

References

1. Carty, D.M., Siwy, J., Brennand, J.E., Zürbig, P., Mullen, W., Franke, J., McCulloch, J.W., North, R.A., Chappell, L.C., Mischak, H., et al.: Urinary proteomics for prediction of preeclampsia. Hypertension **57**(3), 561–569 (2011)
2. Casal, J., Mateu, E.: Tipos de muestreo. Rev. Epidem. Med. Prev **1**(1), 3–7 (2003)
3. Duran, M.E.M., García, O.E.P., CArey, A.C., Bonilla, H.Q., Espitia, N.C.C., Barros, E.C.: Protocolo de vigilancia en salud pública morbilidad materna extrema
4. Farran, B., Channanath, A.M., Behbehani, K., Thanaraj, T.A.: Predictive models to assess risk of type 2 diabetes, hypertension and comorbidity: machine-learning algorithms and validation using national health data from kuwaita cohort study. BMJ Open **3**(5), e002457 (2013)
5. Haaga, J.G., Wasserheit, J.N., Tsui, A.O., et al.: Reproductive Health in Developing Countries: Expanding Dimensions, Building Solutions. National Academies Press, Washington, D.C. (1997)
6. Mariño Martínez, C.A., Fiesco, V., Carolina, D., et al.: Caracterización de la morbilidad materna extrema en el Instituto Materno Infantil-Hospital la Victoria/Characterization of extreme morbidity disease in the Instituto Materno Infantil-Hospital la Victoria. Ph.D. thesis, Universidad Nacional de Colombia
7. Morales-Osorno, B., Martínez, D.M., Cifuentes-Borrero, R.: Extreme maternal morbidity in Clinica Rafael Uribe Uribe, Cali, Colombia, from January 2003 to May 2006. Revista Colombiana de Obstetricia y Ginecología **58**(3), 184–188 (2007)
8. Nanda, S., Savvidou, M., Syngelaki, A., Akolekar, R., Nicolaides, K.H.: Prediction of gestational diabetes mellitus by maternal factors and biomarkers at 11 to 13 weeks. Prenat. Diagn. **31**(2), 135–141 (2011)
9. Neocleous, C.K., Anastasopoulos, P., Nikolaides, K.H., Schizas, C.N., Neokleous, K.C.: Neural networks to estimate the risk for preeclampsia occurrence. In: International Joint Conference on Neural Networks, IJCNN 2009, pp. 2221–2225. IEEE (2009)
10. Organization, W.H., UNICEF.: Revised 1990 estimates of maternal mortality: a new approach. World Health Organization (1996)

11. Park, F.J., Leung, C.H., Poon, L.C., Williams, P.F., Rothwell, S.J., Hyett, J.A.: Clinical evaluation of a first trimester algorithm predicting the risk of hypertensive disease of pregnancy. Aust. N. Z. J. Obstet. Gynaecol. **53**(6), 532–539 (2013)

12. Pedregosa, F., Varoquaux, G., Gramfort, A., Michel, V., Thirion, B., Grisel, O., Blondel, M., Prettenhofer, P., Weiss, R., Dubourg, V., Vanderplas, J., Passos, A., Cournapeau, D., Brucher, M., Perrot, M., Duchesnay, E.: Scikit-learn: machine learning in Python. J. Mach. Learn. Res. **12**, 2825–2830 (2011)

13. Poon, L.C., Kametas, N.A., Maiz, N., Akolekar, R., Nicolaides, K.H.: First-trimester prediction of hypertensive disorders in pregnancy. Hypertension **53**(5), 812–818 (2009)

14. Rojas, J.A., Cogollo, M., Miranda, J.E., Ramos, E.C., Fernández, J.C., Bello, A.M.: Morbilidad materna extrema en cuidados intensivos obstétricos. Cartagena (Colombia) 2006–2008 maternal near miss in obstetric critical care. Cartagena, Colombia, 2006–2008. Revista Colombiana de Obstetricia y Ginecología **62**(2), 131–140 (2011)

15. de la Salud, O.P.: Clasificación estadística internacional de enfermedades y problemas relacionados con la salud: décima revisión: CIE-10. Pan American Health Org (1995)

16. de Vigilancia, S.: Control en salud pública (sivigila). Informe de Intoxicaciones por plaguicidas. Instituto Nacional de Salud, INS. Bogotá, Colombia (2012)

Big Data, Knowledge Discovery and Data Mining

Collaborative Filtering
with Semantic Neighbour Discovery

Bruno Veloso[1,2], Benedita Malheiro[2,3](✉), and Juan C. Burguillo[1]

[1] EET/UVigo – School of Telecommunication Engineering,
University of Vigo, Vigo, Spain
[2] INESC TEC, Porto, Portugal
[3] ISEP/IPP – School of Engineering, Polytechnic Institute of Porto, Porto, Portugal
mbm@isep.ipp.pt

Abstract. Nearest neighbour collaborative filtering (NNCF) algorithms
are commonly used in multimedia recommender systems to suggest
media items based on the ratings of users with similar preferences. However, the prediction accuracy of NNCF algorithms is affected by the
reduced number of items – the subset of items co-rated by both users –
typically used to determine the similarity between pairs of users. In this
paper, we propose a different approach, which substantially enhances the
accuracy of the neighbour selection process – a user-based CF (UbCF)
with semantic neighbour discovery (SND). Our neighbour discovery
methodology, which assesses pairs of users by taking into account all
the items rated at least by one of the users instead of just the set of co-
rated items, semantically enriches this enlarged set of items using linked
data and, finally, applies the Collinearity and Proximity Similarity metric
(CPS), which combines the cosine similarity with Chebyschev distance
dissimilarity metric. We tested the proposed SND against the Pearson
Correlation neighbour discovery algorithm off-line, using the HetRec data
set, and the results show a clear improvement in terms of accuracy and
execution time for the predicted recommendations.

Keywords: User-based collaborative filtering · Semantic neighbour
discovery · Semantic enrichment

1 Introduction

Nowadays viewers are unable to search, find and choose their preferred content
in real time due to the size of the search space both in terms of the number
of offers and of the diversity of sources. Consequently, media content personalisation poses a significant challenge to media content distributors, *i.e.*, how to
choose the most appropriate media content for each viewer from a vast list of
sources and candidates per source. This information overload problem is typically addressed using Information Filtering technologies. A possible approach is
to use Collaborative Filtering (CF), which relies on information retrieval and
data mining techniques to provide recommendations based on the ratings of

© Springer International Publishing AG 2016
M. Montes-y-Gómez et al. (Eds.): IBERAMIA 2016, LNAI 10022, pp. 273–284, 2016.
DOI: 10.1007/978-3-319-47955-2_23

users with similar preferences. CF has three variants – user-based, item-based and tag-based – and two alternative implementations – memory-based, *e.g.*, k-Nearest Neighbour (k-NN), and model-based, *e.g.*, Singular Value Decomposition (SVD) [3]. Our research is focussed on a specific CF memory-based technique called user-based k-NN. This technique is influenced by several factors: (i) the similarity metrics used to determine the nearest neighbours; (ii) the sparsity of the existing user ratings, which emerge from the fact that each user usually rates only a very small percentage of the represented items; and (iii) the set of items used to determine the similarity between two users, which is based solely on co-rated items, *i.e.*, items rated by both users, ignoring all items rated by just one of the users. These weaknesses can lead to the selection of unrelated neighbours, resulting in the inability to provide interesting recommendations. Current nearest neighbour CF assessment methodologies are unable to identify these inefficiencies.

In this paper, we propose a new user-based k-NN filtering approach, which performs a semantic neighbour discovery (SND) using all items rated at least by one of the users, instead of just the subset of co-rated items, to select the most related neighbours and, thus, improve the accuracy of the final recommendations. For each user, it collects semantic data regarding all rated items, *e.g.*, the set of writers, actors, countries or directors, builds multiple vectors and applies the Collinearity and Proximity Similarity (CPS) metric, a similarity metric proposed by [16] to select the best neighbours. CPS combines the well known cosine similarity with Chebyschev distance dissimilarity, which determines the greatest differences between two related feature vectors.

Experiments were conducted with the HetRec 2011 data set, which contains movies, user ratings and tags, to evaluate the quality and accuracy of the recommendations produced. The metrics used were the Global Mean Absolute Error (GMAE), the Global Root Mean Square Error (GRMSE), the Global F1-measure (GF1) as well as the execution time. The comparison between the SND and the standard Pearson Correlation UbCF recommendations show a clear improvement in terms of GF1, comparable results in terms of GMAE and execution times related with the number of LOD features used.

The main contributions of this paper are: (i) the development of a new user-based k-NN collaborative filter supported by semantic neighbour discovery, *i.e.*, which uses linked data to improve the selection of the nearest neighbours; and (ii) the application of the Collinearity and Proximity Similarity (CPS) metric to k-NN.

The rest of the paper is organized as follows. Section 2 presents an overview of collaborative recommendation with semantic enrichment. Section 3 describes the proposed collaborative filter. Section 4 describes the data set, presents the adopted evaluation metrics as well as the tests, and discusses the results. Finally, Sect. 5 provides the conclusions and presents future development lines.

2   Semantically Enriched Collaborative Recommendation

Recommendation systems are valuable support tools for users or applications to find items of interest whenever the search space is too large. They are frequently used in personalisation and data retrieval tasks, and rely on different techniques, *e.g.*, information filtering and semantic enrichment. The information filtering approaches for recommendation systems are organised in Content-based Filters (CbF), Collaborative Filters (CF) and Hybrid Filters (HF). This paper is focussed on k-NN CF.

2.1 Collaborative Filtering

CF algorithms recommend items by making predictions based on the history of items rated by the user. There are two main approaches: (*i*) memory-based algorithms, which calculate the similarity or correlation among users or items; and (*ii*) model-based algorithms, which learn predictive models. Collaborative filtering depends on the availability of the individual viewer records of watched, rated and/or tagged items. CF promotes diversity when dealing with different types of content and creates an effect of controlled serendipity. However, these filters suffer from: (*i*) data sparsity and scalability issues; (*ii*) the first-rater problem, *i.e.*, an item cannot be recommended without being rated by a user; (*iii*) the grey sheep problem, *i.e.*, users who never rate items; and (*iv*) the cold-start problem, *i.e.*, new users have no rated items. Memory and model-based approaches make predictions by analysing users (user-based), items (item-based) or tags (tag-based). Multiple examples can be found in the literature regarding both approaches. Whereas [10, 12, 13] are instances of user, item and tag memory-based approaches, [14, 17] are instances of user and item model-based approaches, respectively. This paper presents a UbCF with semantic enrichment.

2.2 Semantic Enrichment

Semantic enrichment is supported by public interconnected knowledge bases – the Linked Open Data (LOD) repositories. Together they are the fabric of the Semantic Web [2], where each *datum* is described and interlinked through Resource Description Framework (RDF) triplets. The subject-predicate-object triplets include uniform resource identifiers (URI) for resource identification and look up, using open standards. They provide a valuable source of information that can improve conventional recommender systems, if properly exploited [11]. Linked data can be used to enrich the profiles of users and items with new features, allowing the identification of new relationships. In this paper, we enrich the profiles of the users, using complementary features extracted from the LOD, and determine a multiple feature neighbour similarity, using CPS, to choose the nearest neighbours.

Martin *et al.* propose the usage of semantic reasoning techniques to improve the neighbour discovery. It is based on the products to recommend rather than on user similarity, explores hierarchical similarity [8]. The authors propose a pre-filtering approach limiting the neighbour search subspace at the level of the product class hierarchy, followed by the assessment of the ratings and semantic features to identify common interests. Our approach is more general since it is applicable, not only, to the complete set of LOD features, but also to any subset of LOD features. Moreover, we rely on semantically enriched data to choose neighbours at the term level rather than at the concept level.

Kaveh *et al.* propose a weighted k-NN neighbour selection, taking into account the semantic distances between neighbours. The selection mechanism adjusts the weight of these distances to enhance or decrease the importance of the best and worst categories, respectively [6]. While this approach determines the similarity between users based on the discrimination of the relative importance between the categories, we equally consider the selected LOD properties. When compared with our approach, this category discrimination may lead to overspecialisation.

The hybrid film recommender designed by Kushwaha *et al.*, [7] first uses semantic data from DBpedia to minimise the sparsity problem, then applies the Levenshtein distance to map films to DBpedia RDF links and, finally, determines the cosine similarity to find the best neighbours to generate predictions. While this recommender calculates the similarity between neighbours based exclusively on the features of the top co-rated items (5-star items), our proposal contemplates the features of all items rated by the users under analysis, including both liked and disliked items.

The collaborative recommender system reported by Martín-Vicente *et al.* uses semantic reasoning to address the matrix sparsity problem and improve the neighbour selection. User similarity is calculated using the relations and the number of hops between concepts represented in the system's ontology [9]. While this proposal determines similarity between users at the concept level, we choose the nearest neighbours by calculating the similarity between users at the (semantically enriched feature) term level. Consequently, our approach is fully automated, *i.e.*, does not require any previous semantic annotation.

This related work review shows that semantic based neighbour discovery is becoming increasingly popular. Our SND approach, when compared with these works, does not require prior concept annotation as the semantic concept level approaches, uses the LOD to enrich all items rated by any of the users under comparison with complementary features and determines user similarity with the CPS metric.

3   UbCF with Semantic Neighbour Discovery

User-based collaborative filtering systems suggest interesting items by relying on like-minded people called neighbours [1]. The quality of these NNCF depends largely on the similarity metrics used and, in particular, on the data used to determine the set of nearest neighbours. Typically, users have rated quite a small number of items and are compared on the basis of an even smaller subset – the set of co-rated items – regardless of the adopted similarity metric. To overcome this problem, we, rather than comparing two users based solely on the set of co-rated items (the intersection set), use all items rated by at least one of the users (the union set). Although this idea was first proposed by [15], we adopt a diverse approach herein called Semantic Neighbour Discovery (SND). Our UbCF with SND, which is represented in Fig. 1, encompasses three stages: (i) semantic enrichment; (ii) neighbour discovery; and (iii) item prediction.

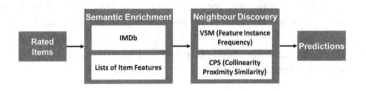

Fig. 1. Semantic UbCF architecture.

In the semantic enrichment stage the system collects, for each user and based on the full set of user rated items, multiple lists of semantic feature instances from the IMDb via the OMDb API[1], *i.e.*, the lists of genres, actors, directors, writers, languages, countries, types and years of the user rated items. These lists are created beforehand off-line, using historical data and are updated on-line to include newly rated user items. Depending on the feature list, each element, representing a specific genre, actor, director, writer, type, year, language or country, holds the corresponding frequency count in the user rated items. During the neighbour discovery stage, the system, based on the semantic feature lists gathered, compares the active and current users to select the k users with the highest similarity to the active user. First, it creates a vector for each semantic feature and user with a dimension equal to the union of instances of the corresponding active and current user lists. Then, it calculates, for each semantic feature instance vector, the normalised Feature Instance Frequency (FIF) according to Eq. 1:

$$FIF_{f,i,u} = \frac{freq_{f,i,u}}{\sum_i^n freq_{f,i,u}} \tag{1}$$

where $freq_{f,i,u}$ is the frequency of the instance i of the semantic feature f in the items rated by the user u and n is the number of instances of the semantic feature f present in the items rated by the user u. The similarity between users is determined through the Collinearity and Proximity Similarity (CPS) using the Eq. 2:

$$CPS_{a,c} = \beta \frac{\sum_{f=1}^m CS_{f,a,c}}{m} + (1-\beta)\frac{\sum_{f=1}^m CDD_{f,a,c}}{m} \tag{2a}$$

$$CS_{f,a,c} = \frac{\sum_{i=1}^n \hat{A}_{f,i}\hat{C}_{f,i}}{\sqrt{\sum_{i=1}^n \hat{A}_{f,i}^2}\sqrt{\sum_{i=1}^n \hat{C}_{f,i}^2}} \tag{2b}$$

$$CDD_{f,a,c} = (1 - \max_i(|\hat{A}_{f,i} - \hat{C}_{f,i}|)) \tag{2c}$$

where $CS_{f,a,c}$ is the cosine similarity between the active user a and the current user c regarding the semantic feature f, $CDD_{f,a,c}$ is the Chebyshev distance

---

[1] http://omdbapi.com.

dissimilarity between the active user a and the current user c regarding the semantic feature f, $\hat{A}_f$ is the vector of the active user a regarding the semantic feature f, $\hat{C}_f$ is the vector of the active user c regarding the semantic feature f, m is the number of semantic features used, i is the index of each semantic feature instance, n is the number of semantic feature instances and β is a linear combining parameter. We used $\beta = 0.5$, i.e., attributed equal weights to the collinearity and the proximity metrics. In the final prediction stage, the algorithm computes the active user predictions using the Eq. 3:

$$P_{a,i} = \bar{r}_u + \frac{\sum_{j=1}^{k} r_{j,i} CPS_{a,c}}{\sum_{j=1}^{k} |CPS_{a,c}|} \tag{3}$$

where $P_{a,i}$ is the item i prediction for the active user a, $\bar{r}_u$ is the average rating for user u, $r_{j,i}$ is the rating attributed by user j to the movie i, $CPS_{a,c}$ is the similarity between the active user a and the current user c and k is the number of neighbours.

4   Tests and Results

The following subsections present the data set used, the evaluation metrics and the results obtained. The recommendation service holds two user-based collaborative filters with the following neighbour discovery algorithms: (i) Pearson correlation (PC); and (ii) semantic neighbour discovery (SND). Both algorithms select the best 50 neighbours ($k = 50$). The SND algorithm can be configured to include, for each rated film, any combination of actor (A), country (C), director (D), genre (G), language (L), type (T), writer (W) and year (Y) features extracted from the LOD. The PC algorithm is the base algorithm for all comparisons. Our aim is to compare the accuracy, prediction errors and run time of the two neighbour discovery algorithms: SND and PC. The experiments were performed with an Intel i7-2600 3.4 GHz Central Processing Unit (CPU) and 16 GB DDR3 Random Access Memory (RAM) platform.

4.1   Data Set

The recommendation service was evaluated off-line with HetRec 2011 v. 2.0[2]. This data set was chosen due to lower data sparsity and data set size. It contains information about users (2113) and movies (10197), including user movie ratings (average of 405 ratings per user) and movie tag assignments (average of 23 tags per user) together with the corresponding timestamps. In terms of semantic data, the data set covers 6404 actors, 106 countries, 4196 directors, 27 genres, 166 languages, 5 types, 16471 writers and 97 years.

The data set was divided into the training set (80 %) and the test set (20 %), ensuring that, regardless of the number of ratings, each user has 80 % of their

---

[2] http://www.grouplens.org/datasets/hetrec-2011/.

ratings in the training set and the remaining 20 % are in the testing set. Since the data set used contains sparse information, only items that were actually rated by users are recommended. This way, the system increases the accuracy of the predictions, but, since the suggested items belong to the group of items seen by the user, the calculated precision value is always one [5].

4.2   Evaluation Metrics

The metrics for evaluating recommender systems can be divided into predictive and classification accuracy metrics.

Predictive Accuracy Metrics. Measure the error between the predicted rating and the rating assigned by the user. According to [4], there are two important metrics: the Mean Absolute Error (MAE) and Root Mean Square Error (RMSE). The normalised MAE and $RMSE$ values range from 0 (best) to 1 (worst). $RMSE$, when compared with MAE, emphasizes the largest errors.

The Global MAE (GMAE), which corresponds to the average MAE for all users, measures the average absolute deviation between the predicted ratings and the ratings assigned by each user. The Global $RMSE$ (GRMSE), which is the average $RMSE$ taking into account all users, measures the root mean square deviation between the predicted ratings and the ratings assigned by each user.

Classification Accuracy Metrics. The *Recall*, *Precision* and *F1-measure* are classification accuracy metrics and, as such, can be used to measure the frequency with which a recommendation system provides correct recommendations. *F1-measure* is the harmonic mean of the *Precision* and *Recall*, where *Precision* is the number of relevant items recommended from the total number of items and *Recall* is the number of relevant items recommended from the total number of relevant items available. These metrics range from 0 (worst) to 1 (best).

The Global *F1-measure* (GF1), which corresponds to the average *F1-measure* for all users, was used to evaluate the performance of both algorithms.

4.3   Results

In this section, we present the set of tests performed to compare the accuracy, predicting errors and execution time of the two neighbour discovery algorithms: SND and PC. These analyses are done by calculating the *GF1*, the *GMAE* and the *GRMSE*, respectively. We expect SND to have improved accuracy – it uses all items rated by the users (not just the co-rated) and semantically enriches the items to identify additional similarities among users – and run time – at least in the cases where the volume of the LOD data used allows it. In terms of predicting errors, we expect SND to exhibit a similar behaviour.

The off-line performance was determined separately for the two neighbour discovery algorithms, using a 60 % similarity threshold. The training data set was submitted to each filter, and the individual user's recommendations were

Table 1. Average CPS results for different β values

β	0.1	0.2	0.3	0.4	0.5	0.6	0.7	0.8	0.9	1.0
GF1	0.739	0.747	0.752	0.756	<u>0.758</u>	0.761	0.762	0.763	0.764	**0.765**
GMAE	0.133	0.133	0.133	0.133	**0.133**	0.133	0.133	0.134	0.134	0.134
GRMSE	**0.170**	**0.170**	**0.170**	0.171	<u>0.171</u>	0.171	0.172	0.172	0.172	0.172
TIME (s)	616	614	609	612	**604**	**604**	614.71	620	626	621

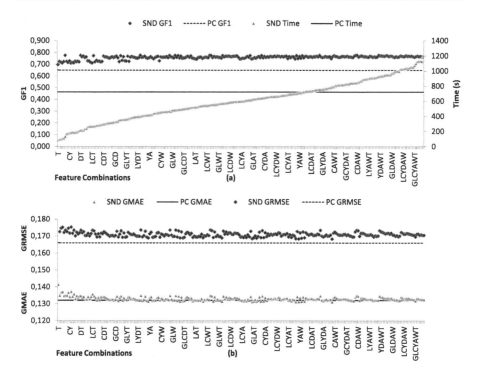

Fig. 2. SND *versus* PC: (a) GF1 and time; (b) GMAE and GRMSE

generated. Finally, these recommendations were evaluated using the test data set and calculating the corresponding global metrics.

Table 1 presents the results using different β values for the combination of CS and CDD in CPS. We can observe that $\beta = 0.5$ provides improved average results in terms of time, prediction errors and good results in terms of accuracy.

Figure 2 compares the results of both algorithms – SND and PC – for all semantic feature combinations. While Fig. 2a compares the *GF1* and the run time for all semantic feature combinations, Fig. 2b contrasts their normalised *GMAE* and *GRMSE*. The horizontal axis represents the 255 feature combinations involving the eight IMDb features – actor (A), country (C), director (D), genre (G), language (L), type (T), writer (W) and year (Y). Figure 2a shows that SND is always more accurate – the black diamond data series *versus* the dashed black line – with an average improvement of 17 %. In terms of execution time, the SND algorithm – grey triangle data series – was faster than the base algorithm – the continuous black line – in 67 % of the cases. Figure 2b displays the negligible statistical errors that SND introduces. In average, the *GMAE* increases 1 % – grey triangle data series *versus* the continuous black line – and the *GRMSE* increases 3 % – black diamonds *versus* the dashed black line.

These results also show the relevance of selecting an appropriate set of LOD features for semantic enrichment. The analysis of the global metrics results for SND-GLCYDAWT, which corresponds to the version that uses all semantic features, and SND-DAT, which uses three features, illustrates this issue. Both combinations improve the recommendation accuracy 18 %, but SND-GLCYDAWT increases the run time 67 % while SND-DAT decreases the run time 25 %. In terms of the prediction errors, SND-DAT maintains the base algorithm errors and SND-GLCYDAWT maintains *GMAE* and increases *GRMSE* 3 %. In this case, the combination of director (D), actor (A) and type (T) provides better results than the combination of the eight IMDb features.

Figure 3 displays the top 8 results in terms of run time (a) and *GF1* (b). The horizontal axis represents the semantic feature combinations – actor (A), country (C), director (D), genre (G), language (L), type (T), writer (W) and year (Y).

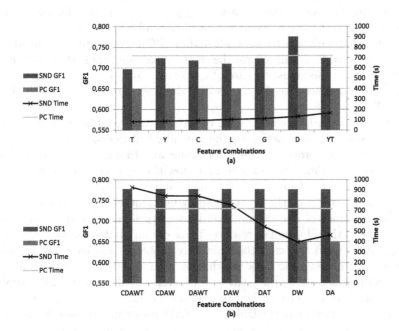

Fig. 3. Top 8 fastest (a) and most accurate (b) SND results.

We can observe in Fig. 3a that the best feature in terms of run time and, then, *GF1* is director (D) since it reduces the run time 82 %, improves the recommendation accuracy 19 % and increases the recommendation errors by 2 % and 4 % for the *GMAE* and *GRMSE*.

SND-T, which corresponds to a semantic enrichment based solely on the type (T), reduces the run time by 89 %, improves the recommendation accuracy by 8 % and increases the recommendation errors by 7 % and 9 % for the *GMAE* and *GRMSE*. In Fig. 3b, SND-DAW, which has a run time comparable to the base algorithm (5 % increase), improves the recommendation accuracy by 20 % and increases the recommendation errors by 2 % and 5 % for the *GMAE* and *GRMSE*, respectively.

As we had anticipated, the UbCF with SND generates more accurate predictions, is faster for the majority of the 255 feature combinations tested and introduces negligible prediction errors.

5   Conclusions

This paper describes an alternative neighbour discovery method for collaborative filters, which relies on semantic enrichment to expand the information available and explore additional relations among neighbours.

The SND algorithm, which uses all items rated by each user, and not only the co-rated items to discover the k nearest neighbours, shows considerable improvements in the quality of the recommendations in all SND feature combinations, when compared with the baseline algorithm (PC). The results obtained show that not all semantic features have the same impact and, thus, the set of semantic features to be used must be carefully selected. Our experiments showed that the best IMDb features are, for the HetRec data set, director (D), actor (A) and writer (W).

A straightforward implementation of the semantic neighbour discovery algorithm, *i.e.*, one that does not chose the semantic features to use, will face scalability and run time problems. These problems are, to some extent, shared with the baseline algorithm (PC). Our results show that it is possible to find a good compromise between the amount of semantic data used, the algorithm run time and the improvement of the recommendations accuracy. Nevertheless, in cases where the accuracy of the recommendations is crucial and the data set is huge, we recommend the adoption of resourceful execution platforms, *e.g.*, cloud computing platforms, to minimise the scalability problem together with the off-line pre-processing of semantic data enrichment prior to launching the on-line service. This approach allows the creation for all registered users of the corresponding data enrichment vectors (directors, actors, writers, genres, *etc.*) beforehand. Once the service is on-line, the UbCF with SND has to update the existing vector when new user feedback arrives and generate new semantic enrichment vector instances for new users.

As future work, we wish to apply this approach to item-based collaborative filters and assess the results to find the impact on the "new item problem".

Furthermore, we intend to explore the use of other semantic sources to improve the subset of relevant semantic features.

Acknowledgements. This work was partially financed by the ERDF – European Regional Development Fund through the Operational Programme for Competitiveness and Internationalisation - COMPETE 2020 Programme within project «POCI-01-0145-FEDER-006961»and by National Funds through the FCT – Fundação para a Ciência e a Tecnologia (Portuguese Foundation for Science and Technology) as part of project UID/EEA/50014/2013.

References

1. Bellogín, A., Castells, P., Cantador, I.: Neighbor selection and weighting in user-based collaborative filtering: a performance prediction approach. ACM Trans. Web (TWEB) **8**(2), 12 (2014)
2. Berners-Lee, T., Hendler, J., Lassila, O.: The semantic web. Sci. Am. **284**(5), 28–37 (2001)
3. Bobadilla, J., Ortega, F., Hernando, A., Gutiérrez, A.: Recommender systems survey. Knowl.-Based Syst. **46**, 109–132 (2013)
4. Herlocker, J.L., Konstan, J.A., Borchers, A., Riedl, J.: An algorithmic framework for performing collaborative filtering. In: Proceedings of 22nd Annual International ACM SIGIR Conference on Research and Development in Information Retrieval, pp. 230–237. ACM (1999)
5. Herlocker, J.L., Konstan, J.A., Terveen, L.G., Riedl, J.T.: Evaluating collaborative filtering recommender systems. ACM Trans. Inf. Syst. (TOIS) **22**(1), 5–53 (2004)
6. Kaveh-Yazdy, F., Zare-Mirakabad, M.R., Xia, F.: A novel neighbor selection approach for knn: a physiological status prediction case study. In: Proceedings of the 1st International Workshop on Context Discovery and Data Mining, p. 2. ACM (2012)
7. Kushwaha, N., Vyas, O.: Semmovierec: extraction of semantic features of dbpedia for recommender system. In: Proceedings of the 7th ACM India Computing Conference, p. 13. ACM (2014)
8. Martín-Vicente, M.I., Gil-Solla, A., Ramos-Cabrer, M., Blanco-Fernández, Y., López-Nores, M.: A semantic approach to avoiding fake neighborhoods in collaborative recommendation of coupons through digital tv. IEEE Trans. Consum. Electron. **56**(1), 54–62 (2010)
9. Martín-Vicente, M.I., Gil-Solla, A., Ramos-Cabrer, M., Pazos-Arias, J.J., Blanco-Fernández, Y., López-Nores, M.: A semantic approach to improve neighborhood formation in collaborative recommender systems. Expert Syst. Appl. **41**(17), 7776–7788 (2014)
10. Melville, P., Sindhwani, V.: Recommender systems. In: Encyclopedia of Machine Learning, pp. 829–838. Springer (2010)
11. Ostuni, V.C., Di Noia, T., Di Sciascio, E., Mirizzi, R.: Top-n recommendations from implicit feedback leveraging linked open data. In: Proceedings of the 7th ACM Conference on Recommender Systems, pp. 85–92. RecSys 2013, NY, USA. ACM, New York (2013)
12. Papagelis, M., Plexousakis, D.: Qualitative analysis of user-based and item-based prediction algorithms for recommendation agents. Eng. Appl. Artif. Intell. **18**(7), 781–789 (2005)

13. Rey-López, M., Díaz-Redondo, R.P., Fernández-Vilas, A., Pazos-Arias, J.J.: T-learning 2.0: A personalised hybrid approach based on ontologies and folksonomies. In: Computational Intelligence for Technology Enhanced Learning, pp. 125–142. Springer (2010)

14. Sarwar, B.M., Karypis, G., Konstan, J., Riedl, J.: Recommender systems for large-scale e-commerce: Scalable neighborhood formation using clustering. In: Proceedings of the Fifth International Conference on Computer and Information Technology, vol. 1 (2002)

15. Symeonidis, P., Nanopoulos, A., Papadopoulos, A.N., Manolopoulos, Y.: Collaborative filtering: Fallacies and insights in measuring similarity. In: Berendt, B., Hotho, A., Mladenic, D., Semeraro, G. (Chairs) Proceedings of the 17th European Conference on Machine Learning and 10th European Conference on Principles and the Practice of Knowledge Discovery in Databases Workshop on Web Mining, pp. 56–67 (2006)

16. Veloso, B., Malheiro, B., Burguillo, J.C.: A multi-agent brokerage platform for media content recommendation. Int. J. Appl. Math. Comput. Sci 25(3) (2015)

17. Vozalis, M.G., Margaritis, K.G.: Applying SVD on item-based filtering. In: Proceedings of the Fifth International Conference on Intelligent Systems Design and Applications, pp. 464–469. IEEE (2005)

Distributed and Parallel Algorithm for Computing Betweenness Centrality

Mirlayne Campuzano-Alvarez[(⊠)] and Adrian Fonseca-Bruzón

Center for Pattern Recognition and Data Mining, Santiago de Cuba, Cuba
{mirlayne,adrian}@cerpamid.co.cu

Abstract. Today, online social networks have millions of users, and continue growing up. For that reason, the graphs generated from these networks usually do not fit into a single machine's memory and the time required for its processing is very large. In particular, to compute a centrality measure like betweenness could be expensive on those graphs. To address this challenge, in this paper we present a parallel and distributed algorithm to compute betweenness. Also, we develop a heuristic to reduce the overall time, which accomplish a speedup over 80x in the best of cases.

Keywords: Online social network · Betweenness · MPI · Distributed computing

1 Introduction

Nowadays, online social networks such as Facebook, Twitter and Instagram have gain a huge popularity among people all around the world. These networks have become important resources to share information. For that reason, social networks have reach a growing interest among the scientific community; motivated, mainly, by the possibility of extract useful knowledge about people behavior. In this context, Social Networks Analysis (SNA) arises as a set of methods and tools for the analysis of the patterns in the connections established among the actors of the social network. One important task in SNA is to discover the most prominent users into a social network. To reach this goal, many researchers use centrality measures.

Among the different centrality measures, betweenness is one of the most used. This measure determines the frequency that a node or an edge acts as a bridge in the shortest paths in the network. In other words, betweenness is an indicator of the amount of information that can pass through a node or an edge.

Many applications have been found for betweenness in a wide variety of study fields such as Energy Management Systems, Chemical Networks Analysis and SNA. A practical application of betweenness in Energy Management Systems was proposed in [6] to make a power grid contingency analysis. In this work, the measure was used to identify high impact components in a power grid topology by finding the most traverse edges. In the field of Chemical Networks Analysis,

M. Montes-y-Gómez et al. (Eds.): IBERAMIA 2016, LNAI 10022, pp. 285–296, 2016.
DOI: 10.1007/978-3-319-47955-2_24

betweenness has been used, among other things, for the generation of skeletal mechanism and its introduction into a combustion system [13]. In SNA, in addition to determine the most important users, betweenness has been used also in other tasks, such as Communities Detection, as was shown in [11], where the authors proposed a method to detect communities by removing edges with the highest betweenness.

However, in spite of the massive importance of betweenness, its computation is expensive in large networks, because it involves to determine all shortest paths from all vertexes to each others. For that reason, several strategies have been proposed in the literature to accelerate this process using different strategies. Some approaches make a preprocessing of the graph for identifying vertexes with *strategical positions* which can be removed or merged, thus reducing the number of nodes to visit when the shortest paths are calculated [12]. Other authors establish that is not necessary to compute exact betweenness of all vertexes, so they process only a subset of them [4].

However, due to the fast grow up of online social networks, which have millions of users, nowadays the techniques previously mentioned are not enough, because the computation time often can reach several days and the main memory of a single machine is not enough to support the network representation. A good strategy to face this situation was proposed in [5], where was presented an algorithm to distribute the data into several machines, thus reducing the memory consumption. Another good alternative to address this problem is to parallelize the computation and thus reduce the execution time. Some authors proposed to exploit the power of GPU to speed up the computation [9]. Nevertheless, this algorithm requires to have the available hardware needed to make the work in the graphic card, which can imply more power consumption and cost.

Another option to parallelize computation is taking advantage of the multiple cores of the modern machines. In this context, it is appropriate to use MPI (Message Passing Interface), which allows that information be shared among processors of one or several machines. This characteristic make MPI suitable to distribute data and thus reduce the memory consumption, which is very important to process large social networks. Also, as the computations can be made in parallel, the speedup of algorithms can be increased. However, might be difficult to reduce the overall time of algorithms because one of MPI drawbacks is the delay in communications, which makes the time spent in the interchange of messages time be greater than the processing time. For that reason, it is recommended its use when the computation in greater than communications or when is possible to get a strategy to overlap computation and communications.

Taking into account these characteristics, in this paper we propose a parallel distributed algorithm for computing betweenness using MPI. Also, we propose a modification that allows our algorithm speeds up when it is used in sparse networks. The experimental results show that our proposal is suitable to process large networks and it can be adapted according to the available resources.

The rest of the paper is organized as follows. First, we expose some definitions related with graph and betweenness concepts. Then, is given a brief description of previous works, including an explanation of the sequential algorithms used. Next, we explain our proposal. Finally, we describe the experimental environment, present the results and make an analysis of them.

2 Definitions and Notations

Commonly, social networks are represented by a graph $G = (V, E)$ where V is a set of nodes, $n = |V|$ and E a set of edges, $m = |E|$. In this research, we work with unweighted and undirected graphs.

Let $d_s(v)$ be the distance of the shortest path from the source vertex s to v. We will denote $\sigma_{st}(v)$ as the number of the shortest paths from s to t that pass through v and σ_{st} the number of the shortest paths from s to t, $\forall s, t \in V$.

Betweenness is considered a medial centrality measure, because all walks passing through a node are considered [1]. This measure is used for computing the frequency that a vertex lies in the shortest path between all nodes in the graph; which means that it ranks nodes according the amount of information that can pass through them. Formally, it is defined as follows:

$$Cb(v) = \sum_{\substack{s \neq v, v \neq t \\ s,v,t \in V}} \frac{\sigma_{st}(v)}{\sigma_{st}} \tag{1}$$

Computing betweenness involves two main steps:

- Compute the shortest paths: This step implies to determine a Directed Acyclic Graph (DAG) for a selected source node s.
- Compute betweenness: At this step we compute the betweenness value for each node in the DAG previously computed.

3 Previous Work

According to Eq. 1, betweenness is costly to compute because requires $\mathcal{O}(n^3)$ run-time. In large networks this implies that compute exact betweenness of all nodes can take hours, even days. For that reason, some researchers have proposed algorithms and methods to reduce the time required for its computing. Some of them will be described below.

3.1 Sequential Algorithms

The best known sequential algorithm was proposed by Brandes in [3]. The base of this algorithm is that the betweenness value of a node depends of their sons in the DAG. Following this idea, Brandes defined the pair-dependency as follows:

$$\delta_{st}(v) = \frac{\sigma_{st}(v)}{\sigma_{st}} \tag{2}$$

Also, he introduced the vertex dependency from s as:

$$\delta_{s\bullet} = \sum_{t \in V} \delta_{st}(v) \tag{3}$$

And he defined the recursive formula:

$$\delta_{s\bullet}(v) = \sum_{w:v \in P_s(w)} \frac{\sigma_{sv}}{\sigma_{sw}} * (1 + \delta_{s\bullet}(w)) \tag{4}$$

where $P_s(w)$ is the set of predecessors of w in the DAG rooted by s.

Using the formulas showed above, the running time of the algorithm was reduced to $\mathcal{O}(nm)$ for unweighted graphs. But this reduction in the time cost is not enough for large social networks.

In [2] was proposed a new algorithm which speed up the Brandes's one by removing iteratively vertexes of degree 1. This method takes advantage of the sparsity of social networks, identifying these key nodes, for which is not necessary to compute the betweenness value for each source node. They compute the betweenness for those vertexes once in a preprocessing step and transform a bit Brandes's algorithm to keep correctness. The authors point out that nodes with degree equal to one had none value of betweenness, however their presence is influential in the betweenness value of their only neighbor. Thus, those "hanging" nodes can be removed, and this process can be repeated with the rest of the graph until there are not more nodes with degree 1. For computing the betweenness value of the removed nodes the authors used the following formula:

$$Cb(v) = Cb(v) + (n - p(v) - p(u) - 2) * (p(u) + 1) \tag{5}$$

where p is an array which store in the position $p(i)$ the number of nodes removed from the sub-tree rooted i. By means:

$$p(v) = p(v) + p(u) + 1 \tag{6}$$

where u is a node of degree 1 at the current iteration.

To keep correctness of Brandes's algorithm they modified the dependency formula as is shown below:

$$\delta_{s\bullet}(v) = \sum_{w:v \in P_s(w)} \frac{\sigma_{sv}}{\sigma_{sw}} * (1 + \delta_{s\bullet}(w) + p(w)) \tag{7}$$

Finally, the betweenness value of the remaining nodes of the graph is computed as follows:

$$Cb(w) = Cb(w) + \delta(w) * (p(s) + 1), w \neq s \tag{8}$$

3.2   Parallel and Distributed Algorithms

In [8] was presented the first parallel algorithm to compute betweenness for the massively multi-threaded Cray XMT system. To achieve this goal, the authors proposed to storage successors instead of predecessors because with that it is possible to obtain a lock free parallel algorithm. They keep the correctness employing the following equation:

$$\delta_{s\bullet}(v) = \sum_{w:w \in S_s(v)} \frac{\sigma_{sv}}{\sigma_{sw}} * (1 + \delta_{s\bullet}(w)) \tag{9}$$

where $S_s(v)$ is the set of successors of v from the source node s. Furthermore, using this relation the algorithm can achieve more asymptotically efficient compute of betweenness, as was noted on this work. Although, they observed that the algorithm is more cache-friendly as well, as the updates are applied to the vertex that is currently being processed.

Besides, in [5] was noted that successor sets yield better locality than predecessor sets for unweighted betweenness centrality. Using this idea, they proposed a distributed algorithm to compute betweenness. In the step of the shortest path traversal they used the algorithm Δ-stepping [10], allowing use their proposal for weighted and unweighed graphs as well. This is a space efficient approach incorporated at present in the Parallel Boost Graph Library. However, as the authors themselves point out on the experimental results, the algorithm is faster than the sequential version when are available 16 processors or more. Besides, they do not propose any heuristic to distributed nodes, so they need to search across processors where are store vertexes every time they send a message.

4   Our Proposal

As was mentioned before, computing betweenness involves two main steps: determining the shortest paths and computing betweenness. The first of these is the most expensive, for that reason we concentrate our efforts in accelerate this step. To achieve this goal, first we distribute the nodes of the graph using a Round Robin strategy across the available processors. Then, we determine the DAGs rooted by the subset of nodes which belong to every processor. Next, we compute the number of the shortest paths that pass through the nodes that lie in the DAGs previously determined. Finally, we compute the dependency values and the betweenness contribution of the nodes reached in the DAGs. Our algorithm is named DBB (Distributed Brandes Betweenness).

In our case, we opted for distributing the nodes of the network across the np available processors. Each node is tagged with a unique id ($0 \leq id < np$) and they will be allocated to the processor r if $id \mod np = r$. Using this strategy, each processor stores a subset of nodes of the graph and we are able to know in which processor the nodes are allocated.

When the shortest paths are computed, it is necessary to visit all the reachable nodes. Due to the vertexes are distributed in multiple processors, it is inevitable to send messages requesting the needed nodes and receive those nodes and their neighbors. This situation might imply a lot of communications and thus an increase of the time cost compared with the sequential version. To avoid that

---

Algorithm 1. Compute the shortest paths (thread 1)

Data: $G_r = (V_r, E_r)$;
V_r: set of nodes corresponding to processor r;
E_r: set of edges adjacent to each $v_r \in V_r$;
np: number of processors; n: number of nodes;
Result: S_s: Successors set

1 Divide V_r in chunks C_r of size ch;
2 **foreach** C_r **do**
3   **for** $s \in C_r$ **do**
4   **for** $i = 0$ to n **do**
5           $d[s][i] = -1$;
6        $d[s][s] = 0$;
7        $S[s].push(s)$ // S: array of stacks;
8   **for** *neighbor* v *of* s **do**
9           $d[s][v] = 1$;
10     $level = 0$;
11   **while** *True* **do**
12   **for** $i = 0$ to np **do**
13           $N_i = \{v \mid d[s][v] = level + 1, \forall s \in C_r \text{ and } v \mod np = i\}$;
14   **foreach** *neighbor* w *of* $v \in N_i$ **do**
15   **if** $d[s][w] = level$ **then**
16                 $S_s[s][w].push(v)$;
17                 $S[s].push(v)$;
18   **if** $d[s][w] = -1$ *and* $d[s][v] = level + 1$ **then**
19                 $d[s][w] = level + 2$;
20   **if** $\bigcup N_i = \emptyset$ **then**
21   **go to** line 23;
22        $level = level + 1$;
23   Calculate betweenness using Algorithm 2;

---

problem, we propose the use of two threads on each processor. The first one do the real computation, and also it demands and receives the needed nodes. The second thread attends requests, builds the array of neighbors and send the nodes in demand along its neighbors to the processor which need them. Thus, we overlap some computations and communications allowing that overall time be reduced. The work corresponding to first thread will be described in the algorithms shown below. In the Algorithm 1 is presented the computing of the shortest paths step and in Algorithm 2 the steps required for betweenness computing.

As we model the network as an unweighted and undirected graph, we use the BFS algorithm to determine the shortest paths. This algorithm performs a breadth search starting at a root node s, by means it visits the nodes by levels

---

Algorithm 2. Computing betweenness

Data: S_s: Successors set;
  S: array of stacks;
  Result: Cb: Partial betweenness value for each node of the graph G
1 **foreach** $s \in S_s$ **do**
2 | $\sigma[s] = 1$;
3 | $R_v.push(s)$ // R_v: queue of reachable nodes from s;
4 | **while** $R_v \neq \emptyset$ **do**
5 | | $v \leftarrow R_v.pop()$;
6 | | **foreach** $w \in S_s[v]$ **do**
7 | | | $\sigma[w] = \sigma[w] + \sigma[v]$;
8 | | | $R_v.push(w)$;
9 | **while** $S[s] \neq \emptyset$ **do**
10 | | $w \leftarrow S[s].pop()$;
11 | | **foreach** $v \in S_s[w]$ **do**
12 | | | $\delta[w] = \delta[w] + \sigma[w] * (1 + \delta[v])/\sigma[v]$;
13 | | **if** $v \neq s$ **then**
14 | | | $Cb[v] = Cb[v] + \delta[v]$;

---

and do not pass to next level until are visited all nodes at the current level. This process must be repeated taking each node of the graph as root. As we have the graph distributed across multiple processors, we are able to discover in parallel the DAG corresponding to a source node s on each processor. However, we notice that various s_i at the same processor might need the same nodes in the same level of BFS algorithm. For that reason, we propose to expand the DAGs corresponding to several s_i at the same time by introducing a parameter ch to indicate the number of s_i. Although we can not guarantee that scenario is going to take place, we can deduce that while greater be the value of ch, more advantage we can take of reducing communications and thus a thrift time.

Nevertheless, we need to store the DAGs corresponding to each source node and this implies keep loaded in memory a lot of information, because apart from the $S_s(v)$, we need the σ and d arrays and the stack S, which stores the nodes in non-increased order from the root, for each source node. Also, we have to keep the queue Q, which stores the nodes that are going to be visited in the BFS algorithm, for each s_i.

Due to the amount of information we need to store, it is necessary to minimize the number of structures used in our algorithms. For that reason, we opted for obtaining the number of the shortest paths σ later, because this structure is not needed to discover the shortest paths and it can be deduced from the successors array. Also, we decided to dispense with the queue Q because the information of the array d can be used to determine the nodes that will be visited at next level of the following way: The first thread only requests those nodes which are not visited yet and that are going to be reachable at next level. This is made by checking the arrays d, which are updated before a node is visited (Algorithm 1 lines 18 and 19). This strategy allows us to know exactly the vertexes that are going to be on some shortest path at next level. To send a request, the thread first searches all nodes needed from one processor (Algorithm 1 line 13), and send a package of those nodes into one message. That package is received by the second thread of the processor owner of the vertexes and sends the requested nodes and their neighbors in one package for each node. When a node and its neighbors are received, the distance arrays d are updated (Algorithm 1 line 19) and the nodes are stored in the stack S in non-increasing order from the root (Algorithm 1 line 17). And this process is repeated until there are no more vertexes to process.

Once there have been determined the DAGs, it is computed the dependency value of each node as is shown in Algorithm 2. For each source node separately, we compute the values of σ using the successor array (Algorithm 2 line 7). Then, we compute the dependency value (Algorithm 2 line 12) and finally a partial betweenness. That procedure is made in each processor in parallel.

Finally, when the contribution to betweenness of all nodes of the graph has been computed and stored in the δ arrays, we are able to compute the final score. To get the betweenness value of each node, we perform a reduction of the partial values.

4.1   Modification for Sparse Networks

As social networks often are distinguished by their sparsity, this property can be taken in advantage to speed up the betweenness computation. This was made by using the idea of the SPVB algorithm, in [2]. In our approach, named DSPVB (Distributed Shortest Paths Vertex Betweenness), we modified our previous distributed algorithm by adding a preprocessing step. First, we look up, in parallel, for all connected components, the number of vertex in each component and which of them belongs to each vertex. Then, we determine, in parallel the nodes of degree 1 for each connected component. For those nodes, we compute the betweenness value that they provide to its unique neighbor and finally those nodes are removed. With the residual graph, we repeat this step, until there are no more nodes of degree 1, following the idea of SPVB algorithm. Also, we removed the nodes of degree 0. When is finished the pre-processing step, we compute the shortest paths using the Algorithm 1 and the same strategies to determine the number of the shortest paths. Then we substitute the lines 12 and 14 of Algorithm 2 by the Eqs. 7 and 8 respectively. We expect that this algorithm offers a reduction in time and memory consumption.

Although our algorithm is designed for processing undirected and unweighted graphs, it can be extended for weighted graphs by changing the algorithm for computing the shortest paths by another one that consider the weight. Also, if the graph, which represent the social network, is directed, we should change the DSPVB algorithm taking into account the direction.

5   Experiments

For validating our proposal, we conducted several experiments with different datasets corresponding to real networks. We analyze the performance of our proposal in comparison with Brandes's algorithm. Also, we make an analysis of the rate between computing time and communications time in our method in relation to the number of processors employed.

To implement our algorithms we use the C++ programming language. To run the experiments, we employed a cluster of 13 machines, each of them with: 4GB of RAM DDR2, 2 processors DualCore Opteron 2.66 GHz.

We use four well known datasets from Stanford Large Network Dataset Collection [7]. These datasets were selected attending their features, especially the density of the network, as is shown in Table 1. The column Average indicates the average degree of each node. As can be observed, the datasets present different density of edges per nodes. The column % Hang indicates the percentage of nodes that can be removed from the graph when the modification for sparse networks is applied.

To evaluate our proposal, we use the speedup as measure. Speedup is defined as the ratio between the sequential time and the parallel time. In Table 2 we show the results, we tagged as DBB (Distributed Brandes Betweenness), our distributed algorithm based in the Brandes one; and DSPVB (Distributed the Shortest Paths Vertex Betweenness) is the modification based in the idea of the SPVB algorithm. As can be observed, the speedup of our DSPVB proposal always gets better behavior than the sequential algorithm and the speedup increases when we augment the number of processors in all cases.

In other hand, we noted that when the graph, which represent the social network, is sparse, it is better to used the DSPVB algorithm, because it can achieve a speed up over 80x, as happened with Email-EuAll dataset. It is important to highlight this result, because a special feature of social networks is their sparsity. However, a better performance depends of the network sparsity as well as the number of nodes eliminated on preprocessing step. Also, we conducted experiments in order to compare our algorithm with the one propose in [5] but we do not show those results because the Boost algorithm was executed in several hours using the Email-Enron dataset, which is the smallest one, and for each number of processors.

Table 1. Datasets from standford large network collection

Dataset	Nodes	Edges	Average	% Hang
soc-Slashdot0922	82168	948464	11.54	2.19
Email-Enron	36692	183831	10.02	31.09
soc-Epinions	75879	508837	6.71	50.84
Email-EuAll	265214	420045	1.58	86.34

Table 2. Speedup

Datas	soc-Slashdot0922		Email-Enron		soc-Epinions		Email-EuAll	
Procs	DBB	DSPVB	DBB	DSPVB	DBB	DSPVB	DBB	DSPVB
2	0.75	2.06	0.99	1.86	0.57	2.78	0.38	20.92
3	1.07	2.99	1.36	2.65	0.80	4.00	0.58	26.82
4	1.41	3.98	1.74	3.39	1.00	5.08	0.64	33.60
5	1.68	4.78	2.14	4.10	1.20	6.17	0.70	39.51
6	1.99	5.75	2.49	5.01	1.36	7.58	0.77	45.43
7	2.22	6.50	2.86	5.60	1.55	8.59	0.83	53.45
8	2.59	7.49	3.16	6.25	1.71	9.61	0.92	60.29
9	2.87	8.55	3.56	6.99	1.92	11.01	0.99	61.59
10	3.15	9.28	3.86	7.60	2.10	11.89	1.03	71.94
11	3.41	10.19	4.14	8.36	2.33	12.94	1.16	77.83
12	3.66	10.99	4.48	8.99	2.50	14.11	1.27	80.58
13	3.96	11.72	4.86	9.59	2.68	15.07	1.34	86.29

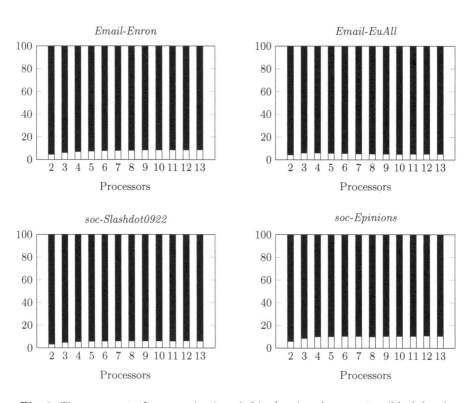

Fig. 1. Time per cent of communications (white bars) and processing (black bars)

We also study the behavior of our proposal with respect to time consuming when it is processing and when it is communicating. We set, a priori, the number of source nodes, from which we expanded the DAGs at the same time in each processor, to 50. In each chunk of source nodes, we measured both times and we computed the average per chunk and the results are shown in the Fig. 1. To establish a comparison, we only show the result of our first proposal, because the other one has a similar behavior. As we can observe for all the datasets, the processing time is bigger than the communication time, because in all cases the average time that processors wait for send o receive messages is never greater than 15 %. That proves that our strategies for reducing the number of messages is effective, also proves that we reach to overlap communications and processing by using two threads.

6 Conclusions and Future Work

In this paper we presented a distributed and parallel algorithm to compute betweenness centrality. Also, we proposed a version which takes advantage of the sparsity of social networks to remove iteratively vertexes of degree 1 and thus reduce the amount of computation. In the experimental results we showed that our proposal is more efficient than the sequential Brandes algorithm. In addition, we made achieve to overlap communication and processing, thus reducing the delay in the messages interchange.

Besides, we proposed to work in a strategy to reduce the memory consumption due to the use of the successor arrays for each source node and extend our work to process weighted and directed graphs. In addition, we believe that it is possible to reduce the processing time employing a global array for each processor to indicate, during the computing the shortest paths step, the nodes that are going to be visited at next level, instead of traverse the distance arrays.

References

1. Aggarwal, C.C. (ed.): Social Network Data Analytics. Springer, Berlin (2011)
2. Baglioni, M., Geraci, F., Pellegrini, M., Lastres, E.: Fast exact computation of betweenness centrality in social networks. In: Proceedings of the 2012 International Conference on Advances in Social Networks Analysis and Mining (ASONAM 2012), ASONAM 2012, pp. 450–456. IEEE Computer Society, Washington, D.C. (2012)
3. Brandes, U.: A faster algorithm for betweenness centrality. J. Math. Sociol. **25**, 163–177 (2001)
4. Brandes, U., Pich, C.: Centrality estimation in large networks. Int. J. Bifurcat. Chaos **17**(7), 2303–2318 (2007)
5. Edmonds, N., Hoefler, T., Lumsdaine, A.: A space-efficient parallelalgorithm for computing betweenness centrality in distributedmemory. In: HiPC, pp. 1–10. IEEE Computer Society (2010)
6. Jin, S., Huang, Z., Chen, Y., Chavarría-Miranda, D., Feo, J., Wong, P.C.: A novel application of parallel betweenness centrality to power grid contingency analysis. In: 2010 IEEE International Symposium on Parallel and Distributed Processing (IPDPS), pp. 1–7. IEEE (2010)

7. Leskovec, J., Krevl, A.: SNAP Datasets: Stanford large network dataset collection, June 2014. http://snap.stanford.edu/data
8. Madduri, K., Ediger, D., Jiang, K., Bader, D.A., Chavarra-Miranda, D.G.: A faster parallel algorithm and efficient multithreaded implementations for evaluating betweenness centrality on massive datasets. In: IPDPS, pp. 1–8. IEEE (2009)
9. McLaughlin, A., Bader, D.A.: Scalable and high performance betweenness centrality on the GPU. In: Damkroger, T., Dongarra, J. (eds.) SC, pp. 572–583. IEEE (2014)
10. Meyer, U., Sanders, P.: Δ-stepping: a parallel single source shortest path algorithm. In: Bilardi, G., Pietracaprina, A., Italiano, G.F., Pucci, G. (eds.) ESA 1998. LNCS, vol. 1461, p. 393. Springer, Heidelberg (1998)
11. Newman, M.E., Girvan, M.: Finding and evaluating community structure in networks. Phys. Rev. E **69**(2), 026113 (2004)
12. Sariyüce, A.E., Saule, E., Kaya, K., Çatalyürek, Ü.V.: Shattering and compressing networks for betweenness centrality. In: SIAM Data Mining Conference (SDM). SIAM (2013)
13. Zhao, P., Nackman, S.M., Law, C.K.: On the application of betweenness centrality in chemical network analysis: computational diagnostics and model reduction. Combust. Flame **162**(8), 2991–2998 (2015)

Principal Curves and Surfaces
to Interval Valued Variables

Jorge Arce G.[1,2(✉)] and Oldemar Rodríguez R.[1]

[1] University of Costa Rica, San José, Costa Rica
oldemar.rodriguez@ucr.ac.cr
[2] National Bank of Costa Rica, San José, Costa Rica
jarceg@bncr.fi.cr
http://www.cimpa.ucr.ac.cr/

Abstract. In this paper we propose a generalization to symbolic interval valued variables, of the Principal Curves and Surfaces method proposed by Hastie in [6]. Given a data set X with n observations and m continuous variables, the main idea of Principal Curves and Surfaces method is to generalize the principal component line, providing a smooth one-dimensional curved approximation to a set of data points in $\mathbb{R}^m$. A principal surface is more general, providing a curved manifold approximation of dimension 2 or more. In our case we are interested in finding the main principal curve that approximates better symbolic interval data variables. In [3,4], authors proposed the Centers Method and the Vertices Method to extend the well-known principal components analysis method to a particular kind of symbolic objects characterized by multi-valued variables of interval type. In this paper we generalize both, the Centers Method and the Vertices Method, finding a smooth curve that passes through the middle of the data X in an orthogonal sense. Some comparisons of the proposed method regarding the Centers and the Vertices Methods are made, this was done with the RSDA package using Ichino data set, see [1,10]. To make these comparisons we have used the correlation index.

Keywords: Interval-valued variables · Principal curves and surfaces · Symbolic data analysis

1 Symbolic Data

Symbolic data was first introduced by [12]. In the classical data analysis a variable takes a single value, in the symbolic analysis the variable may take a finite or an infinite set of values. A type of symbolic variable may take an infinite set of numerical values ranging from a low to a high value (interval).

As the Principal Component Analysis (PCA) is one of the most popular multivariate methods, it is tempting to extend the PCA analysis to symbolic data. Some methods can be found in the literature, among them the vertex

© Springer International Publishing AG 2016
M. Montes-y-Gómez et al. (Eds.): IBERAMIA 2016, LNAI 10022, pp. 297–309, 2016.
DOI: 10.1007/978-3-319-47955-2_25

method and the centers method [1,3,4]. To introduce the vertex method, let $\mathbf{X}$ be a symbolic matrix of $n \times p$, then:

$$\begin{bmatrix} \xi_{11} & \xi_{12} & \xi_{13} & \cdots & \xi_{1p} \\ \xi_{21} & \xi_{22} & \xi_{23} & \cdots & \xi_{2p} \\ \vdots & \vdots & \vdots & \ddots & \vdots \\ \xi_{n1} & \xi_{n2} & \xi_{n3} & \cdots & \xi_{np} \end{bmatrix}. \tag{1}$$

If $\mathbf{X}$ is an intervals matrix, then the matrix $\mathbf{X}$ to the Eq. (1) has the following form,

$$\begin{bmatrix} [a_{11},b_{11}] & [a_{12},b_{12}] & [a_{13},b_{13}] & \cdots & [a_{1p},b_{1p}] \\ [a_{21},b_{21}] & [a_{22},b_{22}] & [a_{23},b_{23}] & \cdots & [a_{2p},b_{2p}] \\ \vdots & \vdots & \vdots & \ddots & \vdots \\ [a_{n1},b_{n1}] & [a_{n2},b_{n2}] & [a_{n3},b_{n3}] & \cdots & [a_{np},b_{np}] \end{bmatrix}, \tag{2}$$

where $a_{ij} \leq b_{ij}$ for all $i = 1,2,\ldots,n$ y $j = 1,2,\ldots,p$.

For the principal component analysis for interval data, the matrix $\mathbf{X}$ is transformed in a vertex matrix $\mathbf{X}^v$. Then, for $i = 1,\ldots,n$,

$$\mathbf{X}_i = ([a_{i1},b_{i1}], [a_{i2},b_{i2}], [a_{i3},b_{i3}], \ldots, [a_{ip},b_{ip}]).$$

Define the vertex matrix for the observation i,

$$\mathbf{X}_i^v = \begin{bmatrix} a_{i1} & a_{i2} & \cdots & a_{ip} \\ a_{i1} & a_{i2} & \cdots & b_{ip} \\ \vdots & \vdots & \vdots & \vdots \\ b_{i1} & b_{i2} & \cdots & a_{ip} \\ b_{i1} & b_{i2} & \cdots & b_{ip} \end{bmatrix}. \tag{3}$$

Define the vertex matrix of $\mathbf{X}$ by

$$X^v = \begin{bmatrix} \begin{bmatrix} a_{11} & a_{12} & \cdots & a_{1p} \\ \vdots & \vdots & \vdots & \vdots \\ b_{11} & b_{12} & \cdots & b_{1p} \end{bmatrix} \\ \vdots \vdots \vdots \vdots \\ \begin{bmatrix} a_{i1} & a_{i2} & \cdots & a_{ip} \\ \vdots & \vdots & \vdots & \vdots \\ b_{i1} & b_{i2} & \cdots & b_{ip} \end{bmatrix} \\ \vdots \vdots \vdots \vdots \\ \begin{bmatrix} a_{n1} & a_{n2} & \cdots & a_{np} \\ \vdots & \vdots & \vdots & \vdots \\ b_{n1} & b_{n2} & \cdots & b_{np} \end{bmatrix} \end{bmatrix}. \tag{4}$$

Then, a classical principal component analysis is applied on the n observations in X^v (4). The kth principal component of X^v is:

$$Y_k^v = X^v v_k^v,$$

where v_k^v denote the kth principal component with associated k^{th} eigenvalues of variance and covariance matrix of X^v. The kth principal component for the observation i is $Y_{ik}^v = [y_{ik}^{lo}, y_{ik}^{up}]$,

$$y_{ik}^{lo} = \min_{\eta \in L_i} y_{\eta k}^v, \tag{5}$$

$$y_{ik}^{up} = \max_{\eta \in L_i} y_{\eta k}^v, \tag{6}$$

L_i is the set of rows X^v describing i, for $N_i = 2^{m_i}$,

$$L_i = \left\{ \sum_{m=1}^{i-1} N_m + 1, \sum_{m=1}^{i-1} N_m + 2, \ldots, \sum_{m=1}^{i-1} N_m + N_i \right\}. \tag{7}$$

Equations 5 and 6 can be calculated by:

$$y_{ik}^{lo} = \sum_{j \in J_c^-} (b_{ij} - \bar{X}^v{}_{(j)}) v_{kj}^v + \sum_{j \in J_c^+} (a_{ij} - \bar{X}^v{}_{(j)}) v_{kj}^v,$$

$$y_{ik}^{up} = \sum_{j \in J_c^-} (a_{ij} - \bar{X}^v{}_{(j)}) v_{kj}^v + \sum_{j \in J_c^+} (b_{ij} - \bar{X}^v{}_{(j)}) v_{kj}^v,$$

where $J_c^- = \{j | v_{kj}^v < 0\}$ and $J_c^+ = \{j | v_{kj}^v \geq 0\}, \bar{X}^v{}_{(j)}$ are the average of jth column.

For more details, see [1].

2   Self-Consistency of a Random Vector

Suppose we want to represent or approximate the distribution of a random vector X by a random vector Y. One measure of how well Y approximates X is the mean squared error $E(\|X - Y\|^2)$, for details of building the best distribution that approximates X, see [2]. Some theorems and definitions are:

Lemma 1. If $E(Y)^2 < \infty$ then $\mu = E(Y)$ exists.

Lemma 2. $E(Y)^2 < \infty$ if and only if $E(Y - c)^2 < \infty \; \forall c$.

Proof.

$$\begin{aligned}
E(Y - c)^2 &= E(Y^2 - 2Yc + c^2) \\
&= E(Y)^2 - 2E(Y)E(c) + E(c^2) \\
&= E(Y)^2 - 2cE(Y) + c^2.
\end{aligned}$$

If $E(Y)^2 < \infty, E(Y) < \infty$ and $c^2 < \infty$ then $E(Y - c)^2 < \infty$.

Lemma 3. $E(Y-c)^2$ *is either* ∞ $\forall c$ *or is minimized uniquely by* $c = \mu = E(Y)$. *In fact, when* $E(Y)^2 < \infty$,

$$E(Y - c)^2 = var(Y) + (c - \mu)^2.$$

Proof. $E(Y)^2 < \infty$ if and only if $E(Y - c)^2 < \infty$ and $E(Y)^2 < \infty$, implies that $\mu = E(Y)$ exists, then

$$(Y - c)^2 = (Y - \mu + \mu - c)^2$$
$$= (Y - \mu)^2 + 2(Y - \mu)(\mu - c) + (\mu - c)^2.$$

Implies

$$E(Y - c)^2 = var(Y) + (c - \mu)^2.$$

For 8, $E((Y - \mu)(\mu - c)) = (\mu - c)E(Y - \mu) = 0$; $E(Y - \mu)^2 = 0$ and $E(\mu - c)^2 = E(E(Y))^2 - 2cE(E(Y)) + c^2 = E(Y)^2 - 2cE(Y) + c^2 = E(Y - c)^2$ then $c = \mu = E(Y)$. We see that $E(Y - \mu)^2$ has a unique minimum in $c = \mu = E(Y)$.

Now we can solve the problem of finding the best MSPE (mean squared prediction error) predictor of Y, given a vector $\mathbf{Z}$; that is, we can find the g that minimizes $E(Y - g(\mathbf{Z}))^2$. By using the substitution theorem for conditional expectations, we have that:

$$E((Y - g(\mathbf{Z}))^2 \mid \mathbf{Z} = \mathbf{z}) = E((Y - g(\mathbf{z}))^2 \mid \mathbf{Z} = \mathbf{z}). \tag{8}$$

Let be

$$\mu(z) = E(Y \mid \mathbf{Z} = \mathbf{z}).$$

Because $g(z)$ is a constant, for 3 we have that:

$$E((Y - g(\mathbf{z}))^2 \mid \mathbf{Z} = \mathbf{z}) = E((Y - \mu(z))^2 \mid \mathbf{Z} = \mathbf{z}) + (g(z) - \mu(z))^2. \tag{9}$$

If we now take expectations of both sides and use the double expectation theorem, we can obtain the following result.

Theorem 1. *If* Z *is any random vector and* Y *any random variable, then either* $E(Y - g(Z))^2 = \infty$ *for every function* g *or*

$$E(Y - \mu(\mathbf{z}))^2 \leq E(Y - g(\mathbf{z}))^2,$$

for every g *with strict inequality holding unless* $g(\mathbf{Z}) = \mu(\mathbf{Z})$. *That is,* $\mu(\mathbf{Z})$ *is the unique best MSPE predictor. In fact, when* $E(Y^2) < \infty$ *we have*

$$E(Y - g(Z))^2 = E(Y - \mu(Z))^2 + E(g(Z) - \mu(Z))^2.$$

Definition 1. *For two jointly distributed random vectors* $\mathbf{X}$ *y* $\mathbf{Y}$, *we can say that* $\mathbf{Y}$ *is self-consistent for* $\mathbf{X}$ *if* $\mathbf{Y} = E(\mathbf{X} \mid \mathbf{Y})$ *almost surely.*

Example 1. Let be X_n a sequence of independent, mean-zero random variables, and let be $S_n = \sum_{i=1}^n X_i$. Then

$$
\begin{aligned}
E(S_{n+k} \mid S_n) &= S_n + E(X_{n+1} + \cdots + X_{n+k} \mid S_n) \\
&= S_n + E(X_{n+1} + \cdots + X_{n+k}) \\
&= S_n.
\end{aligned}
$$

3   Principal Surface

The Principal Surface (Fig. 1(4) is a nonlinear generalization to principal component analysis (Fig. 1(3), providing a smooth one-dimensional curved approximation to a set of data points in $\mathbb{R}^m$. A principal surface is more general, providing a curved manifold approximation of dimension two or more. For details of building the principal surface, see [6]. Some theorems and definitions are as follows.

We denote by $\mathbf{X}$ a random vector in $\mathbb{R}^p$ with density h and finite second moments. Without loss of generality we assume $E(\mathbf{X}) = 0$. Let f denote a smooth (C^∞) unit speed curve $(\|\mathbf{f}'(z)\| = 1)$ in $\mathbb{R}^p$ parametrized over $\Lambda \subset \mathbb{R}$, a closed (possibly infinite) interval, which does not intersect itself $(\lambda_1 \neq \lambda_2 \Rightarrow \mathbf{f}(\lambda_1) \neq \mathbf{f}(\lambda_2))$, and has finite length inside any finite ball in $\mathbb{R}^p$.

Let be

$$
D(\mathbf{x}) = \inf_{\lambda \in \Lambda} d(\mathbf{x}, \mathbf{f}(\lambda)), \forall \mathbf{x} \in \mathbb{R}^p, \tag{10}
$$

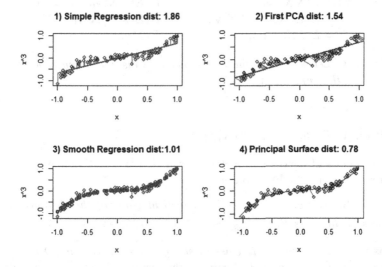

Fig. 1. Distinct methods to aproximate data: (1) The linear regression line minimizes the sum of squared deviations in the dependent variable. (2) The principal components line minimizes the sum of squared deviations in all variables. (3) The smooth regression curve minimizes the sum of squared deviations in the dependent variable, subject to smoothness constraints. (4) The principal surfaces minimizes the sum of squared deviations in all variables, subject to smoothness constraints.

where

$$d(\mathbf{x}, \mathbf{f}(\lambda)) = \|\mathbf{x} - \mathbf{f}(\lambda)\|, \tag{11}$$

the usual euclidean distance between two vectors. Let be

$$M(\mathbf{x}) = \{\lambda \mid \mathrm{d}(\mathbf{x}, \mathbf{f}(\lambda)) = D(\mathbf{x})\}. \tag{12}$$

Since Λ is compact, $M(\mathbf{x})$ is not empty, $\mathrm{d}(\mathbf{x}, \mathbf{f}(\lambda))$ is continuous, $\mathbf{f}$ is continuous, $M^c(\mathbf{x})$ is open, and hence, $M(\mathbf{x})$ is closed.

We define

$$\lambda_{\mathbf{f}} : \mathbb{R}^p \to \mathbb{R}, \text{ with } \lambda_{\mathbf{f}} = \sup M(\mathbf{x}). \tag{13}$$

Definition 2. *The curve $\mathbf{f}$ is called self-consistent or a principal curve of h if $E(\mathbf{X} \mid \lambda_f = \lambda) = \mathbf{f}(\lambda)$ almost surely.*

Theorem 2. *If a straight line $l(\lambda) = \mathbf{u}_0 + \lambda \mathbf{v}_0$ is self consistent and $\mathbf{u}_0 \perp \mathbf{v}_0$, then it is a principal component.*

Proof.

$$\begin{aligned}
0 = E(\mathbf{X}) &= E_\lambda E(\mathbf{X} \mid \lambda_{\mathbf{f}} = \lambda) \\
&= E_\lambda(\mathbf{u}_0 + \lambda \mathbf{v}_0) = \mathbf{u}_0 + \bar{\lambda}\mathbf{v}_0.
\end{aligned}$$

Then $\mathbf{u}_0 = 0$

$$\begin{aligned}
\Sigma \mathbf{v}_0 = E(\mathbf{X}\mathbf{X}^t)\mathbf{v}_0 &= E_\lambda E(\mathbf{X}\mathbf{X}^t \mathbf{v}_0 \mid \lambda_{\mathbf{f}} = \lambda) \\
&= E_\lambda E(\mathbf{X}\mathbf{X}^t \mathbf{v}_0 \mid \mathbf{X}^t \mathbf{v}_0 = \lambda) = E_\lambda E(\lambda \mathbf{X} \mid \mathbf{X}^t \mathbf{v}_0 = \lambda) \\
&= E_\lambda \lambda E(\mathbf{X} \mid \lambda_{\mathbf{f}} = \lambda) \qquad = E_\lambda \lambda^2 \mathbf{v}_0.
\end{aligned}$$

Definition 3. *Let be $d(x, f)$ the usual euclidian distance from a point x to its projection on the curve f $\|x - f(\lambda_f)\|$.*

and we define

Definition 4. $D^2(h, f) = E_h d^2(X, f)$.

Let be $\mathscr{G}$ a class of curves parametrized over Λ. For $g \in \mathscr{G}$ defines $f_t = f + tg$. This creates a perturbed version of f.

Definition 5. *The curve f is called a critical point of the distance function variations in the class $\mathscr{G}$ iff $\frac{dD^2(h, f_t)}{dt} = 0$ if $t = 0, \forall g \in \mathscr{G}$.*

Proposition 1. *Let be $\mathscr{G}_l$ the class of straight lines $g(\lambda) = a + \lambda b$. A straight line $l_0(\lambda) = a_0 + \lambda b_0$ is a critical point of the distance function for variations in $\mathscr{G}_l$ iff b_0 is an eigenvector of $Cov(X)$ and $a_0 = 0$.*

The following algorithm allows to compute the principal curve for classical data.

The principal curve algorithm

Algorithm 1. Principal Surface

Require: X is a matrix of $p{\times}n$, TOL is a tolerance of variations and N
is the maximum number of iterations
Ensure: f principal curve of X
1: $f^{(0)}(\lambda) = v\lambda$ where v is the the first linear principal component of h.
Set $\lambda_0(x) = \lambda_{f^{(0)}}(x)$
2: **while** $|D^2(h, f^{(j)}) - D^2(h, f^{(j-1)})| > TOL$ **AND** $j < N$ **do**
3:    Set $f^{(j)}(.) = E(X \mid \lambda_{j-1}(X) = .)$
4:    $\lambda_j(x) = \lambda_{f^{(j)}}(x)$
5:    $D^2(h, f^{(j)}) = E_{\lambda(j)} E(\|X - f(\lambda_j(X))\|^2 \mid \lambda_j(X))$
6:    $j = j + 1$, $f = f^{(j)}$
7: **end while**
8: **return** f

---

4   The Principal Curves and Surfaces to Interval Valued Variables

Let be I a hyper-rectangle in $\mathbb{R}^n$ and let f nonnegative function $f : \mathbb{R}^n \rightarrow \mathbb{R}^+ \cup \{0\}$ with $\int_{\mathbb{R}^n} f(x)dx = 1$ and μ is a measure.

Definition 6. *If* $p(I) = \int_I f d\mu$, p *is a probability measure.*

Let be $X = (I_1, I_2, ...)$ a random vector of intervals in $\mathbb{R}^n$.

Definition 7.

- *The expectation of a random vector of intervals is the interval* $E(X) = \sum_{i=1}^{\infty} I_i p(I_i)$.
- *The conditional probability of two intervals* A, B *is* $p(A \mid B) = \frac{p(A \cap B)}{p(B)}, p(B) > 0$.
- *The conditional distribution of a random vector of interval* Y *given* $Z = z$ *corresponds to a single probability measure, with* $p(z) > 0$, $p_z = p(Y \mid Z = z)$ *if* $p_z > 0$.

Definition 8. *For two random interval vectors* X *and* Y, *we say that* Y *is self-consistent for* X *if* $Y = E(X \mid Y)$.

Let be $I = [a, b]$ a non-trival and finite interval in $\mathbb{R}$, let $f_I(x) = \begin{cases} 0 & x \notin I \\ \frac{1}{b-a} & x \in I. \end{cases}$

For example, let be $X = \{X_1, X_2\}$ and $Y = \{Y_1\}$, with $X_1 = (0, 10]$, $X_2 = (10, 20]$, $Y_1 = (5, 15]$ y $I = [-10, 10]$

$$
\begin{aligned}
E_I(X \mid Y) &= E_I(X_1 \mid Y_1) + E_I(X_2 \mid Y_1) \\
p_I(Y_1) &= \frac{\mu(Y_1)}{\mu(I)} &&= \frac{10}{20} = \frac{1}{2} \\
p_I(X_1 \cap Y_1) &= \frac{\mu((5, 10])}{\mu(I)} &&= \frac{5}{20} = \frac{1}{4} \\
p_I(X_2 \cap Y_1) &= \frac{\mu((10, 15])}{\mu(I)} &&= \frac{5}{20} = \frac{1}{4} \\
p_I(X_1 \mid Y_1) &= \frac{p_I(X_1 \cap Y_1)}{p_I(Y_1)} &&= \frac{1}{2} \\
p_I(X_2 \mid Y_1) &= \frac{p_I(X_2 \cap Y_1)}{p_I(Y_1)} &&= \frac{1}{2} \\
E_I(X \mid Y) &= \tfrac{1}{2}X_1 + \tfrac{1}{2}X_2 = Y_1.
\end{aligned}
$$

Then Y is self-consistent for X.

The principal curves and surfaces to Interval Valued Variables algorithm

The following algorithm allows to compute the principal curve for interval values data tables.

Algorithm 2. Principal Surface to Interval Valued Variables

Require: X a symbolic matrix of interval data $n \times p$, TOL is a tolerance of variations and N is the maximum number of iterations

Ensure: $\hat{f}$ a matrix of $n \times p$ principal curves of X

1: X^v the vertex matrix of X

2: h is the continuous probability density of X^v.

3: $f^{(0)}(\lambda) = v\lambda$ where v is the the first linear principal component of X^v. Set $\lambda_0(x) = \lambda_{f^{(0)}}(x)$

4: **while** $|D^2(h, f^{(j)}) - D^2(h, f^{(j-1)})| >$ TOL AND $j <$ N **do**

5:     Set $f^{(j)}(.) = E(X \mid \lambda_{j-1}(X) = .)$

6:     $\lambda_j(x) = \lambda_{f^{(j)}}(x)$

7:     $D^2(h, f^{(j)}) = E_{\lambda^{(j)}} E(\|X - f(\lambda_j(X))\|^2 \mid \lambda_j(X))$

8:     $j = j + 1$, $f = f^{(j)}$

9: **end while**

10: $\hat{f}_{iv} = [\min_{k_i \in \Delta_i} f_{vik_i}, \max_{k_i \in \Delta_i} f_{vik_i}]$ **for** $i = 1, \dots, n$ **and** $v = 1, \dots, p$

11: **return** $\hat{f}$

---

5   Experimental Evaluation

All examples presented in this section were processed using the RSDA (R to Symbolic Data Analysis) package developed by this work's authors for applications of the Symbolic Data Analysis, it can be consulted in [10]. To make these comparisons we have used the correlation index.

Example Ichino oils Data. The symbolic surfaces method is illustrated on the oils data of [5]. The variables in the classic data set are (Table 1):

To construct the symbolic data table, use the definition given in the Table 2

In the symbolic data Table 3, each row in this table represents a class of oil described by four variables of quantitative interval.

Step 1: Compute X^v the vertex matrix of X, for example the vertex matrix of the linseed oil (L) is Table 4:

In this case X^v has 128 rows (vertex).

Step 2: Calculate the principal surface to X^v.

For example, to individual linseed (L) in Table 5,

Table 1. Columns Oils data

Variable	Type
Oil	Qualitative
Specific gravity	Quantitive
Freezing point	Quantitive
Iodine value	Quantitive
Saponification	Quantitive

Table 2. Columns oils symbolic data set

Variable	Symbolic type	values
Oil	Concept	linseed (L), perilla (P), cotton (Co), sesame (S),
Oil	Concept	camillia (Ca), olive (O), beef (B), hog (H)
Specific gravity (GRA)	Interval	$[a_{i1}, b_{i1}]$
Freezing point (FRE)	Interval	$[a_{i2}, b_{i2}]$
Iodine value (IOD)	Interval	$[a_{i3}, b_{i3}]$
Saponification (SAP)	Interval	$[a_{i4}, b_{i4}]$

Table 3. Oils symbolic data

	GRA	FRE	IOD	SAP
L	[0.93, 0.94]	[−27, −18]	[170, 204]	[118, 196]
P	[0.9, 0.94]	[−5, −4]	[192, 208]	[188, 197]
Co	[0.9, 0.92]	[−6, −1]	[99, 113]	[189, 198]
S	[0.9, 0.93]	[−6, −4]	[104, 116]	[187, 193]
Ca	[0.9, 0.92]	[−25, −15]	[80, 82]	[189, 193]
O	[0.9, 0.92]	[0, 6]	[79, 90]	[187, 196]
B	[0.9, 0.87]	[30, 38]	[40, 48]	[190, 199]
H	[0.9, 0.86]	[22, 32]	[53, 77]	[190, 202]

Table 4. Vertex matrix to linseed

GRA	FRE	IOD	SAP
0.93	−27	170	118
0.94	−27	170	118
0.93	−18	170	118
0.94	−18	170	118
0.93	−27	204	118
0.94	−27	204	118
0.93	−18	204	118
0.94	−18	204	118
0.93	−27	170	196
0.94	−27	170	196
0.93	−18	170	196
0.94	−18	170	196
0.93	−27	204	196
0.94	−27	204	196
0.93	−18	204	196
0.94	−18	204	196

Table 5. Principal surfaces to vertex matrix of linseed (L)

suf_1	suf_2	suf_3	suf_4	lambda
80.55	**−22.78**	**0.02**	**−67.87**	**249.37**
80.55	−22.78	0.02	−67.87	249.37
80.55	−22.78	0.02	−67.87	249.37
80.55	−22.78	0.02	−67.87	249.37
80.65	−22.73	0.02	−67.61	249.09
80.65	−22.73	0.02	−67.61	249.09
80.65	−22.73	0.02	−67.61	249.09
80.65	−22.73	0.02	−67.61	249.09
60.90	−12.41	0.02	9.04	147.77
60.90	−12.41	0.02	**9.04**	**147.77**
60.51	−12.39	0.02	9.00	147.38
60.51	−12.39	0.02	9.00	147.38
89.05	−11.20	0.03	2.09	177.21
89.05	−11.20	**0.03**	2.09	177.21
89.05	**−11.20**	0.03	2.09	177.21
89.05	−11.20	0.03	2.09	177.21
89.05	−11.20	0.03	2.09	177.21

Table 6. Principal Surfaces: Oils data linseed (L)

	suf_1	suf_2	suf_3	suf_4	lambda
L	[60.5, 89.05]	[−22.78, −11.2]	[−67.87, 9.04]	[0.02, 0.03]	[147.38, 249.37]

Table 7. Principal Surfaces: Oils data

	GRA	FRE	IOD	SAP	lambda
B	[−69.2, −60.27]	[30.38, 37.69]	[6.76, 7.16]	[−0.05, −0.04]	[0, 11.53]
Ca	[−22.7, −20.15]	[−6.05, −4.74]	[3.43, 3.51]	[0, 0.01]	[63.16, 66.06]
Co	[−11.8, 3.61]	[−9.8, −8.34]	[3.46, 4.33]	[0.01, 0.02]	[74.71, 90.23]
H	[−58.6, −41.03]	[12.63, 29.02]	[5.29, 6.68]	[−0.04, −0.02]	[13.65, 37.76]
L	[60.5, 89.05]	[−22.78, −11.2]	[−67.87, 9.04]	[0.02, 0.03]	[147.38, 249.37]
O	[−32.7, −21.72]	[−5.31, 3.5]	[3.47, 4.3]	[−0.01, 0]	[50.18, 64.33]
P	[82.1, 90.39]	[−11.48, −11.22]	[0.67, 5.85]	[0.03, 0.03]	[169.18, 179.16]
S	[−6.3, 6.2]	[−9.97, −9.05]	[3.66, 4.56]	[0.01, 0.02]	[80.25, 92.83]

Step 3: Create symbolic principal surface.

In each case, symbolic principal surface takes the maximum and minimum for each surface. For example, the numbers in bold on step 2 mean the maximum for each surface. Then the symbolic principal surface to linseed (L) is Table 6

Step 4: Repeat **Step 3:** for each individuals

The principal surface for Ichino's oils data (Table 7) are the following:

Correlation

Figure 2 shows:

- In the SAP variable, correlation with the third principal surface is better than the second principal component.
- The FRE variable is highly correlated with principal surface 2 and 4.
- The GRA variable is highly correlated with principal surface 1, 2 and 4.

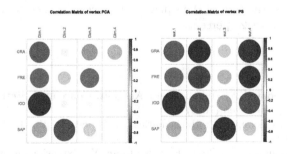

Fig. 2. Vertex correlation matrix: Oils data

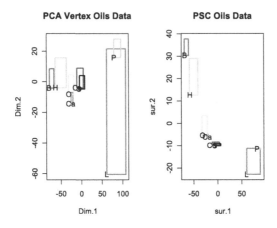

Fig. 3. PS vs PCA: Oils data

Figure 3 shows:

– The principal curves preserve the cluster generated by the PCA.
– The segmentation by individuals has fewer intersections.
– The sum of the squared distances in the method of principal surfaces is 16177.82 while the PCA is 452651.1.

6   Conclusion

The difference between principal components and principal surfaces is the smoothness constraints, some conclusions are:

– The sum of the squared distances in the principal surfaces method is less than principal component.
– Correlations between the original data and the principal surfaces are better than the correlations obtained between the principal components and the original data.
– The segmentation by individuals has fewer intersections.
– In all data sets, the sum of the squared distances obtained by PCA has been reduced by at least 20 percent.

References

1. Billard, L., Diday, E.: Symbolic Data Analysis: Conceptual Statistics and Data Mining. Wiley, Hoboken (2006)
2. Bickel, P.J., Doksum, K.A.: Mathematical Statistics. Prentice Hall, Upper Saddle River (1977)
3. Cazes, P., Chouakria, A., Diday, E., Schektman, Y.: Extension de l'analyse en composantes principales á des données de type intervalle. Rev. Statistique Appliquée **XLV**(3), 5–24 (1997)

4. Douzal-Chouakria, A., Billard, L., Diday, E.: Principal component analysis for interval-valued observations. Stat. Anal. Data Min. **4**(2), 229–246 (2011)
5. Ichino, M.: General metrics for mixed features - the Cartesian space theory for pattern recognition. In: Conference on Systems, Man, and Cybernetics, pp. 494–497. Pergamon, Oxford (1988)
6. Hastie, T.: Principal curves and surface. Ph.D. thesis Stanford University (1984)
7. Hastie, T., Weingessel, A.: Princurve - fits a principal curve in arbitrary dimension (2014). R package version 1.1-12 http://cran.r-project.org/web/packages/princurve/index.html
8. Hastie, T., Stuetzle, W.: Principal curves. J. Am. Stat. Assoc. **84**(406), 502–516 (1989)
9. Hastie, T., Tibshirani, R., Friedman, J.: The Elements of Statistical Learning; Data Mining, Inference and Prediction. Springer, New York (2008)
10. Rodríguez, O. with contributions from Olger Calderon, Roberto Zuñiga and Jorge Arce. RSDA - R to Symbolic Data Analysis (2015). R package version 1.3 http://CRAN.R-project.org/package=RSDA
11. Rodríguez, O.: Classification et Modèles Linéaires en Analyse des Données Symboliques. Ph.D. thesis, Paris IX-Dauphine University (2000)
12. Diday, E.: Introduction a L'approache Symbolique en Analyse des Données. Premieres Journées Symbolic-Numérique, CEREMADE, Université Paris, pp. 21–56 (1987)

In Defense of Online Kmeans for Prototype Generation and Instance Reduction

Mauricio García-Limón, Hugo Jair Escalante[(✉)], and Alicia Morales-Reyes

Instituto Nacional de Astrofísica, Óptica y Electrónica, Puebla, Mexico
{mauricio.garcia,hugojair,a.morales}@inaoep.mx

Abstract. The nearest neighbor rule is one of the most popular algorithms for data mining tasks due in part to its simplicity and theoretical/empirical properties. However, with the availability of large volumes of data, this algorithm suffers from two problems: the computational cost to classify a new example, and the need to store the whole training set. To alleviate these problems instance reduction algorithms are often used to obtain a new condensed training set that in addition to reducing the computational burden, in some cases they improve the classification performance. Many instance reduction algorithms have been proposed so far, obtaining outstanding performance in mid size data sets. However, the application of the most competitive instance reduction algorithms becomes prohibitive when dealing with massive data volumes. For this reason, in recent years, it has become crucial the development of large scale instance reduction algorithms. This paper elaborates on the usage of a classic algorithm for clustering: K-means for tackling the instance reduction problem in big data. We show that this traditional algorithm outperforms most state of the art instance reduction methods in mid size data sets. In addition, this algorithm can cope well with massive data sets and still obtain quite competitive performance. Therefore, the main contribution of this work is showing the validity of this often depreciated algorithm for a quite relevant task in a quite relevant scenario.

1 Introduction

Nowadays the availability of massive repositories of information is ubiquitous across many domains and applications. This type of repositories are known as big data when computational resources are not enough to store or process them. The existence of these type of data sets has challenged the fields of pattern recognition and machine learning, whose communities have had to revolutionize the way they process information in order to cope with it. These volumes of information are particularly challenging for the nearest-neighbor rule (NN), one of the most effective and popular classifiers in data mining [28]. In NN the classification of a new instance is done by assigning it the label of the most similar (usually, that with the closest Euclidean distance) instance in the training set. This form of classification requires storing the entire set of instances (as opposed to learning a model) and estimating the distance with each stored instance every time a

© Springer International Publishing AG 2016
M. Montes-y-Gómez et al. (Eds.): IBERAMIA 2016, LNAI 10022, pp. 310–322, 2016.
DOI: 10.1007/978-3-319-47955-2_26

new observation has to be classified. Clearly, this procedure becomes prohibitive when facing big data sets [25].

Instance reduction (IR) algorithms comprise methods that condense, clean, or synthesize a data set with the goal of reducing the computational/storage cost for NN and, in some cases, they even enhance the classification performance of the classifier [9,24]. There are two approaches for IR: **instance selection**: choose a subset of the training set. **prototype generation**: generate a new synthetic set of instances. Two important aspects are considered when evaluating an IR method: its generalization performance (e.g., ratio of correctly classified instances by NN when using the reduced set), and its reduction rate (w.r.t the original data set), there are a vast number of methods available out there for IR, see e.g. [9,24]. However, whereas existing methods have obtained satisfactory results in both criteria, almost all of these methods have been only evaluated in mid size data sets with few thousands of data at the most. Therefore, its application in massive data sets is not viable for most existing methods [23,24].

For the previous reason methods for IR in massive data sets are receiving increasingly more attention [3,25]. In this regard, there are two main schemes for facing large scale IR, those based on batch and online learning. In the batch fashion (e.g., methods like CNN [6], DROP3 [26], etc.[1]), the training set is processed in its entirely for several times, clearly, this way of processing becomes prohibitive when they are dealing with a massive datasets (for which a even single pass through the data set may be computing demanding). In addition, these methods can provide a solution only after the IR algorithm finishes. On the other hand, online-learning based IR methods (e.g., Online LVQ [8]) model the training set as a data stream, in which a single instance is processed at once [16]; where in each iteration the reduced set of instances is updated and the training set is processed a single time. Compared to batch-based methods, online techniques are advantageous because they do not need to store the training set as a whole and they can provide a reduced data set at any time during the learning process. Despite the existence of very efficient IR methods based on online learning for facing massive data sets, most of them are still limited in that they are not able to provide satisfactory results in terms of both, generalization and reduction performance.

This paper elaborates on the usage of a traditional clustering method, K–means, for tacking the IR problem for NN in massive data sets. K–means is a widely know clustering method, and even it has been used as initialization technique by some IR methods. Nevertheless, its performance has not been benchmarked with state of the art IR methods, not even in small/mid size data sets. We show, that this algorithm performs as well as the state of the art methods in IR (comprising both instance selection and prototype generation) in small and mid size data sets. What is more, we show that the online version of this algorithm outperforms online based methods designated for IR in massive data sets and obtains competitive performance with batch techniques that are far more computationally expensive.

---

[1] In [9,24] a complete taxonomy of instance selection and prototype generation algorithms are presented, respectively.

The contributions of this work are twofold: (1) we show that K–means, an old known algorithm is as effective as more elaborate and sophisticated techniques from the state of the art on IR (to the best of our knowledge no previous study has been reported elsewhere); (2) we show that the online version of K–means can be used for IR in the context of NN for massive data sets and that it can obtain better performance than existing online techniques, at the same time, we show the performance of K–means is even competitive with batch methods. Therefore, this paper can be considered as a position paper, supported by empirical evidence, in defense of K–means for IR in the context of NN. Our work will pave the way for the development of IR methods for big data based in this classical algorithm.

2   Related Work

This section reviews related work on IR for NN. We start reviewing those methods based on batch learning (requiring the repeating processing of the whole data set) and then those based on online learning (that aim at processing instances sequentially and do not need to store the whole training set). For comprehensive surveys on IR (yet not related to big data) we refer the reader to [9,24].

2.1   Batch Learning Based Methods for IR

There is a wide range of IR methods that work under the batch learning scheme, in fact, this type of methods include the vast majority of existing methods for IR (including classical and recent techniques). In general terms these methods have been classified into the following families:

- **Condensation**. Methods that aim to determine a subset of instances that classify the training set correctly [2,12].
- **Editing**. Methods that edit or delete the noisy instances according to the consistency of its closest neighbors, commonly they have been used to eliminate the overlap between classes [27].
- **Graph-based**. Methods that represent instances as vertexes of a graph, whose edges meet any criteria of proximity (e.g. RNG, Gabriel, NN, or Delaunay triangulation) [21,23].
- **Evolutionary**. Methods where a criterion is optimized, commonly classifier accuracy, with heuristic optimization techniques [10,15].
- **Centroid or clustering based**. Methods that synthesize instances using the centroid (arithmetic mean) obtained with, e.g., clustering techniques [18,22].
- **Positioning adjustment**. Methods which select a subset of the initial set, and by an optimization process adjust the position of the instances [14,17].
- **Space partitioning**. Methods that divide the space into regions where a single prototype representing each region [20].

Although these methods have reported impressive improvements in NN, most of them cannot be applied in large scale data sets (in fact, these methods have not been evaluated in data sets with more than a few dozens of instances).

On the other hand, when dealing with massive datasets, batch-based methods have been implemented under the divide and conquer approach, (e.g. parallel computation or stratification process) [5,11,25]. Generally, these methods divide the original set into smaller subsets so that they can be processed individually, using any of the methods reviewed above. Although results in large scale data set have been reported with these methods, they suffer from several limitations, including: The whole training set has to be split, and each batch has to be processed many times (i.e., expensive processes in small data sets); there is no rule of thumb for the partition of these subsets; reduced instances can cover overlapping regions of the input space (as data sets are processed separately); and they require of high performance computing infrastructure.

2.2   Incremental/Online Learning

Incremental methods are characterized by being methods that *incrementally* (either per instance or mini-batch level) process the training set, and this process is often done several times in order to obtain acceptable performance. On the other hand, online methods are those that also process instances incrementally but the distinctive idea is to process each instance a single time, thus giving a single pass to the data set and determining the reduced instances set on the fly. There are methods that are incremental but they can be adapted in an online mode. One of the most representative incremental IR algorithms is LVQ3 [14]. In this algorithm each instance is used to update the position of the prototypes; approaching the prototypes of the same class, and keeping away the prototypes of opposite classes. A problem with this algorithm is that it has to iterate several times over the data set in order to obtain satisfactory results, plus it is sensitive to the initialization. Perhaps the first online algorithm for IR is IB2 [1]. In this method each instance is used only once, at each iteration an instance is added to the reduced set of instances if the nearest prototype belongs to opposite class. One problem of this algorithm is that the order of instances has a great impact on its performance, so it may vary greatly its instance reduction rates. There are algorithms based on clustering as the DRHC algorithm [19]. This online algorithm updates the reduced set after of a window of the most recent instances is filled. One drawback is to define the ideal size of the window at the beginning of the algorithm. Finally, in the literature have been proposed online versions of LVQ such as LVQ-OIL [7]. However, although these methods are an alternative to process massive data sets efficiently, their classification performance is poor. In this paper we benchmark the performance of online K−means against existing methods for IR for massive data sets, we show that a simple and well known algorithm outperforms most techniques reviewed in this section, motivating more research on the development of K−means based methods for IR.

3   Online K-Means for Instance Reduction

The K-means algorithm is one of the best known algorithms for clustering, it tries to find the centroids, in a unsupervised manner, that minimize the distance of the original instance set with respect to the centroids. The standard, batch algorithm for finding K clusters under K–means can be found in any standard text book on machine learning or pattern recognition (e.g., [13]). The loss function of the K–means algorithm is as follows:

$$Q = min_k \frac{1}{2}(x - c_k)^2 \tag{1}$$

where x is an instance of data set, and c_k represents the closest centroid to it. It should be noted that K–Means algorithm works in a batch fashion, i.e. processing the data set more than once. As the data set size increases this process becomes prohibitive. In [4] stochastic update rules were presented for K–Means algorithm, enabling K–means to process the data as a data stream, in an online fashion, this allows to process as an unique way the complete data set and making its application in massive sets possible, these rules are:

(1) $k^* = argmin(x_t - c_k)^2$
(2) $n_{k^*} = n_{k^*} + 1$
(3) $w_{k^*} = w_{k^*} + \frac{1}{n_{k^*}}(x_t - c_{k^*})$

From these rules an online algorithm for IR is derived (SGD K-Means). In general terms, the position of the nearest centroid is updated, this new position converges to a local minimum while the number of synthesized samples is greater [4]. In the equations, k^* is the index of the closest centroid of the sample x in the time t of the datastream, n_{k^*} is the learning rate of k^*, and w_{k^*} represents the closest centroid of the sample x_t. Algorithm 1 presents the online K–means algorithm for IR in NN. In this algorithm centroids play the role of prototypes, so we look for a subset of class-dependent centroids to be used as the reduced set of instances. Unlike the original version of K-Means, this algorithm works as a supervised manner, i.e., it uses the information of the labels or classes. This algorithm preserves the original distribution of instances per class with respect to the original set, so the ratio of new instances of each class is calculated (lines 1–3). Then, the instances in the data set are processed iteratively one by one; at each iteration, if the limit of instances per class has not been reached, then that instance is considered as a new centroid, and the learning rate of this centroid is initialized (lines 7–12). If the x_i instance exceeds the limit of instances per class is then used to update the position of closest centroid (lines 14–16).

This algorithm is a slight modification of the online K–means algorithm, we will see in Sect. 4 that this simple algorithm is able to obtain quite competitive results when compared with state of the art IR techniques designed for both mid size and massive data sets.

In terms of computational complexity, it can be noticed that the algorithm has a single cycle, where each instance is processed one by one; internally, for each instance, a search for the nearest prototype that corresponds to the same class has a complexity of O (C * P * log (P)) (A KD Tree complexity is O(M * log (M)) where M is the number of elements in the structure). It can be concluded that the complexity of this algorithm is O (N*C*P*log(P)), where N represents

Algorithm 1. Online K-Means for Instance reduction, SGD K-Means

---

Require: Instance retention rate P,
    number of classes C,
    Training set $\mathcal{X}$,
    Label set $\mathcal{Y}$.

Ensure: Centroids $\mathcal{C}$
    Labels Centroids $\mathcal{L}$.

    // Centroids initialization
1: **for** $i = 1$ to C **do**
2:     $Total\_Class[i] = length(T) * P$ where $T = |\mathcal{Y} == i|$;
3: **end for**

4: $N = length(\mathcal{X})$;

    // One pass to training set
5: $k = 0$ // Number of centroids
6: **for** $i = 1$ to N **do**
7:     **if** $Total\_Class[\mathcal{Y}[i]] > 0$ **then**
8:       $\mathcal{C}[k] = \mathcal{X}[i]$
9:       $\mathcal{L}[k] = \mathcal{Y}[i]$
10:      $Total\_Class[\mathcal{Y}[i]] = Total\_Class[\mathcal{Y}[i]]\text{-}1$
11:      $m[k] = 0$
12:      $k = k + 1$
13:     **else**
14:       $j^* = \arg\min_j(\mathcal{X}_i - \mathcal{C}_j)^2$ where $\mathcal{L}_j == \mathcal{Y}_i$ // Closest centroid of the same class
15:      $m[j^*] = m[j^*] + 1$ // Learning rate.
16:      $\mathcal{C}[j^*] = \mathcal{C}[j^*] + \dfrac{1}{m[j^*]}(\mathcal{X}_i - \mathcal{C}[j^*])$) // Gradient step.
17:     **end if**
18: **end for**
19: **return** $\mathcal{C}$

---

the total size of the training set, C the number of classes, and P represents the instance reduction rate.

4 Experimentation

Table 1 describes the main characteristics of the data sets considered for experimentation. These data sets have been used in previous studies on IR for NN, so using these data sets will allow us to compare K–means with state of the art techniques [9,24]. The experimental study for benchmarking K–means for IR in NN is divided into three parts: first we evaluate the performance of the algorithm in medium size data sets with at most 20,000 instances; in this experiment we want to compare the performance of online K–means for IR to state of the art techniques in prototype selection/generation algorithms. Also, we compare in that section the performance of batch vs. online versions of K–means. Next, the method was evaluated in data sets with a larger number of instances/features. Due to time constraints, online K-means was compared with standard supervised version of K-means and 1NN. Finally, to assess performance in massive sets, the comparison was made with respect to 1NN and reported in [25]. The evaluation methodology that was adopted for the two first data sets was 10 fold cross validation, as performed in previous works, whereas for experiments with massive data sets we used 5 fold cross validation.

Table 1. Description of data sets

Dataset	Instances	Features	Classes	Dataset	Instances	Features	Classes
Medium data set							
Abalone	4,172	8	28	Ring	7,400	20	2
Banana	5,298	2	2	Satimage	6,435	36	6
Chess	3,196	36	2	Segment	2,310	19	7
Coil2000	9,822	85	2	Spambase	4,597	57	2
Magic	19,020	10	2	Splice	3,190	60	3
Marketing	8,993	13	9	Texture	5,500	40	11
Nursery	12,960	8	5	Thyroid	7,200	21	3
Page-Blocks	5,472	10	5	Titanic	2,201	3	2
Penbased	10,992	10	10	Twonorm	7,400	20	2
Phoneme	5,404	5	2				
Large data set							
Ads	3,279	1,558	2	Madelon	2,600	500	2
Adult	48,842	105	2	Mushroom	8,124	117	2
Basehock	1,993	4,862	2	Musk	6,598	166	2
Dexter	600	20,000	2	Pcmac	1,943	3,289	2
Gina	3,468	970	2	Pendigits	10,992	16	10
Gisette	7,000	5,000	2	Shuttle	58,000	9	7
Hiva	4,229	1,617	2				
Massive data set							
SUSY	5,000,000	18	2	Poker	1,250,010	10	10
KDD-CUP 99	4,856,151	122	2				

4.1 Medium Datasets

First, the batch version of K–Means and the proposed algorithm were evaluated
to measure the accuracy and the computation time in the medium data set. The
goal of this experiment is to determine the detriment in performance that the
online version has when compared with the batch variant. In this experiment
the instance retention rate was varied to 1 %, 5 % and 10 %. Table 2 shows the
accuracy and computation time (in seconds) for each IR method. In terms of
accuracy, the online version outperforms the results obtained by the batch version. On the other hand, the time required by the online version is between 2 %
and 5 % of the time required by the batch version. These results show that there
is no loss (instead we observed a gain) in performance when using the online
version instead of the batch version of K–means, this is very important, because
as we will see below, the online version can cope with massive data sets.

The next experiment consists of comparing the proposed algorithm with other
Incremental/Online IR algorithms, for this experiment we considered the reference methods: IB2, oiGRLVQ, GRLVQ, LVQ-OIL, LVQ3 (with only one iteration

Table 2. Online K-means against Batch K-means

SGD K-Means (1%)		K-Means (1%)		SGD K-Means (5%)		K-Means (5%)		SGD K-Means (10%)		K-Means (10%)	
Average	Std	Average	Std	Average	Std	Average	Std	Average	Std	Average	Std
Accuracy											
0.7963	0.2051	0.7912	0.2143	0.8150	0.2116	0.7952	0.2417	0.8142	0.2160	0.7944	0.2489
Time (seconds)											
21.7170	13.0795	733.2345	845.2200	19.9954	15.7006	531.3199	603.5209	20.6744	17.4235	393.9765	335.9072

over the training set), which to the best of our knowledge comprise the most representative IR methods based on online/incremental learning. Table 3 shows the accuracy, instance reduction and execution time (in seconds) obtained by these algorithms and compared to the online version of K–means; the instance reduction rate used to compare the SGD K-Means with the other algorithms was 1 % (for being the worst performing). In terms of accuracy, it can be observed that the method outperforms the other incremental/online algorithms in the state of the art. Regarding instance reduction, the GRLVQ obtains the highest instance reduction rate compared to the other methods, SGD K-Means obtains a very close performance to GRLVQ, but the accuracy of GRLVQ is much lower than that obtained by online K-Means. Finally, in terms of runtime, the method is the second more expensive (after IB2), while the most efficient is LVQ3 although the difference in terms of accuracy is considerable. In conclusion, the SGD K-means is the algorithm that offers the best tradeoff with regard to the incremental/online methods for IR over medium sets. With respect to NN when using the whole data set, the method obtains competitive performance. From these results we can conclude the online version of K–means is indeed competitive with the state of art of online IR for NN.

Finally, the method was compared against state of the art IR algorithms, both instance selection and prototype generation methods were considered. We would like to emphasize that the reference algorithms include the most effective IR methods developed so far. Although it is important to mention that most of the reference algorithms cannot be applied in massive data sets, because they rely on very computationally expensive procedures.

Table 3. SGD K-Means against incremental/online algorithms (Medium Data Sets)

SGD K-Means (1%)		IB2		oiGRLVQ		GRLVQ		LVQ3		LVQ-OIL		1NN	
Accuracy													
Average	Std	Average	Std	Average	Std	Average	Std	Average	Std	Average	Std	Average	Std
0.7963	0.2051	0.7762	0.2229	0.6801	0.2412	0.6266	0.2244	0.6402	0.2519	0.7036	0.2453	0.806	0.2224
Instance reduction rate													
Average	Std	Average	Std	Average	Std	Average	Std	Average	Std	Average	Std	Average	Std
0.9898	0.0008	0.7555	0.2161	0.9454	0.0031	0.9956	0.0039	0.9395	0.0835	0.9165	0.0433	-	-
Time (seconds)													
Average	Std	Average	Std	Average	Std	Average	Std	Average	Std	Average	Std	Average	Std
21.7170	13.0795	31.8718	39.5393	23.1949	27.4008	0.9466	0.5314	0.8605	0.5512	5.5407	6.2896	-	-

Table 4. Overall performance of SGD K-Means against Instance Selection (left) and Prototype Generation (right) algorithms. In [9,24] are presented in detail each of the algorithms shown above.

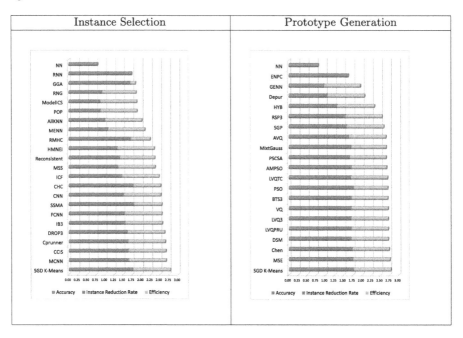

Table 4 shows the overall performance of the reference methods and the proposed one; the left plot compares the performance of the proposed method with instance/prototype selection techniques, whereas the right one shows a comparison with prototype generation methods. For comparison is considered the accuracy (blue part of bar), instance reduction rate (red portion of the bar), and efficiency (green part of bar); the latter was calculated as follows: $1 - (algorithm\_time/maximum\_time)$, where $maximum\_time$ is the time of the most expensive algorithm[2]. Therefore, the larger the bar, the better the combined performance. It can be seen from both plots the K–means based method is the one that obtains the best performance overall, offering the best tradeoff among the three objectives, i.e., the method does not degrade the accuracy of NN while reducing around 99 % the training set, with a high degree of efficiency.

These are very important and interesting results, as we are providing evidence on that K–means, a widely know algorithm (often depreciated because of its simplicity) can obtain performance comparable to state of the art techniques,

---

[2] In the case of instance selection algorithms the RNN algorithm was the most expensive with 24480.0439 s, and with respect to prototype generation the most expensive algorithm was ENPC with 10931.1977 s.

most of which cannot be used to process data sets with more than a few dozens of instances. Our result suggest K–means is in fact, an obligated strong-baseline when developing IR methods and that better IR methods can be developed by extending this algorithm.

4.2   Large Datasets

The purpose of the following experiment was to evaluate the performance of proposed algorithm against Incremental/Online methods in large datasets. The methods of the previous section were used to comparisons. Table 5 reports the accuracy, instance reduction rate and runtime (in seconds).

In terms of accuracy, the proposed algorithm and IB2 outperformed the performance of NN, with the difference that the proposed method obtains a higher reduction rate than IB2, with much more efficiency. Regarding LVQ based methods, they obtain inferior accuracy, with a high instance reduction rate, while being very efficient.

Table 5. SGD K-Means against Incremental/Online algorithms (Large Data Sets)

SGD K-Means (1%)		IB2		oiGRLVQ		GRLVQ		LVQ3		LVQ-OIL		NN	
Accuracy													
Average	Std	Average	Std	Average	Std	Average	Std	Average	Std	Average	Std	Average	Std
0.8642	0.1510	0.8130	0.1719	0.7003	0.1786	0.6438	0.1543	0.6253	0.1824	0.7094	0.1791	0.7561	0.2855
Instance reduction rate													
Average	Std	Average	Std	Average	Std	Average	Std	Average	Std	Average	Std	Average	Std
0.9899	0.0003	0.7977	0.1689	0.9464	0.0076	0.9956	0.0048	0.9585	0.0469	0.9566	0.0575	-	-
Time (seconds)													
Average	Std	Average	Std	Average	Std	Average	Std	Average	Std	Average	Std	Average	Std
81.4419	87.3217	110.1969	113.6915	429.0495	667.6254	94.0000	277.0597	20.3944	46.2390	24.6931	42.4398	-	-

Our results confirm those from the previous section and show that K–means is quite competitive in large data sets as well.

4.3   Massive Datasets

Finally, we evaluate the performance of K–means in massive data sets (with millions of instances). We compare the performance of the proposed method against quite effective state of the art methods based in batch and incremental learning. Table 6 presents the results obtained with the online version of K–means, we compare the performance of this method with SSMA-SFLSDE [25], which, to the best of our knowledge, is the best method proposed so far for prototype generation, we compare to a parallel implementation of the batch algorithm based in the Map-Reduce paradigm. We also compare the performance of the proposed method with LVQ3.

It can be seen from this table that in general the performance of the online K–means method is quite competitive with SSMA-SFLSDE, obtaining performance that is quite similar in terms of accuracy and instance reduction. In terms

Table 6. SGD K-Means against state of the art in massive datasets

Data sets	Algorithms	Accuracy		Instance reduction rate		Time (Seconds)	
		Average	Std	Average	Std	Average	Std
Susy	SGD K-Means	0.7043	0.0007	0.9777	0	55,652	1,442.60
	LVQ3	0.5155	0.0225	**0.9999**	0	**204.86**	17.6681
	SSMA-SFLSDE (Best)	**0.7282**	0.0003	0.9777	0	66,370	4,352.11
	SSMA-SFLSDE (Worst)	0.7188	0.0417	0.9715	0	11,524.57	1,941.27
	SSMA-SFLSDE (Average)	0.723	0.0026	0.9788	0.0073	37,087	24,304
	NN	0.7157	0.0001	0	0	1,167,200.33	-
KDD-CUP 99	SGD K-Means	0.9994	0.0001	0.9993	0	19,487	1,786
	LVQ3	0.44	0.3283	**0.9999**	0	**3.46**	784.73
	Map Reduce (Best)	**0.9994**	0	0.9993	0	8,655.70	149
	Map Reduce (Worst)	0.9989	0.0001	0.9987	0	4,941.77	44.88
	Map Reduce (Average)	0.9991	0.0002	0.9985	0.0008	5,485	2,494.50
	NN	0.9993	0.0001	0	0	2,354,279.87	-
Poker Hand	SGD K-Means	0.4969	0.0004	0.9914	0	8,508	285.67
	LVQ3	0.1318	0.0048	**0.9994**	0	**106.14**	6.07
	Map Reduce (Best)	**0.5181**	0.0015	0.9914	0	14,419.39	209.95
	Map Reduce (Worst)	0.4641	0.001	0.9621	0	585.432	8.45
	Map Reduce (Average)	0.5037	0.0164	0.9758	0.0119	4,310.80	4,993.90
	NN	0.5001	0.0011	0	0	48,760.82	-

of runtime our method is a bit slower, but one should note that the SSMA-SFLSDE method was run in a cluster with 11 nodes, each node having the same computing power as the workstation we used for our experiments. Also, it is remarkable that we are obtaining quite good performance with a very simple and well know algorithm. Therefore, we are showing evidence that k–means is indeed a state of the art method for IR in massive data sets.

5   Conclusions

We presented an empirical study on the usage of a classic algorithm for clustering: K-means, for tackling the instance reduction problem in big data. We show that this traditional algorithm outperforms most state of the art instance reduction methods in mid size data sets. In addition, this algorithm can cope well with massive data sets and still obtain quite competitive performance. We foresee our study will change the perception that the instance reduction community has on this algorithm for IR.

References

1. Aha, D.W., Kibler, D., Albert, M.: Instance-based learning algorithms. Mach. Learn. **6**(1), 37–66 (1991)
2. Angiulli, F.: Fast nearest neighbor condensation for large data sets classification. IEEE Trans. Knowl. Data Eng. **19**(11), 1450–1464 (2007)

3. Arnaiz, A., Diez, F., Rodrguez, J.J., Garca, C.: Instance selection of linear complexity for big data. Knowl.-Based Syst. **107**, 83–95 (2016)
4. Bottou, L.: Stochastic learning. In: Bousquet, O., von Luxburg, U., Rätsch, G. (eds.) Machine Learning 2003. LNCS (LNAI), vol. 3176, pp. 146–168. Springer, Heidelberg (2004)
5. Cano, J.R., Herrera, F., Lozano, M.: Stratification for scaling up evolutionary prototype selection. Pattern Recogn. Lett. **26**(7), 953–963 (2005)
6. Cover, T.M., Hart, P.E.: Nearest neighbor pattern classification. IEEE Trans. Inf. Theor. **13**(1), 21–27 (1967)
7. Cruz-Vega, I., Escalante, H.J.: Improved learning rule for LVQ based on granular computing. In: Carrasco-Ochoa, J.A., Martínez-Trinidad, J.F., Sossa-Azuela, J.H., Olvera López, J.A., Famili, F. (eds.) MCPR 2015. LNCS, pp. 54–63. Springer, Heidelberg (2015)
8. Cruz-Vega, I., Escalante, H.J.: An online and incremental GRLVQ algorithm for prototype generation based on granular computing. Soft Comput. 1–14 (2016)
9. Garcia, S., Derrac, J., Cano, J., Herrera, F.: Prototype selection for nearest neighbor classification: taxonomy and empirical study. IEEE Trans. Pattern Anal. Mach. Intell. **34**(3), 417–435 (2012)
10. García-Limón, M., Escalante, H.J., Morales, E., Morales-Reyes, A.: Simultaneous generation of prototypes and features through genetic programming. In: Proceedings of the Conference on Genetic and Evolutionary Computation, pp. 517–524. ACM (2014)
11. Garcia-Pedrajas, N., de Haro-Garcia, A., Perez-Rodriguez, J.: A scalable approach to simultaneous evolutionary instance and feature selection. Inf. Sci. **228**, 150–174 (2013)
12. Hart, P.: The condensed nearest neighbor rule (corresp.). IEEE Trans. Inf. Theor. **14**(3), 515–516 (1968)
13. Hastie, T., Tibshirani, R., Friedman, J.: The Elements of Statistical Learning. Springer Series in Statistics. Springer, New York (2001)
14. Kohonen, T.: The self-organizing map. Neurocomputing **21**(1), 1–6 (1998)
15. Kuncheva, L.I., Jain, L.C.: Nearest neighbor classifier: simultaneous editing and feature selection. Pattern Recogn. Lett. **20**(11–13), 1149–1156 (1999)
16. Lemaire, V., Salperwyck, C., Bondu, A.: A survey on supervised classification on data streams. In: Zimányi, E., Kutsche, R.-D. (eds.) eBISS 2014. LNBIP, vol. 205, pp. 88–125. Springer, Heidelberg (2015)
17. Nanni, L., Lumini, A.: Particle swarm optimization for prototype reduction. Neurocomputing **72**(4), 1092–1097 (2009)
18. Olvera-López, J.A., Carrasco-Ochoa, J.A., Martínez-Trinidad, J.F.: A new fast prototype selection method based on clustering. Pattern Anal. Appl. **13**(2), 131–141 (2010)
19. Ougiaroglou, S., Evangelidis, G.: RHC: a non-parametric cluster-based data reduction for efficient k-NN classification. Pattern Anal. Appl. **19**, 1–17 (2014)
20. Raicharoen, T., Lursinsap, C.: A divide-and-conquer approach to the pairwise opposite class-nearest neighbor (POC-NN) algorithm. Pattern Recogn. Lett. **26**(10), 1554–1567 (2005)
21. Sánchez, J.S., Pla, F., Ferri, F.: Prototype selection for the nearest neighbour rule through proximity graphs. Pattern Recogn. Lett. **18**(6), 507–513 (1997)
22. Lozano, M., Sotoca, J.M., Sanchez, J.S., Pla, F.: An adaptive condensing algorithm based on mixtures of gaussians. Recent Adv. Artif. Intell. Res. Dev. **113**, 225 (2004)
23. Toussaint, G.T.: Proximity graphs for nearest neighbor decision rules: recent progress. In: Interface-2002, 34th Symposium on Computing and Statistics (2002)

24. Triguero, I., Derrac, J., Garcia, S., Herrera, F.: A taxonomy and experimental study on prototype generation for nearest neighbor classification. Trans. Syst. Man Cybern. Part C **42**(1), 86–100 (2012)
25. Triguero, I., Peralta, D., Bacardit, J., García, S., Herrera, F.: MRPR: a mapreduce solution for prototype reduction in big data classification. Neurocomputing **150**, 331–345 (2015). Part A
26. Wilson, D.R., Martinez, T.: Reduction techniques for instance-based learning algorithms. Mach. Learn. **38**(3), 257–286 (2000)
27. Wilson, D.L.: Asymptotic properties of nearest neighbor rules using edited data. IEEE Trans. Syst. Man Cybern. **3**, 408–421 (1972)
28. Wu, X., Kumar, V., Quinlan, R., Ghosh, J., Yang, Q., Motoda, H., McLachlan, G., Ng, A., Liu, B., Yu, P., Zhou, Z., Steinbach, M., Hand, D., Steinberg, D.: Top 10 algorithms in data mining. Knowl. Inf. Syst. **14**(1), 1–37 (2007)

Computer Vision and Pattern Recognition

In Search of Truth: Analysis of Smile Intensity Dynamics to Detect Deception

Michal Kawulok[(✉)], Jakub Nalepa, Karolina Nurzynska, and Bogdan Smolka

Silesian University of Technology, Gliwice, Poland
{michal.kawulok,jakub.nalepa,
karolina.nurzynska,bogdan.smolka}@polsl.pl

Abstract. Detection of deceptive facial expressions, including estimating smile genuineness, is an important and challenging research topic that draws increasing attention from the computer vision and pattern recognition community. The state-of-the-art methods require localizing a number of facial landmarks to extract sophisticated facial characteristics. In this paper, we explore how to exploit fast smile intensity detectors to extract temporal features. This allows for real-time discrimination between posed and spontaneous expressions at the early smile onset phase. We report the results of experimental validation, which indicate high competitiveness of our method for the UvA-NEMO benchmark database.

Keywords: Affective computing · Face analysis · Facial expressions · Smile genuineness · Deception detection · Support vector machines

1 Introduction

Smiles are very likely the most common and straightforward-to-recognize facial expressions. However, they may convey various underlying (positive or negative) emotions, depending on the context. Although the subtle differences between these types of smiles can be relatively easily perceived by humans in most cases[1], it is a demanding computer vision and pattern recognition task. Recognizing genuine (spontaneous) and deliberate (posed) smiles, along with understanding which facial features exhibit overwhelmingly different human intensions became a vital topic and attracted attention in many domains, ranging from machine learning to clinical research [2].

There are a plethora of real-life applications of such recognition systems. They encompass surveillance engines detecting deceptive facial expressions, medical solutions for recognizing genuine pain, responsive human-machine interaction interfaces, animation frameworks, and many others [3].

In the work reported here, we exploit the dynamics of smile intensity, estimated from the entire face and from coarsely located regions of eyes and mouth.

---

[1] This appears not trivial for patients with mental disorders, e.g., schizophrenia [1].

© Springer International Publishing AG 2016
M. Montes-y-Gómez et al. (Eds.): IBERAMIA 2016, LNAI 10022, pp. 325–337, 2016.
DOI: 10.1007/978-3-319-47955-2_27

Our contribution consists in: (i) relying exclusively on a face detector without the necessity to localize and track facial landmarks, (ii) analysis of the smile intensity time series using regressed trend lines, which does not require high accuracy of smile detectors, (iii) exploiting evolutionary training set selection for support vector machines (SVMs) [4] to improve training from weakly-labeled datasets. Consequently, (iv) we utilize basic and fast-to-compute *uniform local binary pattern* (LBP) features to train the smile detectors. It is worth noting that these features are extracted in the spatial domain at first (independently for every frame), and the temporal features are captured later from the smile intensity series. This is in contrast to many methods [5,6] that employ the spatial-temporal features, which are sensitive to landmark displacements in consecutive frames.

In Sect. 2, we outline the state-of-the-art on smile genuineness detection. Our algorithm is described in detail in Sect. 3 and the obtained experimental results are reported in Sect. 4. The paper is concluded in Sect. 5.

2   Related Literature

Analysis and recognition of facial expressions has been intensively studied [7], and the current state of the art is thoroughly summarized in an interesting survey [8] that has been published recently. Since smiles can reveal various feelings and moods of humans (not necessarily expressed honestly), recognizing spontaneous and posed smiles, together with understanding the underlying smile psychophysics has become an active research topic [9,10].

Most of the state-of-the-art algorithms for recognizing spontaneous and posed facial behaviors are focused on the temporal analysis of various facial features [11]. In [12], the authors investigated changes in the smile onset amplitudes and durations, and found a strong evidence that spontaneous smiles are characterized by smaller amplitudes and significantly more stable relations between those two measures. An interesting approach, in which the asymmetry of facial expressions is exploited, was presented in [13]. Extracting distance-based and angular features from eyelid movements was proposed in [14]. Also, deep learning was investigated to determine the discriminative features [15]—it occurred quite effective here, however it suffers from low interpretability, which remains relevant when the psychological aspects are considered.

In a very recent work [5], it is demonstrated that the eyelid features are most discriminant, but they should be coupled with other facial characteristics (encompassing, e.g., cheeks and/or lip corner features) in order to boost the classification performance. This finding proved that different facial regions can contribute differently to the classification of smiles. While the majority of approaches rely on extracting geometry-based facial features, they are very vulnerable to inaccurate localization, which may easily affect their behavior. The appearance-based techniques are therefore being intensively developed.

Various texture operators were shown efficient in recognizing facial expressions [16,17] and they have been explored to distinguish between posed and spontaneous smiles as well. In [6], the authors utilize the completed local binary pattern (CLBP)—the standard LBP is complemented with textural features from three orthogonal planes, which creates an appearance-based local spatial-temporal descriptor (CLBP-TOP). First, the entire image sequence is divided into blocks in both spatial and temporal domains. Five facial landmarks (eyes, lip corners, and nose tip) are detected and tracked to retrieve facial sub-region volumes, and each sub-region volume is divided into three blocks in the temporal domain, reflecting three smile phases: *onset*, *apex*, and *offset*. Finally, an adaptive learning procedure is applied to extract an optimal subset of all CLBP features (termed disCLBP-TOP). Although this algorithm retrieved high classification scores, inaccurate detection of facial landmarks can jeopardize its performance.

Multi-modal techniques are based on the observation that people communicate by the means of language, facial expressions, head movement, gestures and poses [18]. To fully exploit the information coming from different sources, the multi-modal methods fuse them to improve the classification performance. This fusion may be performed at various abstraction levels (they are often referred to as *early*, *mid-level*, and *late* fusion strategies), e.g., across different smile phases, or for various facial regions. In [19], three different facial regions are used to extract features (eyes, cheeks, and mouth). Then, SVMs are trained for each region separately, and they are used to classify the feature vectors. The algorithm which fuses head, face, and shoulder modalities was proposed in [20] (different landmark trackers were employed for each modality). The authors efficiently combined these modalities, and highlighted which of them carry discriminative information. Another interesting research direction is to exploit thermal imaging—the heat radiated from the face is used to recognize deception [21].

3   Proposed Method

Our approach towards determining smile genuineness consists in analysing the dynamics of smile intensity estimated from the area of (i) eyes, (ii) mouth, and from (iii) the entire face. Importantly, the facial subregions are only roughly localized—in contrast to many existing techniques, we require only the face to be detected rather than a larger set of facial landmarks. Furthermore, we estimate the smile intensity from fairly simple features—as we focus on capturing the smile intensity trends, the effectiveness of the smile classifier is not critical here.

A flowchart of our algorithm is presented in Fig. 1 (the operations performed independently for each frame are indicated with stacked blocks). First of all, we detect faces in every input frame to crop and normalize the facial region. Afterward, the LBP mapping followed by principal components analysis (PCA) is employed to extract the features from the entire face and from the subregions of eyes and mouth. As a result, we obtain three feature vectors from each frame, which are classified using SVMs to assess the smile intensity. In addition, we train another SVM to distinguish between spontaneous and posed smiles (this step is omitted in some variants of our method, hence it is rendered with a dotted line in the flowchart, as an optional step). The frame-wise responses of these classifiers are later analyzed to extract the temporal features capturing the dynamics of smile intensity across the entire sequence. Finally, these temporal feature vectors are classified using SVM to determine whether the smile is spontaneous or posed.

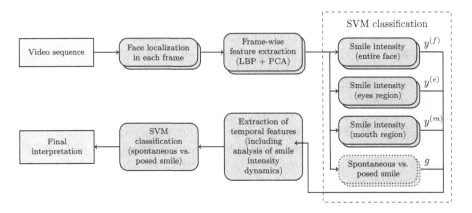

Fig. 1. Flowchart of the proposed algorithm.

3.1 Extracting Frame-Wise Basic Features

In order to extract the frame-wise features, every facial region, located using the multi-level face detector [22], is normalized to the size of 112×150 pixels, and divided into 7×10 sub-images. We extract the features from the entire face and from two roughly-estimated regions, encompassing eyes and mouth (see Fig. 2).

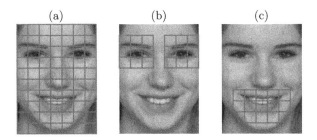

Fig. 2. Sub-images determined in: (a) entire face, (b) eyes region, and (c) mouth region.

The uniform-pattern LBP (with the radius $r_{LBP} = 1$, thus 8 neighboring pixels are considered for the pixel being analyzed) is used to extract the LBP codes. A histogram of these codes becomes the descriptor for the corresponding sub-image. As we exploit the uniform patterns [16], each histogram is composed of 59 bins and all of the non-uniform patterns are assigned to a single label. Finally, the feature vector representing the investigated facial region includes the histograms retrieved for all of its sub-images (e.g., the mouth-region feature vector is a concatenation of 13 independent histograms—see Fig. 2(c)).

The LBP codes are highly-dimensional (4130 features for the entire face), so we reduce their dimensionality with PCA. We determined the numbers of intrinsic dimensions on the experimental basis—we use 100 principal components for the entire face, and 30 components for the regions of eyes and mouth. As a result, we obtain three *frame-wise feature vectors*.

3.2   Estimating Frame-Wise Smile Intensity and Genuineness

The frame-wise feature vectors are classified using SVMs (with radial basis functions) to assess the smile intensity and genuineness. SVM is a binary classifier, which is trained to determine a decision hyperplane separating the data from the opposite classes. The absolute value of the SVM decision function expresses the vector's distance from the hyperplane, which corresponds to the smile intensity or genuineness here. In the former case, the SVM is trained using images with and without the smile, while in the latter—with posed and spontaneous smiles.

A potential drawback of SVMs lies in their high $O(n^3)$ time and $O(n^2)$ memory complexities of the training (where n is the number of vectors in the training set). Here, the feature vectors are extracted from every frame in a video sequence, which results in large training sets. In addition, the data suffer from weak labeling—a video sequence presenting a posed smile may contain a number of neutral frames without any features indicating the posed expression, so the feature vectors extracted from these frames are not helpful in training the classifier. Similarly, it is hard to determine the exact moment when the smile appears, which translates to uncertainty of the ground-truth "smile" labels.

To effectively deal with these problems, we exploited our memetic algorithm for selecting training data for SVMs—PCA^{2}MA [4]. It consists in evolving a "good" training sample, whose size is adaptively increased during the evolutionary optimization. A population of individuals (i.e., training samples) is evolved using genetic operators enhanced with a memetic *education* operator to improve the individuals using a pool of vectors determined as support vectors (SVs) earlier during the evolution. Fitness of each individual is assessed based on how well the trained SVM classifies the entire training set—this is measured with the area under the receiver operating characteristic curve. Importantly, our memetic algorithm makes it possible to train SVMs from weakly labeled data [23]—as discussed earlier, such data are inherent to the analyzed scenario.

3.3   Extracting Temporal Features to Capture Smile Dynamics

After applying SVMs to estimate the smile genuineness (g) and intensity from the region of eyes $(y^{(e)})$, mouth $(y^{(m)})$ and the entire face $(y^{(f)})$, we obtain four time series of frame-wise classification outcomes for each video sequence. To capture the smile intensity dynamics, we analyze each series with a sliding window, in which we apply linear regression for subseries of $\omega = 9$ subsequent scores $(\{y_i\})$. We consider the slope of the trend line (a) and the regression coefficient (r) independently for $y^{(e)}$, $y^{(m)}$ and $y^{(f)}$:

$$a = S_{ty}/S_t, \quad r = S_{ty}/\sqrt{S_t S_y}, \tag{1}$$

$$S_{ty} = \sum_{i=1}^{\omega} (t_i - \bar{t})(y_i - \bar{y}), \quad S_t = \sum_{i=1}^{\omega} (t_i - \bar{t})^2, \quad S_y = \sum_{i=1}^{\omega} (y_i - \bar{y})^2, \tag{2}$$

where t is the frame capture timestamp (for a constant frame rate, the frame number may be used as t), while $\bar{y}$ and $\bar{t}$ are the mean values of y and t inside the window. The regression coefficient $r \in [-1; 1]$ indicates how well the linear trend fits the data (the higher its absolute value is, the more linear they are).

We compute a and r for every frame in the sequence, moving the sliding window with a unitary step—for the first and the last frame, we regress the trend relying on the $\lceil \omega/2 \rceil$ first or last frames, respectively, and then we increase the window width up to ω for the inner frames. To extract the smile intensity dynamics for an i-th frame, we adjust the slope value relying on the regression coefficient: $\delta_i = a_i |r_i|$. These features are obtained independently for the region of mouth (δ^m), eyes (δ^e), and for the whole face (δ^f).

This procedure is illustrated in Fig. 3—the smile intensity for the entire sequence is rendered on the right hand side and a relevant part of the sequence is magnified on the left. Here, the trend line is regressed for frames 7–15 to determine the slope value for the 11-th frame—it may be observed that the smile intensity increases indeed, as estimated by the classifier. The graph of the obtained slope values (before the adjustment) is also rendered in the Figure.

Frame-Level Temporal Features: In the work reported here, we explore *frame-level* and *sequence-level* temporal features, and we consider only those frames, for which the smile intensity is increasing ($\delta_i^{(f)} > 0$). At the frame level, we extract 6 features for every i-th frame to capture the relations between the facial regions in terms of the smile intensity growth:

$$\boldsymbol{\nu}_i = \left[\delta_i^{(f)}, \delta_i^{(f)} - \delta_i^{(e)}, \delta_i^{(f)} - \delta_i^{(m)}, \frac{\delta_i^{(m)}}{\delta_{\max}^{(e)}}, \frac{\delta_i^{(f)}}{\delta_{\max}^{(e)}}, \frac{\delta_i^{(f)}}{\delta_{\max}^{(m)}} \right], \tag{3}$$

where $\delta_{\max} = \max_i (\delta_i)$. From these feature vectors ($\{\boldsymbol{\nu}_i\}$), we train an SVM to differentiate between posed and spontaneous smiles (PCA^2MA is employed to

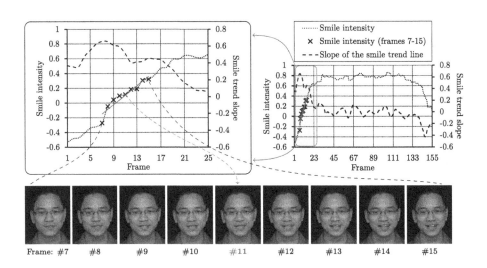

Fig. 3. Example of extracting features related to smile intensity dynamics.

select a training sample). So as to determine the smile genuineness for a sequence, we classify every frame at first, and then we pick 10 % of the frames with the strongest positive SVM responses (i.e., classified as most "posed")—an average SVM response from these selected frames is used as the classification outcome for the entire sequence. This approach is motivated by the hypothesis that only in some frames in a sequence, presenting a posed smile, the relevant features are manifested. As the training set is classified in the same way to obtain the fitness in PCA^{2}MA, it is possible to select appropriate feature vectors from the entire training set during the evolutionary training set optimization.

Sequence-Level Temporal Features: Another approach we explored consists in extracting the features from the entire sequence. Here, the feature vectors are composed of 13 determinants:

$$v = \left[\overline{g}, A^{(f)}, A^{(m)}, A^{(e)}, \overline{\delta}^{(f)}, \overline{\delta}^{(m)}, \overline{\delta}^{(e)}, \delta_{\max}^{(f)}, \delta_{\max}^{(m)}, \delta_{\max}^{(e)}, \tau(f,e), \tau(f,m), \tau(m,e) \right], \quad (4)$$

where $A = (y_{max} - y_{min})$ is the amplitude of the smile intensity, $\overline{\delta}$ is the average intensity dynamics, and $\tau(p,q) = \left[\arg\max_i \delta_i^{(p)} - \arg\max_i \delta_i^{(q)} \right]$ expresses the time difference between the maximum growth observed in two time series.

Illustrative Example: An example of the smile intensities estimated for posed and spontaneous smile of the same individual is shown in Fig. 4. In order to visualize the responses of several SVM classifiers, their outcomes were normalized to the range [0;1]. We also show the outcome of two genuineness classifiers, trained from (i) frame-wise LBP+PCA features, and (ii) from 6 temporal features at the frame level. In the latter case, the frames with decreasing smile intensity are ignored, hence there are some gaps in the rendered graphs. It can be observed that for both cases the SVM responses grow during the smile onset phase, however the peak value is much larger for the posed smile, which overall makes these cases easily separable. This example also clearly indicates that our temporal features are discriminative mainly during the onset phase of a smile, which has been confirmed by the experimental results reported later in Sect. 4. The SVM learned from the frame-wise features returns slightly higher results for the apex phase, but overall its distinctiveness is much lower here.

4   Experimental Validation

We have validated the proposed algorithm using the UvA-NEMO benchmark database, which contains 1240 video sequences of posed and spontaneous smiles (643 and 597 sequences, respectively). All of the reported results were obtained with 10-fold cross-validation following the official experimental protocol [24] (the cross-validation scheme has been applied to train the SVM classifiers). As the PCA^{2}MA evolutionary optimization is a randomized process, we have run 50 tests

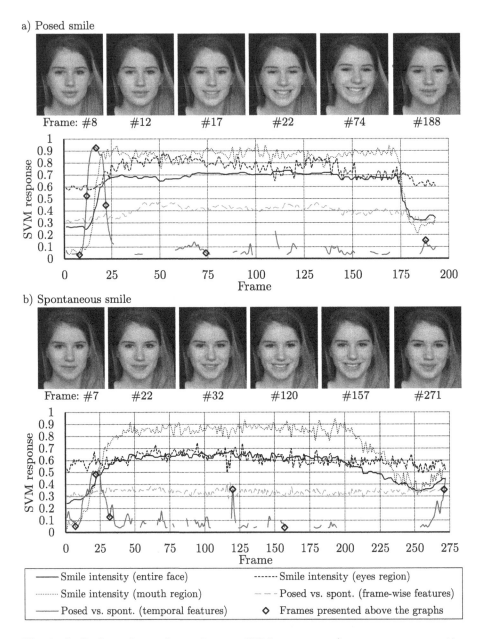

Fig. 4. Smile intensity and genuineness SVM responses for sequences presenting posed (a) and spontaneous (b) smiles. Selected frames are shown above the graphs.

for each scenario. The algorithms were implemented in C++ using LIBSVM and run on a computer equipped with an Intel Core i7-3740QM 2.7 GHz (16 GB RAM) processor. We process 30 frames per second including all the necessary preprocessing operations (face detection being the most time consuming one).

First of all, we trained the SVM classifiers to measure the smile intensity. The achieved classification scores along with the number of obtained SVs (with the standard error) and classification time of a single frame are reported in Table 1. The values of SVM parameters (γ and C), found using a grid search with an exponential step, are also given in the Table.

Table 1. Results of the frame-wise smile intensity estimation using SVMs.

	Accuracy	Number of SVs	Time per frame	γ	C
Entire face	$95.47 \pm 0.123\%$	793.3 ± 14.34	0.14 ms	22.94	3.17
Mouth region	$92.44 \pm 0.068\%$	1218.8 ± 28.57	0.09 ms	68	16.18
Eyes region	$69.08 \pm 0.510\%$	1199.9 ± 20.33	0.09 ms	40.96	10.24

We have investigated several variants of our algorithm, outlined in Table 2. Three basic configurations rely on (i) the frame-wise features (F-static) and the temporal features extracted at (ii) the sequence (S13-inc) or (iii) frame (F6-inc) levels. For the temporal features, we consider processing all the frames (-all), as well as only those with increasing smile intensity (-inc)—here, we focus only on the smile onset phase, detected as the longest interval with $\delta^{(f)} > 0$. Also, we report the results obtained when the SVM is trained using the entire training set (-full). For the sequence-level features, we created yet another variant (S12) which excludes the frame-wise classification outcome ($\bar{g}$).

The obtained classification accuracy and the number of SVs along with the standard error, and the mean time of classifying a whole sequence are shown

Table 2. Investigated variants of the proposed algorithm.

Name	Feature extraction	#features	Frame filter	SVM training
S13-inc	temporal (sequence)	13	$\delta^{(f)} > 0$	PCA^{2}MA
S13-all	temporal (sequence)	13	all	PCA^{2}MA
S13-inc-full	temporal (sequence)	13	$\delta^{(f)} > 0$	full SVM
S12-inc	temporal (sequence)	12 (no $\bar{g}$)	$\delta^{(f)} > 0$	PCA^{2}MA
S12-all	temporal (sequence)	12 (no $\bar{g}$)	all	PCA^{2}MA
S12-inc-full	temporal (sequence)	12 (no $\bar{g}$)	$\delta^{(f)} > 0$	full SVM
F6-inc	temporal (frame)	6	$\delta^{(f)} > 0$	PCA^{2}MA
F6-all	temporal (frame)	6	all	PCA^{2}MA
F6-inc-full	temporal (frame)	6	$\delta^{(f)} > 0$	full SVM
F-static	frame-wise (LBP+PCA)	100	all	PCA^{2}MA

Table 3. Results of smile genuineness estimation using SVMs.

Variant	Accuracy	Number of SVs	Time per sequence	γ	C
S13-inc	84.68 ± 0.096%	340.3 ± 48.61	0.014 ms	0.64	10.24
S13-all	84.04 ± 0.064%	443.0 ± 36.75	0.032 ms	0.64	10.24
S13-inc-full	84.10 ± 0.000%	508.2 ± 0.00	0.024 ms	0.64	10.24
S12-inc	84.60 ± 0.088%	432.5 ± 12.53	0.022 ms	0.64	10.24
S12-all	83.82 ± 0.146%	194.7 ± 18.52	0.009 ms	0.64	10.24
S12-inc-full	84.19 ± 0.000%	548.9 ± 0.00	0.029 ms	0.64	10.24
F6-inc	82.92 ± 0.262%	525.2 ± 64.13	2.205 ms	1.28	0.01
F6-all	82.02 ± 0.252%	434.7 ± 32.09	4.008 ms	1.28	0.01
F6-inc-full	79.66 ± 0.000%	72475.9 ± 0.00	328.1 ms	1.28	0.01
F-static	66.89 ± 0.452%	877.8 ± 21.90	26.68 ms	80.28	96.66

in Table 3. In Table 4, we report the levels of the statistical significance of the differences between the variants employing PCA^{2}MA, measured with the two-tailed Wilcoxon test. Only the differences between S13-inc and S12-inc are not statistically important, so in fact $\bar{g}$ may be excluded from the sequence-level features in Eq. (4). In general, classifying the sequence-level features leads to better scores than relying on the frame-level ones, and it is clearly evident that exploiting PCA^{2}MA to train SVM is beneficial in terms of the accuracy and classification time (resulting from a smaller number of SVs). Furthermore, as noted earlier when discussing Fig. 4, the proposed temporal features capture the differences between posed and spontaneous smiles manifested during the onset phase, and extending it to all the frames deteriorates the accuracy.

In Table 5, we compare the obtained classification scores with several state-of-the-art methods validated using the UvA-NEMO database. Our algorithm delivers the best results, when only the onset phase is considered. If all the phases are analyzed, it is the method proposed by Wu et al. [6] which gains the most and outperforms the others, when disCLBP-TOP features are used. It is

Table 4. The level of statistical significance obtained using the two-tailed Wilcoxon tests. The differences which are statistically important (at $p < 0.05$) are boldfaced.

	S12-inc	S12-all	S13-inc	S13-all	F6-inc	F6-all
F-static	**<0.0001**	**<0.0001**	**<0.0001**	**<0.0001**	**<0.0001**	**<0.0001**
S12-inc	—	**<0.0001**	0.7039	**0.0056**	**0.0031**	**<0.0001**
S12-all		—	**<0.0001**	**<0.0001**	**<0.0001**	**<0.0001**
S13-inc			—	**0.0045**	**0.0033**	**<0.0001**
S13-all				—	**<0.0001**	**0.0006**
F6-inc					—	**<0.0001**

Table 5. Comparison with the state-of-the-art methods (best scores are boldfaced)

Smile phase	Pfister [25]	Dibeklioğlu [5]	Wu [6]		Proposed algorithm
			CLBP-TOP	disCLBP-TOP	
Onset	N/A	84.52 %	N/A	78.20 %	**84.68 %**
All phases	73.06 %	86.37 %	83.03 %	**91.40 %**	84.04 %

worth noting, however, that this method requires a number of facial landmarks to be precisely located and the facial regions to be properly aligned in subsequent frames so as to enable correct extraction of the disCLBP-TOP features in the spatial-temporal domain. Unfortunately, the authors do not quote the processing time, however the required operations may be computationally expensive. It is an important advantage of our algorithm that it relies on simple features which are extracted from images downsampled to a rather small size of 112×150 pixels. This, along with high effectiveness in the onset phase, makes it appropriate for real-time applications, when early smile classification does matter. An interesting fact is that the accuracy of our method deteriorates, when it is applied to all the phases. Possibly, extracting different features when the smile intensity is stable and/or decreasing, followed by their proper fusion could improve the final score, however this leads us towards our future work.

5   Conclusions and Outlook

In this paper, we focused on capturing the dynamics of smile intensity, estimated from multiple facial parts, namely from the region of eyes, mouth, as well as from the entire face. While many researchers put a lot of effort into locating a number of facial landmarks to extract sophisticated frame-wise features, we demonstrated that even relying on simple features extracted from coarsely located eyes and mouth regions, our approach to analyze the dynamics of smile intensity makes it possible to distinguish between posed and spontaneous expressions with the accuracy competitive with the state-of-the-art methods.

In the research reported here, the temporal features, which capture the smile dynamics, discriminate between the posed and spontaneous expressions primarily during the smile onset phase. In our ongoing work, we aim at combining the proposed frame-level and sequence-level features, as well as preparing a set of classifiers optimized for specific smile phases and for different categories of subjects (depending on age, sex, etc.)—ensembles of such classifiers were reported to increase the classification score [5,6]. Also, we will explore whether adopting our face relevance maps optimization technique [26] could be beneficial here. Furthermore, we plan to include other facial dynamics indicators (such as eye blink detector or facial action units classifiers) to provide more discriminative information for our algorithm. Finally, it may be worthwhile to consider extracting whole-body dynamics rather than analysing exclusively the facial region.

Acknowledgements. This work has been supported by the Polish National Science Centre (NCN) under the Grant: DEC-2012/07/B/ST6/01227.

References

1. Barkhof, E., de Sonneville, L.M., Meijer, C.J., de Haan, L.: Specificity of facial emotion recognition impairments in patients with multi-episode schizophrenia. Schizophr. Res.: Cogn. **2**(1), 12–19 (2015)
2. Ross, E.D., Pulusu, V.K.: Posed versus spontaneous facial expressions are modulated by opposite cerebral hemispheres. Cortex **49**(5), 1280–1291 (2013)
3. Trutoiu, L.C., Carter, E.J., Pollard, N., Cohn, J.F., Hodgins, J.K.: Spatial and temporal linearities in posed and spontaneous smiles. ACM Trans. Appl. Percept. **11**(3), 12:1–12:15 (2014)
4. Nalepa, J., Kawulok, M.: Adaptive memetic algorithm enhanced with data geometry analysis to select training data for SVMs. Neurocomputing **185**, 113–132 (2016)
5. Dibeklioğlu, H., Salah, A.A., Gevers, T.: Recognition of genuine smiles. IEEE Trans. Multimedia **17**(3), 279–294 (2015)
6. Wu, P., Liu, H., Zhang, X.: Spontaneous versus posed smile recognition using discriminative local spatial-temporal descriptors. In: Proceedings of the IEEE ICASSP, pp. 1240–1244 (2014)
7. Calder, A.J., Young, A.W.: Understanding the recognition of facial identity and facial expression. Nature Rev.: Neurosci. **6**, 641–651 (2005)
8. Martinez, B., Valstar, M.F.: Advances, challenges, and opportunities in automatic facial expression recognition. In: Kawulok, M., Celebi, M.E., Smolka, B. (eds.) Advances in Face Detection and Facial Image Analysis, pp. 63–100. Springer, Heidelberg (2016)
9. Krumhuber, E.G., Likowski, K.U., Weyers, P.: Facial mimicry of spontaneous and deliberate Duchenne and non-Duchenne smiles. J. Nonverbal Behav. **38**(1), 1–11 (2014)
10. Girard, J.M., Cohn, J.F., Jeni, L.A., Sayette, M.A., De la Torre, F.: Spontaneous facial expression in unscripted social interactions can be measured automatically. Behav. Res. Methods **47**(4), 1136–1147 (2015)
11. Valstar, M.F., Pantic, M., Ambadar, Z., Cohn, J.F.: Spontaneous vs. posed facial behavior: automatic analysis of brow actions. In: Proceedings of the ICMI, pp. 162–170 (2006)
12. Cohn, J.F., Schmidt, K.L.: The timing of facial motion in posed and spontaneous smiles. Int. J. Wavelets Multiresolut. Inf. Process. **02**(02), 121–132 (2004)
13. Sénéchal, T., Turcot, J., el Kaliouby, R.: Smile or smirk? Automatic detection of spontaneous asymmetric smiles to understand viewer experience. In: Proceedings of the IEEE FG, pp. 1–8 (2013)
14. Dibeklioğlu, H., Valenti, R., Salah, A.A., Gevers, T.: Eyes do not lie: spontaneous versus posed smiles. In: ACM International Conference on Multimedia, pp. 1–4 (2010)
15. Gan, Q., Wu, C., Wang, S., Ji, Q.: Posed and spontaneous facial expression differentiation using deep boltzmann machines. In: 2015 International Conference on Affective Computing and Intelligent Interaction (ACII), pp. 643–648. IEEE (2015)
16. Zhao, G., Pietikäinen, M.: Dynamic texture recognition using local binary patterns with an application to facial expressions. IEEE Trans. Pattern Anal. Mach. Intell. **29**(6), 915–928 (2007)

17. Smolka, B., Nurzynska, K.: Power LBP: a novel texture operator for smiling and neutral facial display classification. Procedia Comp. Sci. **51**, 1555–1564 (2015)
18. Abouelenien, M., Pérez-Rosas, V., Mihalcea, R., Burzo, M.: Deception detection using a multimodal approach. In: Proceedings of the ICMI, pp. 58–65. ACM, New York (2014)
19. Cohn, J., Reed, L., Moriyama, T., Xiao, J., Schmidt, K., Ambadar, Z.: Multimodal coordination of facial action, head rotation, and eye motion during spontaneous smiles. In: Proceedings of the IEEE FG, pp. 129–138 (2004)
20. Valstar, M.F., Gunes, H., Pantic, M.: How to distinguish posed from spontaneous smiles using geometric features. In: Proceedings of the ICMI, pp. 38–45. ACM (2007)
21. Rajoub, B.A., Zwiggelaar, R.: Thermal facial analysis for deception detection. IEEE Trans. Inf. Forensics Secur. **9**(6), 1015–1023 (2014)
22. Kawulok, M., Szymanek, J.: Precise multi-level face detector for advanced analysis of facial images. IET Image Process. **6**(2), 95–103 (2012)
23. Kawulok, M., Nalepa, J.: Towards robust SVM training from weakly labeled large data sets. In: Proceedings of the ACPR, pp. 464–468 (2015)
24. Dibeklioğlu, H., Salah, A.A., Gevers, T.: Are you really smiling at me? Spontaneous versus posed enjoyment smiles. In: Fitzgibbon, A., Lazebnik, S., Perona, P., Sato, Y., Schmid, C. (eds.) ECCV 2012, Part III. LNCS, vol. 7574, pp. 525–538. Springer, Heidelberg (2012)
25. Pfister, T., Li, X., Zhao, G., Pietikäinen, M.: Differentiating spontaneous from posed facial expressions within a generic facial expression recognition framework. In: Proceedings of the IEEE ICCV, pp. 868–875 (2011)
26. Kawulok, M., Wu, J., Hancock, E.R.: Supervised relevance maps for increasing the distinctiveness of facial images. Pattern Recogn. **44**(4), 929–939 (2011)

Sign Languague Recognition Without Frame-Sequencing Constraints: A Proof of Concept on the Argentinian Sign Language

Franco Ronchetti[1(✉)], Facundo Quiroga[1], César Estrebou[1],
Laura Lanzarini[1], and Alejandro Rosete[2]

[1] Instituto de Investigación en Informática LIDI, Facultad de informática,
Universidad Nacional de La Plata, La Plata, Argentina
{fronchetti,fquiroga,cesarest,laural}@lidi.info.unlp.edu.ar
[2] Instituto Superior Politécnico Jose Antonio Echeverría, Havana, Cuba
rosete@ceis.cujae.edu.cu

Abstract. Automatic sign language recognition (SLR) is an important topic within the areas of human-computer interaction and machine learning. On the one hand, it poses a complex challenge that requires the intervention of various knowledge areas, such as video processing, image processing, intelligent systems and linguistics. On the other hand, robust recognition of sign language could assist in the translation process and the integration of hearing-impaired people, as well as the teaching of sign language for the hearing population.

SLR systems usually employ Hidden Markov Models, Dynamic Time Warping or similar models to recognize signs. Such techniques exploit the sequential ordering of frames to reduce the number of hypothesis. This paper presents a general probabilistic model for sign classification that combines sub-classifiers based on different types of features such as position, movement and handshape. The model employs a bag-of-words approach in all classification steps, to explore the hypothesis that ordering is not essential for recognition. The proposed model achieved an accuracy rate of 97 % on an Argentinian Sign Language dataset containing 64 classes of signs and 3200 samples, providing some evidence that indeed recognition without ordering is possible.

Keywords: Sign language recognition · Bag-of-words · Argentinian sign language

1 Introduction

1.1 Background

Automatic sign recognition is a complex, multidisciplinary problem that has not been fully solved. While recently there has been some lateral progress through

---

F. Ronchetti and F. Quiroga—Contributed equally.

© Springer International Publishing AG 2016
M. Montes-y-Gómez et al. (Eds.): IBERAMIA 2016, LNAI 10022, pp. 338–349, 2016.
DOI: 10.1007/978-3-319-47955-2_28

gesture recognition, driven mainly by the development of new technologies, there is still a long road ahead before accurate and robust applications are developed that allow translating and interpreting the signs performed by an interpreter [1]. The complex nature of signs draws effort in various research areas such as human-computer interaction, computer vision, movement analysis, machine learning and pattern recognition.

The full task of recognizing a sign language involves a multi-step process [1], which can be simplified as:

1. Tracking the hands of the interpreter
2. Segmenting the hands and creating a model of its shape
3. Recognizing the shapes of the hands
4. Recognizing the sign as a syntactic entity.
5. Assigning semantics to a sequence of signs.
6. Translating the semantics of the signs to the written language

While these tasks can provide feedback to each other, they can be carried out mostly independently, and in different ways. For example, there are several approaches for tracking hand movements: some use 3D systems [4,5], such as MS Kinect, and others simply use a 2D image from an RGB camera [1,10]. Most older systems employed movement sensors such as special gloves, accelerometers, etc., but recent approaches generally focus on video, such as the one presented here.

There are numerous publications dealing with the automated recognition of sign languages, a field that started mostly in the 90s. Von Agris [10] and Cooper et al. [1] both present a general view of the state of the art in sign language recognition.

Sign language recognition employs different types of features, usually classified as manual and non-manual.

Non-manual features such as pose, lip-reading or face expressions are sometimes included to enhance the recognition process, since some signs cannot be differentiated from manual information only [10]. In this regard, the tracking of the head is mostly solved [9], but its segmentation with respect to an arbitrary background or in the presence of hand-head occlusions is still an unsolved problem, as is the robust recognition of such non-manual features [10]. Manual information, on the other hand, generally conveys most of the information in a sign.

For tracking and segmentation of the hands, there is much interest in creating skin color models to detect and track the hands of an interpreter on a video [8], and adding the possibility of segmenting the hands [2], even in the presence of hand-hand occlusions [11].

The handshape information of a sign is composed by a sequence of hand poses. After segmenting the hand, it must represented in a convenient way for handshape recognition. However, turning a hand pose into another requires a non-rigid transformation of the hand, which must be also modeled, and capturing non-rigid 3D transformations with occlusions using a 2D RGB camera is a hard task. While the best possible output from this step would be a full 3D model of the hand, this is generally hard to do without multiple cameras, special sensors or markers [5]. In most cases the handshape is instead represented as a combination of more abstract features based on geometric or morphologic properties of its shape or texture [10].

Some researchers focus on fingerspelling [5], which is essentially a static handshape recognition task. While some signs do indeed present a static handshape in one or both hands, and no movement, most involve many handshapes and their transitions (i.e., non-rigid body transformations of the hand), or rigid body transformations of a single handshape (i.e., rotation and translation), and a certain movement of the hands. To deal with these dynamic signs, SLR systems are based usually on Hidden Markov Models (HMMs), Dynamic Time Warping (DTW) or similar models, whether to recognize segmented signs or a continuous stream [1,10]. These techniques attempt to model the sequence of positions and handshapes, therefore exploiting the sequential ordering of frames to reduce the number of hypothesis to test.

Finally, SLR techniques have traditionally employed sign-level models to recognize sign language. Recently, there have been many efforts to move from sign-level models to sub-unit level models, analogously to the transition from word-level to phoneme-level models in speech recognition [2]. The main problem found in these attempts has been the difficulty to find a useful definition of sub-units of a sign. Unlike speech, signing is very much multimodal and there is little standardization of both languages and specification languages, therefore a promising approach is to infer such sub-units from training data [10].

1.2   Presented Work

This paper presents a general classification model for sign language recognition that focuses on step 4, that is, the recognition of signs as a syntactic entity on the sign-level (i.e., a correspondence between a video containing a sign and a word). The model is composed of a set of subclassifiers, each employing a bag-of-words approach that ignores sequence information. This setup allows us to explore the hypothesis that a classifier can still achieve high recognition accuracies under such a constraint.

While models that exploit the ordering of frames such as HMMs should theoretically achieve greater recognition accuracy, such constraint complicates the inference of sub-units, since by definition they must too be ordered. The bag of words approach can ease the task of determining sub-units, since their sequence or transitions do not need to be specified or learned.

To test the model, we performed experiments on the LSA64 dataset, which consists of 64 signs of the Argentinian Sign Language (LSA) and was recorded with normal RGB cameras.

The document is organized as follows. Section 2 describes the LSA64 dataset, the image processing and feature extraction. Section 3 defines the classification model. Section 4 details the experiments carried out, and finally Sect. 5 presents the general conclusions.

2   Dataset and Features

2.1   Argentinian Sign Language Dataset (LSA64)

The sign dataset for the Argentinian Sign Language[1], includes 3200 videos where each of the 10 non-expert subjects performed 5 repetitions of 64 different types of signs.

To simplify the problem of hand segmentation within an image, subjects wore fluorescent-colored gloves, as can be seen in Fig. 1. The glove substantially simplifies the problem of recognizing the position of the hand and performing its segmentation by removing all issues associated to skin color variations and hand-head occlusions, while fully retaining the difficulty of recognizing the handshape.

The dataset contains 22 two-handed signs and 42 one-handed ones, which were selected among the most commonly used ones in the LSA lexicon.

Fig. 1. Snapshots of the performance of four different signs of the LSA64 database. (Color figure online)

2.2   Preprocessing and Features

The pre-processing and feature extraction activities carried out with the database's videos consist of extracting hand and head positions, along with images of the hand, segmented.

The detection of the hand, and generation of features based on the hand images for each video frame is based on the work of [7]. Additionally, the head of the subjects is tracked via the Viola-Jones's face detector [9]. The 2D position of each hand is then transformed to be relative to that of the head. The positions are normalized by dividing by the arm's length of the subject, measured in centimeters/pixels. In this way, the transformed positions represent displacements from the head, in units of centimeters.

The result of this process is a sequence of frame informations, in which for each frame we calculate the normalized position of both hands, and we extract an image of each hand with the background segmented.

---

[1] More information about this dataset can be found at http://facundoq.github.io/unlp/lsa64/.

3   Classification Model

The model combines the output of two subclassifiers, one for each hand. The subclassifier for each hand combines as well the output of three other subclassifiers that each use position, movement and handshape information (Fig. 2).

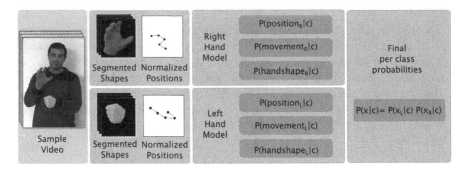

Fig. 2. Diagram of the model.

From the sample video, the sequence of segmented hand images and hand positions is extracted. For each hand, the position info is fed to the position and movement subclassifiers, while the segmented hand images are used as input for the handshape subclassifier. The right hand and left hand subclassifiers each output probabilities for each class, and the final output combines those per class.

Since the proposed model classifies each hand separately, we define the probability of a sample sign x given a class of sign c as:

$$P(x|c) = P(x^l|c)P(x^r|c) \tag{1}$$

where x^l and x^r is the sample sign information in the left and right hand respectively. Being able to split the probabilities in this way depends on the (naive) assumption that the synchronization between the hands is not important for the recognition of the sign, or at least that such information is not crucial for recognition.

The classifier for a single hand depends on several subclassifiers that focus on specific parts of the sign. Since the same type of subclassifiers are employed for each hand, in the following we assume that h can be either l or r, for *left* and *right*.

The three subclassifiers for a hand h use the information of the sequence of positions of the hand of the sample, x_p^h, the movement it performs, x_m^h, and the sequence of handshapes it goes through, x_s^h. Therefore the probability for a hand can be written as:

$$P(x^h|c) = P(x_p^h|c)P(x_m^h|c)P(x_s^h|c) \tag{2}$$

As before, Eq. 2 assumes independence between x_p^h, x_m^h and x_s^h, that is, that the position of the hand in no way restricts the types of possible movements or configurations, etc. Since signers are usually restricted to move their hands inside an imaginary square centered at their torso, this assumption holds in most practical cases.

Hence, to classify a sample sign x we pick the class of sign c with the maximum probability $P(x|c)$, that expands into:

$$P(x|c) = P(x^l|c)P(x^r|c) = P(x^l_p|c)P(x^l_m|c)P(x^l_s|c)P(x^r_p|c)P(x^r_m|c)P(x^r_s|c) \quad (3)$$

Table 1 describes symbols of the notation. In the following subsections, we describe how we calculate the probability for each subclassifier, and how we extend this model to deal with the absence of a hand and signs in which a hand does not move.

Table 1. Notation reference. The variable x refers always to sample information. Subscripts indicate type of information. Superscripts indicate the hand. Variable a refers to a parameter of the model.

c	Class	x	Sample	h	Hand h (generic)	$(\cdot)^h$	Hand h info
$(\cdot)^l$	Left hand info	$(\cdot)^r$	Right hand info	x^h_p	Position	x^h_s	Handshape
x^h_m	Movement	x^h_{tm}	Trajectory	x^h_{am}	Amount of movement	x^h_a	Absence of hand h in testing
a^h_c	Hand h is not used in class c	$a^h_{c,m}$	Absence of movement in training				

3.1 Position-Based Probability

From the set of absolute positions a hand goes through in the execution of the sign, we only employ the first (fp) and last (lp). We performed a 2D Kolmogorov-Smirnov normality test on the first and last positions and found that with a confidence level of 95 % there is enough evidence to reject the hypothesis of normality in 30 % of the classes. However, when fitting the positions with a Gaussian Mixture Model we found that for 78.5 % of the models a single component provided the best Bayesian Information Criteria score, and that performance on the test set was lower than with a single gaussian, possibly due to overfitting avoidance. Therefore, we chose to model the positions for each class using a single 2D normal distribution.

From the training set data we calculate the means $\mu_{fp,c}$ and $\mu_{lp,c}$ and covariances $\Sigma_{fp,c}$ and $\Sigma_{lp,c}$ of the first and last positions, for each class c. The probability for a new sample with position information x^p given class c is computed as:

$$P(x^h_p|c) = g_{fp,c}(x^h_p)g_{lp,c}(x^h_p) \quad (4)$$

where $g_{fp,c}$ is a 2D gaussian probability density function with mean $\mu_{fp,c}$ and covariance $\Sigma_{fp,c}$. The gaussian $g_{lp,c}$ is defined analogously for the last position lp.

3.2   Movement-Based Probability

We consider two factors for the movement-based probability, based on trajectory (x_{tm}) and amount of movement (x_{am}) information, so that:

$$P(x_m^h|c) = P(x_{tm}^h|c)P(x_{am}^h|c) \qquad (5)$$

Amount of Movement. We calculate the amount of movement for a hand h and a sign c as the maximum distance between two positions of the hand. In the training phase, we compute the mean amount of movement $\mu_{am,c}^h$ along with standard deviation $\sigma_{am,c}^h$. During testing, we penalize classes for which the movement of testing sample x differs greatly from $\mu_{am,c}^h$, so that:

$$P(x_{am}^h|c) = g_{am,h}(x_{am}^h) \qquad (6)$$

where $g_{am,h}$ is a 1D Gaussian probability density function with mean and standard deviation $\mu_{am,c}^h$ and $\sigma_{am,c}^h$.

Trajectory. We calculate the probability for the trajectory of the hand, $P(x_{tm}^h|c)$, adapting a classifier which was employed succesfully in [6] for action recognition. In this model, a movement of the hand is described as a discrete path in space, i.e., a list of positions in space. The model calculates the set of directions of that discrete path in space, which is formed by the set of (normalized) vector difference between hand positions in successive frames. By quantizing the possible directions, the classifier computes a distribution of directions of a sample movement, which is then compared to the distribution of directions of training samples to determine the probability of a sample for each class.

This subclassifier employs some sequence information, since to compute the directions of the sign, we need to compute the vector difference between positions of consecutive frames. However, after computing this difference, the order of the directions is irrelevant for the subclassifier. Moreover, if we consider each direction as an estimate of the instantaneous unit velocity of the hand at each frame, we can see that this scheme is simply a proxy for computing the velocities in each frame.

Signs with No Movement in One or Both Hands. Signs in which a hand does not move present a problem. If a class c has very little or no movement, then the trajectory probability $P(x_{tm}^h|c)$ is not useful and will probably penalize classes randomly. To avoid this situation, we employ again the per-class amount of movement means $\mu_{am,c}^h$ computed at training time, and we set a parameter:

$$a_{c,m}^h = \begin{cases} 1 & \text{if } \mu_{am,c}^h > 5cm \\ 0 & \text{if } \mu_{am,c}^h \leq 5cm \end{cases} \qquad (7)$$

where we determined the threshold as 5 cm experimentally. We can use this parameter as an exponent for $P(x_{tm}^h|c)$ to neutralize that factor for classes with no or little movement by redefining Eq. 5 as:

$$P(x_m^h|c) = P(x_{tm}^h|c)^{a_{c,m}^h} P(x_{am}^h|c) \tag{8}$$

In this way, if a sign does not have movement in a hand, by setting $a_{c,m}^h = 0$ the model can ignore the trajectory information.

3.3 Handshape-Based Probability

To obtain the probabilities of each class for the sequence of handshapes of a sign x, we first calculate the probability for each handshape on each frame of a sign, based on the segmented 2D image of the hand, using the classifier described in [7]. Then, we use the probabilities for all frames to model the sequence of handshapes and transformations of the hand.

To calculate the handshape-based probability for the whole sign, we also employ a similar approach as in the trajectory-based subclassifier, but now instead of quantizing over directions, we quantize over the set of vectors that indicate the probability that the hand is in a given hand pose.

3.4 One-Handed Signs

Ignoring the Information of an Unused Hand. Some classes of signs only use one hand. It is possible, however, that at recognition time the other hand is not off camera but present in the video with random positions, movements or handshapes. It is desirable not to consider the other hand's information for one-handed signs, since such random information could be interpreted by the classifier as a genuine attempt to perform part of a sign. To avoid those situations, we can turn Eq. 1 into:

$$P(x|c) = P(x^l|c)^{a_c^l} P(x^r|c)^{a_c^r} \tag{9}$$

where a_c^l and a_c^r have value either 1 or 0. Setting $a_c^h = 0$ allows the model to ignore information for hand h with respect to sign c, and is equivalent to setting the probability for that hand to 1.

We can calculate the parameters a_c^h in various ways; the simplest approach, used in this paper, is to set these parameters according with the annotations provided for the dataset specifying which hands are employed for each class of signs.

Exploiting the Absence of a Hand in the Video. When a hand h is missing from a sample video, we can assume that there is no possibility that the sign in the video uses that hand. If from the annotations mentioned in the previous subsection we know that hand h is used in the signs of class c (i.e., $a_c^h = 1$), but that hand is missing from the video, then the probability for a class c in that case should be set to 0.

We can compute x_a^h, a boolean variable with value 1 when hand h is present in a sign when testing. We consider a hand as present when it can be detected in more than 50 % of the frames of the video. With it we can define:

$$P(x_a^h|c) = \begin{cases} 0 & \text{if } a_c^h = 1 \text{ and } x_a^h = 0 \\ 1 & \text{otherwise} \end{cases} \tag{10}$$

We can then add two factors (one for each hand) to Eq. 9 to penalize this situation, such that:

$$P(x|c) = P(x^l|c)^{a_c^l} P(x_a^l|c) P(x^r|c)^{a_c^r} P(x_a^r|c) \tag{11}$$

4 Results

We performed experiments with the proposed model, setting parameters as described in the previous sections. We employed a stratified random subsampling cross-validation methodology, with an 80/20 training/testing split and 30 independent runs for each experiment. In the following subsections we present and analyze the results of the experiments.

4.1 Subject-Dependent Experiments

Table 2 shows the result of the subject-dependent experiments, where the model obtains an accuracy of 96.2 %. The table also shows the accuracies obtained by the model when using a subset of the features, to measure how much each information each of the corresponding subclassifiers is providing. All subclassifiers seem to provide non-redundant information, since the decreases in accuracy after removing a feature are all significant. Nonetheless, the position subclassifier can classify correctly a large amount (76.1 %) of signs by itself, which could be an effect due to the distribution of positions in the dataset.

The ALL-HMM column also shows the mean accuracy when replacing the trajectory and handshape subclassifiers with Hidden Markov Models with Gaussian Mixture Models output probabilities (HMM-GMM). We trained a model for each class and feature, using EM. Each model is a left-to-right HMM with skip transitions and 4 states in all cases. The handshape subclassifier also uses as input the output probabilities of the static handshape classifier described in [7].

We also performed subject-dependent experiments with the binary features described in [3], that include the same kind of information our classifier considers (position, movement and handshape). The features were calculated for each sample, and in each case the resulting $features \times frames$ matrix was resampled so that the number of frames was the same for all samples (32 frames in our experiments). The resulting $features \times 32$ matrices were used as input to a

Table 2. Subject-dependent experiments, with various combinations of features. ALL: all features. HS: Handshape features. MOV: Movement features. POS: Position features. ALL-HMM: All features with HMM subclassifiers. ALL-BF-SVM: All features with Binary Features and a SVM classifier.

	ALL	HS	MOV	POS	HS-POS	HS-MOV	POS-MOV	ALL-HMM	ALL-BF-SVM
μ	97.44	52.97	54.03	76.05	94.91	83.59	84.84	95.92	95.08
σ	0.59	1.74	1.71	0.62	0.52	0.87	0.90	0.95	0.69

one-versus-all multiclass SVM with a linear kernel. The results of these experiments, that followed the same cross-validation scheme, are shown in column ALL-BF-SVM.

To test the confusion of the model between one and two handed signs, we performed independent experiments dividing the dataset in two subsets: one with the one-handed signs (1H, 42 classes) and another with the two-handed ones (2H, 22 classes) (Table 3). Since the weighted mean between the accuracies of this two experiments does not differ significantly from the subject-dependent mean from Table 2 (97.44 %), this provides evidence that the model is discriminating very well between one and two-handed signs.

Table 3. Accuracy of the model on the one-handed (1H) and two-handed (2H) subsets of the LSA64 dataset. The last column shows the mean accuracy between 1H and 2H, weighted by the percentage of signs in each subset.

Subset	1H	2H	Mean
μ	95.93	99.09	97.01
σ	1.31	0.77	-

Figure 3 shows the confusion matrix for all signs of the dataset. There are few visible patterns in the matrix, providing some evidence that in general the model does not suffer from biases. There are however some exceptions. For example, the model confuses classes 24 and 26, since both posses similar movements and positions, and have only a subtle difference in handshape. Figure 4 shows the same confusion matrix after removing the handshape classifier from the model. Here the confusion between classes 24 and 26 increases dramatically, since without handshape information it is impossible to distinguish the two.

4.2 Subject-Independent Experiments

To evaluate how well the model generalizes with an unseen subject, we performed subject-independent experiments, where we trained the model with nine of the ten subjects and tested it with the remaining one. For each subject left out, we performed 30 runs of the experiment, whose results are show in Table 4. As expected, the accuracy decreases on these experiments, but not overmuch.

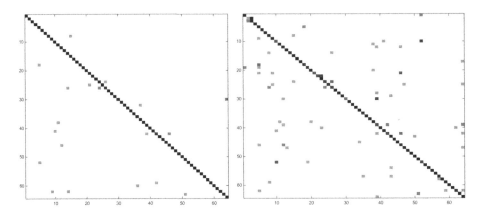

Fig. 3. Confussion Matrix of the LSA64 database.

Fig. 4. Confussion Matrix of the LSA64 without HandShapes Subclassifier.

Table 4. Subject-independent experiments on the LSA64 database. Each column shows the mean accuracy when testing with a subject. The final column shows total mean.

Subject	1	2	3	4	5	6	7	8	9	10	Mean
μ	94.5	93.8	87.7	93.8	91.8	92.6	89.1	90.3	88.4	94.6	91.7
σ	0.66	0.83	1.05	0.79	0.65	0.41	0.91	0.70	0.85	0.66	0.75

5   Conclusion

We have presented a sign language recognition model that does not employ frame-sequence information and still achieves low classification error for both subject-dependent and independent tasks. We tested the model on a medium-sized Argentinian Sign Language dataset. We also compared the model to a sequence dependent one by replacing the trajectory and handshape subclassifiers with HMM-GMM models, and found little difference in the accuracy of both models, providing further evidence of the validity of the sequence-agnostic model. This approach could point the way to new ways of defining sub-units for sign language recognition. The model could also provide advantages for real-time recognition, specially for dealing with out-of-order or missing frames.

In future work, we plan on testing the model on continuous sign language tasks with a sentence-level dataset, as well as determining its suitability for real time tasks. We also plan on introducing new sub-classifiers to improve the model's performance.

References

1. Cooper, H., Holt, B., Bowden, R.: Sign language recognition. In: Moeslund, T.B., Hilton, A., Krüger, V., Sigal, L. (eds.) Visual Analysis of Humans: Looking at People, Chap. 27, pp. 539–562. Springer, Heidelberg (2011)
2. Cooper, H., Ong, E.J., Pugeault, N., Bowden, R.: Sign language recognition using sub-units. J. Mach. Learn. Res. **13**, 2205–2231 (2012)
3. Kadir, T., Bowden, R., Ong, E., Zisserman, A.: Minimal training, large lexicon, unconstrained sign language recognition. In: British Machine Vision Conference, pp. 96.1–96.10 (2004)
4. Matuck, G.R., Moreira, G.S.P., Saotome, O., da Cunha, A.M.: Recognizing the Brazilian Signs language alphabet with neural networks over visual 3D data sensor. In: Bazzan, A.L.C., Pichara, K. (eds.) IBERAMIA 2014. LNCS, vol. 8864, pp. 637–648. Springer, Heidelberg (2014)
5. Pugeault, N., Bowden, R.: Spelling it out: real-time ASL fingerspelling recognition. In: 1st IEEE Workshop on Consumers Depth Cameras for Computer Vision, in Conjunction with ICCV 2011 (2011)
6. Ronchetti, F., Quiroga, F., Lanzarini, L., Estrebou, C.: Distribution of action movements (DAM): a descriptor for human action recognition. Front. Comput. Sci. **9**(6), 956–965 (2015)
7. Ronchetti, F., Quiroga, F., Lanzarini, L., Estrebou, C.: Handshape recognition for argentinian sign language using probsom. J. Comput. Sci. Technol. **16**(1), 1–5 (2016)
8. Roussos, A., Theodorakis, S., Pitsikalis, V., Maragos, P.: Hand tracking and affine shape-appearance handshape sub-units in continuous sign language recognition. In: Kutulakos, K.N. (ed.) ECCV 2010 Workshops, Part I. LNCS, vol. 6553, pp. 258–272. Springer, Heidelberg (2012)
9. Viola, P., Jones, M.J.: Robust real-time face detection. Int. J. Comput. Vis. **57**(2), 137–154 (2004)
10. Von Agris, U., Zieren, J., Canzler, U., Bauer, B., Kraiss, K.F.: Recent developments in visual sign language recognition. Univers. Access Inf. Soc. **6**(4), 323–362 (2008)
11. Zieren, J., Kraiss, K.-F.: Robust person-independent visual sign language recognition. In: Marques, J.S., Pérez de la Blanca, N., Pina, P. (eds.) IbPRIA 2005. LNCS, vol. 3522, pp. 520–528. Springer, Heidelberg (2005)

Computational Intelligence Soft Computing

Automatic Generation of Type-1 and Interval Type-2 Membership Functions for Prediction of Time Series Data

Andréia Alves dos Santos Schwaab[(⊠)], Silvia Modesto Nassar,
and Paulo José de Freitas Filho

Federal University of Santa Catarina, Florianópolis, Brazil
{andreia.schwaab,silvia.nassar,Freitas.filho}@ufsc.br

Abstract. The use of type-1 or type-2 membership functions in fuzzy systems offers a wide range of research opportunities. In this aspect, there are neither formal recommendations, methods that can help to decide which type of membership function should be chosen nor has the process of generating these membership functions been formalized. Against this background, this paper describes a study comparing the results of employing both a Genetic Algorithm and a Simulated Annealing for automatic generation of type-1 and interval type-2 membership functions. The paper also describes tests with different degrees of uncertainty inherent both to the input data and the fuzzy system rules. Experiments were conducted to predict the Mackey-Glass time series and the results were verified using statistical tests. The data obtained from statistical analysis can be used to determine which type of membership function is most appropriate for the problem.

Keywords: Genetic algorithms · Interval Type-2 fuzzy sets · Membership functions · Prediction of time series data · Simulated annealing

1 Introduction

Type-1 fuzzy sets [1] have been used for a wide range of applications [2–7]. Although it is common for this type of set to be employed in domains characterized by uncertainty, in Klir and Yuan [8], the authors say that type-1 membership functions, which describe type-1 fuzzy sets, are often too precise. Type-1 functions are appropriate when there is enough information to determine elements' membership of the set [9], once the shape and the parameters of the function have already been defined.

The need to represent uncertainty with relation to definition of the fuzzy set itself motivated Zadeh to propose type-2 sets [10], which have membership functions that are capable of representing uncertainty in their parameters. Applications of this type of membership function have expanded over recent years [11–15]. Type-2 sets are an approach with higher numbers of degrees of freedom, since the number of parameters that must be defined is larger. Type-2 sets can themselves be subdivided into General sets, which are three-dimensional, or Interval sets, which are fully described by a two-dimensional region of uncertainty, known as the footprint of uncertainty (FOU).

© Springer International Publishing AG 2016
M. Montes-y-Gómez et al. (Eds.): IBERAMIA 2016, LNAI 10022, pp. 353–364, 2016.
DOI: 10.1007/978-3-319-47955-2_29

Regarding the design of membership functions, there is no formal mechanism to aid in the process of deciding which type should be chosen in a fuzzy system modeling. Mendel [16] conducted an introductory study of the subject, the results of which included an equation designed to compare the results of type-1 and type-2 fuzzy systems, but there is still a need for practical applications of the results of that research, which was based on the underlying principle of generating type-2 functions from type-1 functions. Other authors have conducted experiments designed to compare the utilization of type-1 and type-2 sets [17–19]. These authors considered that using type-2 fuzzy sets is more promising than employing type-1 fuzzy sets. However, a series of adjustments to the parameters can prove necessary. In view of this, using an automatic method could be a good strategy to adopt.

Automatic methods for generation of membership functions are often based on using algorithms to conduct optimization of the function parameters [20–22]. Among other types of optimization, Genetic algorithms and the Simulated Annealing algorithm are two types of metaheuristic [23] that have been used to achieve this end in previous research. Application of genetic algorithms is a common method for generating membership functions [24–29]. In contrast, applications of Simulated Annealing are limited for this purpose and, in the case of type-2 membership functions, are a recent phenomenon [19]. Almaraashi [19] conducted a study employing Simulated Annealing and reported good results for parametrization of both type-1 and type-2 systems for predicting temporal series, with type-2 systems exhibiting similar and even superior results. In this context, this article describes a study comparing application of Genetic Algorithms and Simulated Annealing to the generation of type-1 and interval type-2 membership functions. This paper also describes evaluation of the performance of type-1 membership functions and interval type-2 functions when faced with different sources of uncertainty: data and rules. The experiments were conducted using data from the Mackey-Glass [30] time series.

2 Type-1 and Type-2 Fuzzy Sets and Systems

A type-1 fuzzy set is described by a membership function that maps the elements of the domain to a degree of membership that varies from zero to one. In general, the mapping can be defined by Eq. (1):

$$\mu_A : X \rightarrow [0, 1], \tag{1}$$

where A is a type-1 fuzzy set, X represents the domain and μ_A represents the degree of membership of x in the interval from 0 to 1. This means that the elements of the domain have a fixed value for their degree of association with the fuzzy set.

One alternative to deal with the need to represent uncertainty within the definition of membership functions is to employ type-2 sets and, in a more specific approach, interval type-2 sets [31]. A general type-2 set can be seen as a three-dimensional set with a primary degree of membership that, in turn, leads to a secondary degree of membership. In an interval type-2 set, the secondary degree of membership is always 1.

As such, their characteristics can be given by the union of the primary degrees of membership. Equation (2) is the mathematical expression of an interval type-2 set.

$$\tilde{A} = \{(x, u), 1 | \forall x \in X, \forall u \in J_x, J_x \subseteq [0, 1]\} \tag{2}$$

In Eq. (2) x is an element of the domain X, u is the primary degree of membership and J_x is the domain of the secondary degree of membership and the image of the primary degree of membership. An interval type-2 fuzzy set is represented by its primary membership, reflected by its FOU. A FOU is the area delimited by two membership functions, which are known as the upper and lower functions of the type-2 function.

When modeling fuzzy systems, type-1 or type-2 sets may be used. A type-1 fuzzy inference system (or type-1 fuzzy logic system) has type-1 membership functions. In turn, a type-2 fuzzy inference system has at least one type-2 membership function. Additionally, the type-2 inference system includes one more step than the type-1 system, which is the fuzzy system type reduction step. Type reduction is a procedure by which a type-2 set is reduced to a type-1 set. The type reduction method most commonly employed is the centroid technique, which consists of an iterative procedure through which an interval type-1 set is obtained. If it is necessary to obtain a numerical value, it becomes necessary to perform defuzzification, which generally employs the center-of-sets method. For further details on fuzzy inference systems, consult Karnik, Mendel and Liang [32].

3 Generation of Fuzzy Type-1 and Type-2 Membership Functions Using Genetic Algorithms and Simulated Annealing

As part of this study, methods were created to define type-1 and type-2 membership functions. Since the objectives of the study include the employment of type-2 fuzzy sets, an application that is considered appropriate for this context was chosen. The problem chosen was the prediction of the Mackey-Glass chaotic time series, since utilization of type-2 sets can be appropriate for applications in which data vary with time [33]. Furthermore, prediction of the Mackey-Glass series is already a reference application used in other investigations in the area [19, 31, 34–37]. The time series data are obtained using the nonlinear equation shown in Eq. (3).

$$\frac{dx(t)}{dt} = \frac{a * x(t - \tau)}{1 + x^n(t - \tau)} - b * x(t) \tag{3}$$

In Eq. (3), t refers to the current point in time; $a = 0.2$, $b = 0.1$ and $n = 10$ are constants. The parameter $\tau = 17$ is applied because from this value onwards the series exhibits chaotic behavior. These parameters were also used by Almaraashi [19].

Experiments were conducted using this data series with and without the addition of noise. The noise was normally distributed and was added at two different levels, so that the resulting Signal to Noise Ratio (SNR) would be 10 dB or 20 dB. Modeling of the fuzzy systems was accomplished using the Juzzy framework [38], which is a code library developed in the Java® language that makes it possible to develop fuzzy inference systems, both type-1 and type-2. All other algorithms were also implemented in the Java language to facilitate integration with the framework.

Gaussian membership functions were adopted to construct both types of fuzzy system. The system is composed of 4 input variables and 1 output variable, which represents the prediction made by the system. The input variables are the points in the series denoted by x_{t-18}, x_{t-12}, x_{t-6} and x_t. The output of the inference system is point x_{t+6}. This prediction strategy using a 6-point difference between each variable has been adopted by other authors [19, 34, 39]. Each variable is described by 3 (three) fuzzy sets: low, medium and high. Three sets were used in order to prevent the rule base from becoming too large.

The rule base was generated using two methods, one empirical and one automatic. A procedure inspired by the Wang and Mendel method [40, 41] was used for automatic rule generation. The first step was to divide the time series data space into three regions (low, medium and high) and the data in the time series were then classified in each of these regions. The rules were constructed on the basis of the classification of the data. Where rules conflicted, the least frequent were eliminated. The automatic method was used to define two groups of six rules: the most frequent and the least frequent. In addition to automatic extraction, another experimental case was conducted using six rules defined empirically, on the basis of observation of the data.

Two analysis strategies were adopted:

1. Test both the Genetic Algorithm and the Simulated Annealing method for generation of type-1 and type-2 membership functions in the presence of different sources of uncertainty: noise-free data and noisy data (SNR = 20 dB); and empirical rules, least frequent rules and most frequent rules. The objective of this strategy was to evaluate differences between the optimization methods employed.
2. Test the Genetic Algorithm for generation of type-1 and type-2 membership functions considering the following conditions: a noisy data series (SNR = 10 dB), empirical rules, most frequent rules and least frequent rules. The objective of this strategy was to evaluate differences between the results obtained using type-1 and type-2 fuzzy systems.

The performance of each algorithm was tested by calculating the Root Mean Square Error (RMSE), which is represented by Eq. (4).

$$RMSE = \sqrt{\frac{1}{n}\sum_{k=1}^{n}(y_i - \hat{y}_i)^2} \tag{4}$$

In Eq. (4), n is the total number of points analyzed, y_i is the true value observed in the time series and $\hat{y}_i$ is the value predicted by the system. The next section describes the two optimization methods for discovery of membership functions for predicting the Mackey-Glass time series. A Genetic Algorithm and a Simulated Annealing Algorithm were developed, both for definition of type-1 functions and type-2 functions.

3.1 Genetic Algorithm

A genetic algorithm [42] is a metaheuristic that performs population optimization. It is based on the evolutionary processes of living beings. This type of approach works with a sufficiently large number of solutions that pass through many stages until the number of maximum generations of the initial population has been reached. Pseudocode representing the genetic algorithm used for these experiments is shown in Fig. 1. The genetic algorithm generates an initial population at random, made up of N individuals, in which each individual represents one configuration of membership functions (input and output) for the fuzzy system. The fitness of each individual is tested by calculating the RMSE.

---

1. Generate a population with N solutions at random;
2. Calculate the fitness of each individual in the population;
3. For L=1 to *Max_generations* do
   a. Perform genetic recombination (select parents and cross breed), generating two children
   b. Perform mutation with probability α (50%)
   c. If the individual thus generated has lower fitness than the population mean then substitute with the worst member of the existing population

N=2500, *Max_generations*=10000

Fig. 1. Pseudocode for the genetic algorithm.

Two items of information are needed in order to represent type-1 Gaussian functions: the mean and the standard deviation. The chromosome is made up of real values and has size 30, because two genes are needed to represent each set of input and output variables (5 variables composed of 3 sets, each represented by 2 genes).

There is a little more complexity involved in representing type-2 sets, since it is necessary to define two means and two standard deviations, or one mean and two standard deviations. In order to limit this complexity, one mean and two standard deviations were adopted per interval type-2 set. As such, the size of the chromosome is now 45, because one extra gene is needed to represent each set.

3.2 Simulated Annealing

Simulated Annealing [43] is an alternative method of optimization. The method works by starting from a given solution within the search space and exploring neighboring solutions. The strategy is known as single-point optimization. The principal idea underlying the method is to reproduce the movement of molecules during the process of cooling metals (which is continued until cooling is complete and the molecules are no longer able to move). The Simulated Annealing algorithm implemented for this study is summarized in Fig. 2. Certain parameters must be defined before these algorithms can be executed and these are also specified in Fig. 2.

1. Select an initial state *i* within the search space *S*
2. Select a temperature *T >0*
3. Repeat
Calculate the number of iterations to be performed at each temperature *N(T)*
Repeat
        Generate state *j*, which is a neighbor of *i*
        If the cost of *j* is lower than the cost of *i* then
            i←j
        Else, if random(0,1)<exp((-f(j)-f(i))/T) then
            i←j   */* j may be accepted if a given probability is met*/*
            n←n+1
     until *n=N(T)* */* until the number of iterations per
                 temperature is reached */*
t ← t + 1
T ← *f* (T)*/* Cooling function */*
Until stop criterion */* minimum temperature is reached */*

T=500; N(T)=10; *f*(T)=αT, where α=0.999

Fig. 2. Pseudocode for simulated annealing.

The initial solution is generated at random and then neighboring solutions are explored. The algorithm chooses a neighboring solution at random and this is achieved using a step size (0.003). If the neighboring solution has a lower cost than the current state, it is immediately accepted by the algorithm. If not, it may still be accepted at a probability that is higher during the first iterations (when temperature is still high). The inclusion of the capability to accept worse solutions is one of this method's differentiating factors and helps to avoid the algorithm becoming trapped at local minimums.

In common with the genetic algorithm, each solution for type-1 membership functions has size 30 while solutions for type-2 membership functions have size 45. Both the genetic algorithm and the simulated annealing algorithm generate type-2 membership functions independently from type-1 membership functions.

4 Results

Table 1 presents the results (mean RMSE) for the simulations conducted with each of the algorithms implemented, for each type of membership function and for the three different sets of rules (empirical, most frequent and least frequent). In this step, simulations were conducted using data free from noise and using data with an SNR of 20 dB. Each experiment was repeated 10 times and the results shown are the means of these repetitions. The results given in Table 1 for these tests conducted using noise-free data show that Simulated Annealing produced the best results (0.031 and 0.036). Compared with the Genetic Algorithm it produced better results when the most frequent rule base was used. However, when the empirical and least frequent rules were used the genetic algorithm produced superior results. Using the least frequent rules, the results using Simulated Annealing were significantly worse.

Table 1. Results of simulations (mean RMSE) for the noise-free Mackey-Glass series and for the series with a SNR of 20 dB.

Method	Type of rule	Noise-free data		Noisy data (SNR 20 dB)	
		Type-1	Type-2	Type-1	Type-2
Genetic algorithm	Empirical rules	0.0488003	0.0555451	0.0570526	0.06042858
	Most frequent rules	0.0378878	0.0385007	0.0433673	**0.04598173**
	Least frequent rules	0.0376379	0.0448113	0.0455106	0.05238116
Simulated Annealing	Empirical rules	0.0509107	0.0572951	0.0569687	0.06370764
	Most frequent rules	**0.0313965**	**0.0369703**	**0.0403663**	0.04859556
	Least frequent rules	0.0862893	0.0797911	0.0885165	0.07822319

Analysis of the results for noisy data reveals that the best results for type-1 functions were obtained using the Simulated Annealing method. In contrast, the best results for type-2 functions were produced by the Genetic Algorithm. Comparison of the type-1 and type-2 membership function results shows that, in general, the results of using type-1 membership functions were superior.

The ANOVA statistical test was applied in order to understand the influence on the results of each of the factors tested (method, type of membership function and type of rule) and to test whether the differences observed between different sets of results were significant. The objective of this test is to verify whether the mean results are similar or not and to investigate whether there are interactions between the factors tested. The null hypothesis (H_0) tested with ANOVA is that the means are equal. The alternative hypothesis is that the means for different treatments are different (H_1). Table 2 lists the results of this statistical analysis to a significance level of 5%.

The statistical analysis in Table 2 suggests that there are significant differences between results in terms of which method is used (since according to the test p value, the null hypothesis should be rejected). This is because of the influence of the results for the group using the least frequent rules. Similarly, the test also indicates that means are different when different groups of rules are compared (the p value is 0.001). However, the results comparing membership function types were statistically similar. Table 3 lists the results of the test for interactions between factors.

It can be observed from Table 3 that the statistical tests indicate that for both noise-free data and for noisy data there was an interaction between the method of optimization used and the type of rules followed. This has already been noted above, where it was observed that SA was superior to GA when the most frequent rules were followed, but inferior when empirical rules and least frequent rules were followed. None of the other combinations of factors exhibited significant interactions.

Table 2. Statistical analysis of mean RMSE.

	Noise-free data		SNR = 20 dB		Test to 5% significance [a,b]
	mean RMSE	p value	mean RMSE	p value	
Genetic Algorithm	0.04383	0.001	0.05079	0.001	Means are different (H_1)
Simulated Annealing	0.05711		0.06273		
Type-1 MF	0.04879	0.389	0.0553	0.331	Means are similar (H_0)
Interval Type-2 MF	0.05215		0.05822		
Empirical Rules	0.05314	0.001	0.05954	0.001	Means are different (H_1)
Least Frequent Rules	0.06213		0.06616		
Most Frequent Rules	0.03614		0.04458		

[a] The statistical significance level of this test indicates that there is a less than 5% probability that the null hypothesis has been rejected when in fact it was correct.
[b] Conclusions for both tests

Table 3. Statistical test for interactions between factors.

Interaction	p value (noise-free)	Test to 5% significance [a]	p value (SNR 20 dB)	Test to 5% significance [b]
Method and Type of Membership Function	0.679	No significant interaction (H_0)	0.627	No significant interaction (H_0)
Rules and Methods	0.001	**Significant interaction (H_1)**	0.001	**Significant interaction (H_1)**
Type of Membership Function and Rules	0.756	No significant interaction (H_0)	0.429	No significant interaction (H_0)

[a] Result for tests using data free from noise
[b] Results for tests using noisy data

In terms of the time taken for each simulation, the Simulated Annealing algorithm exhibited worse performance, since a longer simulation time was needed to obtain the membership functions for the fuzzy system. The simulation times for optimization of a type-1 system were similar, with the Simulated Annealing algorithm taking around 1.05 times longer. However, the difference in time taken to optimize a type-2 system was larger, with SA taking around 2.88 times longer than GA.

Table 4. Results of simulations (mean RMSE) with the Genetic Algorithm and a data series with a 10 dB SNR.

Type of rules	SNR = 10 dB	
	Type-1	Type-2
Empirical rules	0.070622578	0.071183647
Most frequent rules	0.063050118	0.063079639
Least frequent rules	0.073096869	0.072745525

Another round of tests was conducted using the Genetic Algorithm only. In these experiments more noise was added to the time series data. The results of these simulations are shown in Table 4.

The data in Table 4 show that the means for the results obtained are very similar when one type of membership function is compared with the other. Similarly, the results of the ANOVA statistical test to 5% significance indicate that there were no differences between the means for the two different types of membership function (p value 0.946). Otherwise, the test of different types of rules shows that the means are different (p value 0.001), with the most frequent rules exhibiting the best results and the least frequent rules exhibiting the worst result. The statistical test detected no significant interaction between type of rules and type of membership function (p value 0.95).

Although at this moment the goal of this work is not to achieve the best method for predicting the Mackey-glass time series, the developed methods seem promising when their results are compared to those obtained by Almaraashi [19]. This occurs because of the better results acquired for the series with noise, as can be observed in Table 5. Thus, the Genetic Algorithm and the Simulated Annealing developed in this work revealed to be more suitable in presence of uncertainty.

Table 5. Performance comparison (mean RMSE) for the noise-free Mackey-Glass series and for the series with an SNR of 20 dB.

	Noise free	SNR = 20 dB
Almaraashi [19] - Type-1 MF	0.00704	0.159
Almaraashi [19] - Type-2 MF	**0.003988**	0.1558
Genetic Algorithm – Type-1 MF- this paper	0.0378878	0.0433673
Genetic Algorithm – Type-2 MF- this paper	0.0385007	0.0459817
Simulated Annealing – Type-1 MF- this paper	0.0313965	**0.0403663**
Simulated Annealing – Type-2 MF- this paper	0.0369703	0.0485956

It is observed in Table 5, however, that the results obtained for the noise free data were inferior to those results achieved by Almaraashi [19]. The results presented in Table 5 consider the use of the most frequent rules group for both Simulated Annealing and Genetic Algorithm. In Almaraashi's [19] research, sixteen rules were applied.

5 Conclusion

This article describes implementations of two optimization methods for generating type-1 and type-2 fuzzy membership functions. The optimization algorithms used were the Genetic Algorithm and the Simulated Annealing Algorithm. A range of different levels of uncertainty in the input data and the rule bases were tested. In order to vary the level of uncertainty inherent in the data, the Mackey-Glass time series was predicted with different levels of noise. In order to test different levels of rule uncertainty (varying in terms of subjectivity and frequency), rules defined empirically and rules defined automatically were tested, using two sets of automatically defined rules, most frequent and least frequent.

Analysis of the type of method employed demonstrated that simulated annealing offered good results for situations in which there was less uncertainty in the rules, but although simulated annealing resulted in lower means when the most frequent rules were used, statistical analysis showed that the differences between simulated annealing and the genetic algorithm were significant when all of the experiments were considered. When deciding whether to employ simulated annealing for optimization of fuzzy systems, an analysis should also be conducted to determine whether the computational cost of using SA is justified by the results, since simulation times are longer.

The type of rules followed had a significant influence on the results, with the set of rules made up of the most frequent rules producing results that were closest to the expected results. This shows that definition of the rule base is very important to obtaining satisfactory results and that the results are dependent on both functions and rules. With regard to situations in which uncertainty is present in the rules, type-2 fuzzy systems were not significantly superior to type-1 fuzzy systems.

It was observed that these optimization processes were capable of producing type-1 functions that offered superior results to type-2 functions, which was confirmed by statistical analysis of the means. However, when noise was added to the data, the performance of the system using type-2 functions improved to the extent that the results of the test conducted with an SNR of 10 dB were approximately equal. This confirms the claim that type-2 systems respond well in situations of uncertainty and, more specifically, in situations in which uncertainty is present in the input data. Notwithstanding, since the statistical test showed that the means were similar, type-1 functions are still the better choice because they are less complex.

References

1. Zadeh, L.A.: Fuzzy sets. Inf. Control **8**, 338–353 (1965)
2. Abd, K., Abhary, K., Marian, R.: Development of a fuzzy-simulation model of scheduling robotic flexible assembly cells. J. Comput. Sc. **9**(12), 1761–1768 (2013)
3. Amindoust, A., Ahmed, S., Saghafinia, A., Bahreininejad, A.: Sustainable supplier selection: a ranking model based on fuzzy inference system. Appl. Soft Comput. **12**(6), 1668–1677 (2012)

4. Chrysafiadi, K., Virvou, M.: Evaluating the integration of fuzzy logic into the student model of a web-based learning environment. Expert Syst. Appl. **39**(18), 13127–13134 (2012)
5. Frantti, T., Majanen, M.: An expert system for real-time traffic management in wireless local area networks. Expert Syst. Appl. **41**(10), 4996–5008 (2014)
6. Laasri, E.H.A., Akhouayri, E.-S., Agliz, D., Zonta, D., Atmani, A.: A fuzzy expert system for automatic seismic signal classification. Expert Syst. Appl. **42**(3), 1013–1027 (2015)
7. Onieva, E., Milanés, V., Villagrá, J., Pérez, J., Godoy, J.: Genetic optimization of a vehicle fuzzy decision system for intersections. Expert Syst. Appl. **39**(18), 13148–13157 (2012)
8. Klir, G., Yuan, B.: Fuzzy Sets and Fuzzy Logic: Theory and Applications. Prentice Hall, New Jersey (1995)
9. Ross, T.J.: Fuzzy Logic with Engineering Applications, 3rd edn. Wiley, UK (2010)
10. Zadeh, L.A.: The concept of a linguistic variable and its application to approximate reasoning–I. Inf. Sci. **8**, 199–249 (1975)
11. Sharifian, A., Sharifian, S.: A new power system transient stability assessment method based on type-2 fuzzy neural network estimation. Int. J. Electr. Power Energy Syst. **64**, 71–87 (2015)
12. Pan, Y., Li, H., Zhou, Q.: Fault detection for interval type-2 fuzzy systems with sensor nonlinearities. Neurocomputing **145**, 488–494 (2014)
13. Murthy, C., Varma, K.A., Roy, D.S., Mohanta, D.K.: Reliability evaluation of phasor measurement unit using type-2 fuzzy set theory. IEEE Syst. J. **8**(4), 1302–1309 (2014)
14. Bi, Y., Srinivasan, D., Lu, X., Sun, Z., Zeng, W.: Type-2 fuzzy multi-intersection traffic signal control with differential evolution optimization. Expert Syst. Appl. **41**(16), 7338–7349 (2014)
15. Melin, P., Castillo, O.: A review on type-2 fuzzy logic applications in clustering, classification and pattern recognition. Appl. Soft Comput. **21**, 568–577 (2014)
16. Mendel, J.M.: A quantitative comparison of interval type-2 and type-1 fuzzy logic systems: first results. In: IEEE International Conference on Fuzzy Systems (FUZZ), pp. 1–8. Barcelona (2010)
17. Linda, O., Manic, M.: Comparative analysis of type-1 and type-2 fuzzy control in context of learning behaviors for mobile robotics. In: Proceedings of the 36th Annual Conference of the IEEE Industrial Electronics Society (IECON 2010),. Glendale, AZ, pp. 1092–1098 (2010)
18. Linda, O., Manic, M.: Evaluating uncertainty resiliency of type-2 fuzzy logic controllers for parallel delta robot. In: 4th International Conference on Human System Interaction, pp. 91–97. Yokohama (2011)
19. Almaraashi, M.: Learning of Type-2 Fuzzy Logic Systems using Simulated Annealing. De Montfort University (2012)
20. Sanchez, M.A., Castillo, O., Castro, J.R.: Information granule formation via the concept of uncertainty-based information with interval type-2 fuzzy sets representation and Takagi-Sugeno-Kang consequents optimized with Cuckoo search. Appl. Soft Comput. **27**, 602–609 (2015)
21. Martínez-Soto, R., Castillo, O., Aguilar, L.T., Díaz, A.R.: A hybrid optimization method with PSO and GA to automatically design type-1 and type-2 fuzzy logic controllers. Int. J. Mach. Learn. Cybern. **6**(2), 175–196 (2015)
22. Maldonado, Y., Castillo, O., Melin, P.: A multi-objective optimization of type-2 fuzzy control speed in FPGAs. Appl. Soft Comput. **24**, 1164–1174 (2014)
23. Sörensen, K., Glover, F.: Metaheuristics. In: Gass, S.I., Fu, M.C. (eds.) Encyclopedia of Operations Research and Management Science, 3rd edn. Springer, Boston (2013)
24. Shill, P.C., Akhand, M.A.H., Murase, K.: Simultaneous design of membership functions and rule sets for type-2 fuzzy controllers using genetic algorithms. In: 14th International Conference on Computer and Information Technology, Dhaka, pp. 554–559 (2011)

25. Maldonado, Y., Castillo, O., Melin, P.: Optimization of membership functions for an incremental fuzzy PD control based on genetic algorithms. Stud. Comput. Intell. **318**, 195–211 (2010)
26. Hidalgo, D., Melin, P., Castillo, O.: Optimal design of type-2 fuzzy membership functions using genetic algorithms in a partitioned search space. In: IEEE International Conference on Granular Computing, San Jose, pp. 212–216 (2010)
27. Herman, N.S., Yusuf, I., Shamsuddin, S.M.b.H.: Genetic algorithms and designing membership function in fuzzy logic controllers. In: World Congress on Nature & Biologically Inspired Computing (NaBIC 2009), Coimbatore, pp. 1753–1758 (2009)
28. Ghaemi, M., Akbarzadeh-T, M., Jalaeian-F, M.: Adaptive interval type-2 fuzzy PI sliding mode control with optimization of membership functions using genetic algorithm. In: 2nd International eConference on Computer and Knowledge Engineering (ICCKE), Mashhad, pp. 123–128 (2012)
29. Patri, A., Nayak, A.: A fuzzy-based localization in range-free wireless sensor network using genetic algorithm & Sinc membership function. In: International Conference on Green Computing, Communication and Conservation of Energy (ICGCE), Chennai, pp. 140–145 (2013)
30. Mackey, M.C., Glass, L.: Oscillation and chaos in physiological control systems. Science **197**, 287–289 (1977)
31. Liang, Q., Mendel, J.M.: Interval type-2 fuzzy logic systems: theory and design. IEEE Trans. Fuzzy Syst. **8**(5), 535–550 (2000)
32. Karnik, N.N., Mendel, J.M., Liang, Q.: Type-2 fuzzy logic systems. IEEE Trans. Fuzzy Syst. **7**(6), 643–658 (1999)
33. Mendel, J.M.: Fuzzy Sets for Words: Why Type-2 Fuzzy Sets Should be Used and How They Can be Used. IEEE-FUZZ (2004) tutorial. http://ewh.ieee.org/cmte/cis/mtsc/ieeecis/Mendel.pdf
34. Castellano, G., Fanelli, A.M., Mencar, C.: Design of transparent Mamdani fuzzy inference systems. In: Abraham, A., Köppen, M., Franke, K. (eds.) Design and Application of Hybrid Intelligent Systems, pp. 468–476. IOS Press, The Netherlands (2003)
35. Kasabov, N., Song, Q.: DENFIS: Dynamic evolving neural-fuzzy inference system and its application for time-series prediction. IEEE Trans. Fuzzy Syst. **10**, 144–154 (2002)
36. Angelov, P., Filev, D.: An approach to online identification of Takagi-Sugeno fuzzy models. IEEE Trans. Syst. Man Cybern. **34**, 484–498 (2004)
37. Serir, L., Ramasso, E., Nectoux, P., Zerhouni, N.: E2GKpro: an evidential evolving multi-modeling approach for system behavior prediction with applications. Mech. Syst. Signal Process. **37**, 213–228 (2013)
38. Wagner, C.: Juzzy - a Java based toolkit for type-2 fuzzy logic. In: IEEE Symposium on Advances in Type-2 Fuzzy Logic Systems (T2FUZZ), pp. 45–52. Singapore (2013)
39. Kasabov, N.: Evolving Connectionist Systems: The Knowledge Engineering Approach, 2nd edn. Springer, London (2007)
40. Wang, L.-X.: The WM method completed: a flexible fuzzy system approach to data mining. IEEE Trans. Fuzzy Syst. **11**(6), 768–782 (2013)
41. Wang, L.-X., Mendel, J.M.: Generating fuzzy rules by learning from examples. IEEE Trans. Syst. Man Cybern. **22**(6), 1414–1427 (1992)
42. Holland, J.H.: Adaptation in Natural and Artificial Systems. University of Michigan Press, Ann Arbor (1975)
43. Kirkpatrick, S., Gelatt, C.D., Vecchi, M.P.: Optimization by simulated annealing. Science **220**, 671–680 (1983)

Calibration of Microscopic Traffic Flow Simulation Models Using a Memetic Algorithm with Solis and Wets Local Search Chaining (MA-SW-Chains)

Carlos Cobos[1]([⊠]), Carlos Daza[1], Cristhian Martínez[1],
Martha Mendoza[1], Carlos Gaviria[2], Cristian Arteaga[2],
and Alexander Paz[2]

[1] Universidad del Cauca, Popayán, Colombia
{ccobos, carlosdazar, cristhianm,
mmendoza}@unicauca.edu.co
[2] University of Nevada, Las Vegas, USA
{cgaviria, carteaga, apaz}@unlv.edu

Abstract. Traffic models require calibration to provide an adequate representation of the actual field conditions. This study presents the adaptation of a memetic algorithm (MA-SW-Chains) based on Solis and Wets local search chains, for the calibration of microscopic traffic flow simulation models. The effectiveness of the proposed MA-SW-Chains approach was tested using two vehicular traffic flow models (McTrans and Reno). The results were superior compared to two state-of-the-art approaches found in the literature: (i) a single-objective genetic algorithm that uses simulated annealing (GASA), and (ii) a stochastic approximation simultaneous perturbation algorithm (SPSA). The comparison was based on tuning time, runtime and the quality of the calibration, measured by the GEH statistic (which calculates the difference between the counts of real and simulated links) .

Keywords: Calibration · Local search chaining · Solis and wets · Traffic flow simulation · Single-objective optimization · Memetic algorithm

1 Introduction

The last decade has observed an increase in the use of microscopic traffic flow simulation for a wide range of purposes. Calibration of the simulation models is critical to ensure validity and significance of the analysis. Calibration requires field measurements to be compared to the corresponding simulation outputs [1].

Microscopic traffic flow simulation models include many parameters that are adjusted to capture the required behavior. General principles for calibration of a microscopic traffic flow simulation models are summarized as follows: (1) pre-calibration, (2) selection of parameters to calibrate and the corresponding range, (3) selection of goodness of fit measures, (4) defining of a calibration procedure, and (5) execution of calibration, validation and analysis of results [2]. Optimization is recommended and typically used in step

© Springer International Publishing AG 2016
M. Montes-y-Gómez et al. (Eds.): IBERAMIA 2016, LNAI 10022, pp. 365–375, 2016.
DOI: 10.1007/978-3-319-47955-2_30

4 (calibration procedure) for searching the numeric value to assign to each of the selected parameters to minimize the difference between simulation and field measurements.

Studies in the past have proposed a variety of methods and algorithms for conducting the search described [3]. The single sequence algorithm [4] and methodologies of stochastic approximation have been used for the simultaneous calibration of the model parameters [5, 6]. The two techniques used today include genetic algorithms (GA) [7, 8] and the simultaneous perturbation stochastic approximation algorithm (SPSA) [9, 10]. These techniques do not require information about the characteristics of the microscopic traffic flow simulation model [11]. That is, information about the methods used by the traffic flow simulation model to move vehicles and represent traffic flow dynamics is not required. Hence, these techniques assumed to work well with various types of traffic flow models.

Previous studies indicate that SPSA and GA provide similar results [10, 12]. However, SPSA requires less time and computational resources than GA. A memetic algorithm [13], GASA combines genetic algorithms with simulated annealing, which provides adequate results for low and medium complexity networks. A comparison with SPSA showed promising results [13].

There is no approach in the literature for the calibration of microscopic traffic flow models using memetic algorithms based on local search chains, as proposed in this paper. The first memetic algorithm based on local search chain approach was introduced in 2010 [14] and 2015 [15]. The algorithm showed great performance solving complex problems (continuous).

One of the interpretations of the No-Free-Lunch Theorem for optimization is that it is unknown which meta-heuristic can solve a problem in the best possible manner [16]. Thus, this paper investigates the performance of the MA-SW-Chains algorithm [14] compared to two alternative state-of-the-art algorithms found in the resent literature [11, 13].

The rest of the article is organized as follows: Sect. 2 defines the problem and explains how the MA-SW-Chains algorithm was adapted to solve the proposed calibration problem. Section 3 describes the experiments and results obtained with the MA-SW-Chains algorithm. Section 4 presents the comparison between the MA-SW-Chains, GASA, and SPSA algorithms. Finally, Sect. 5 provides conclusions and directions for future work.

2 Proposal

2.1 Formal Definition of the Problem

This study used the general principles described previously and in detail [2] for the calibration of microscopic traffic flow models. The objective function to minimize is the Root Mean Squared Normalized Error (RMSNE) [10] as shown in Eq. (1). Minimize RMSNE implies the minimization of the difference between actual and simulated volumes (vehicle counts) and actual and simulated vehicle speeds.

$$RMSNE = \frac{1}{\sqrt{N}} * \sum_{t=1}^{T} \left(W * \sqrt{\sum_{i=1}^{N} \left(\frac{V_{i,t} - \tilde{V}_{(\theta)i,t}}{V_{i,t}} \right)^2} + (1 - W) * \sqrt{\sum_{i=1}^{N} \left(\frac{S_{i,t} - \tilde{S}_{(\theta)i,t}}{S_{i,t}} \right)^2} \right) \quad (1)$$

where $V_{i,t}$ is the actual count for link i at time t and $\tilde{V}_{(\theta)i,t}$ is the simulated count for link i at time t. $S_{i,t}$ is the actual speed for link i at time t and $\tilde{S}_{(\theta)i,t}$ is the simulated speed for link i at time t. W is a weight used to assign value to the volume and the speed, N is the total number of links in the model, and T is the total number of time periods t.

Calibration criteria: The Federal Highway Administration (FHWA) guidelines for the calibration of CORSIM models were used in this study. The difference between the count for real and simulated links ought to be less than 5 for all links. The statistical GEH [2] was used, which should be less than 5 in at least 85 % of the links. The formula used to calculate the GEH statistic is provided by Eq. (2).

$$GEH = \sum_{i=1}^{N} \sqrt{\frac{2(V_i - \tilde{V}_{(\theta)i})^2}{V_i + \tilde{V}_{(\theta)i}}} \quad (2)$$

where V_i and $\tilde{V}_{(\theta)i}$ are, respectively, the actual and simulated count for link i.

2.2 Adaptation of the MA-SW-Chains Algorithm

In this paper, each agent (solution or individual) represents a parameter vector containing a solution for the optimization problem. In addition to a measure of fitness (efficacy), a bias vector indicates the direction of the local search. A standard deviation value (rho) is used in the Gaussian Distribution within the Solis and Wets algorithm [17] to generate an offset vector, which helps to change the direction of the bias vector. The representation of an agent is illustrated in Fig. 1.

P_1 P_2	Parameters	P_{n-1} P_n	0 1	Bias	$n-1$ n	rho	Fitness
	...			...		2,5	0,016

Fig. 1. Representation of an agent in the MA-SW-Chains algorithm.

Users (traffic engineers) need to select parameters that are calibrated for a specific model. These selected parameters (or record types) are part of the agent (see [11] for more details). These parameters take integer values based on restrictions defined in the CORSIM user manual [18].

The MA-SW-Chains algorithm uses a steady state GA to explore the solution search space. In this steady state GA, an (or several) offspring is created in every generation and competes with the agents in the population [19] to be a part of the new population. The best agents are used to generate new populations through the genetic operators of the algorithm. The best agent of the population in each generation is selected for the process of exploitation using the Solis and Wets algorithm to refine the search to find better solutions.

The differentiating approach of MA-SW-Chains is the chaining of local searches, assigning a local search intensity (Istr) to each agent and exploiting the most promising agents more intensely. To adapt the local search intensity, local search can be applied several times to the same agent and their final parameters stored. These are used again when the agent is selected to be optimized locally, thereby creating the local search chain [14]. The stopping criterion for MA-SW-Chains is the maximum number of generations (a parameter of the algorithm), or the maximum execution time (an alternative parameter).

The MA-SW-Chains algorithm for supporting the calibration process is defined by the following steps:

Step 1:   Generate the initial population randomly and apply restrictions to the agents to avoid non-permitted values. Each parameter to be calibrated has a previously defined range and type (in this case, integer). Fitness is calculated for all agents of the population based on RMSNE according to Eq. (1). The size of the population is a parameter previously defined by the user

Step 2:   Generate new agents (the number of agents is determined by the algorithm parameters set by the user) using a tournament selection genetic operator, uniform crossover, and multigene mutation. These new agents comply with the restrictions to avoid non-permitted values

Step 3:   Calculate fitness for new agents based on RMSNE and add to the population

Step 4:   Given that the lowest value of RMSNE is sought, agents with high fitness values are removed from the population. As many as required agents are removed to maintain the population size defined by the user

Step 5:   Select the best agent in the population (this agent has the lowest fitness measure because a minimum value of RMSNE is sought)

Step 6:   If local search has been applied previously to this best agent, then the bias vector and standard deviation (rho) of the local search are initialized using the bias vector and standard deviation (rho) values of the agent selected

Step 7:   If local search has not been applied previously to this best agent, the parameters of local search are initialized using default values, with the bias vector equal to the zero vector and the standard deviation (rho) equal to half of the distance from the nearest neighbor to the best agent in the population [20]

Step 8:   Apply the local search algorithm, Solis and Wets, to the best agent of the population with an intensity of Istr, which is the number of iterations the local search will run and is defined previously by the user

Step 9: On completion of the local search, the bias vector and standard deviation (rho) of the local search should be set at the bias vector and standard deviation (rho) of the agent to which the local search was applied. These parameters will be used in future local searches if the agent is selected again as the best of the population

Step 10: If the stop condition is not satisfied, then back to step 2

The pseudocode of the original MA-SW-Chains algorithm can be reviewed in [14]. Local searches are performed with the Solis and Wets algorithm, which is a random Hill-Climber with adaptive step size. The Solis and Wets algorithm works while the local search iterations are less than or equal to the parameter Istr. The pseudocode of the Solis and Wets algorithm is found in [20] and was adapted for the calibration process using the following steps:

Step 1: Initialize the offset vector (same size as the bias vector) with values from the Gaussian distribution, whose parameters are zero for both mean and standard deviation (rho) according to the data sent from MA-SW-Chains

Step 2: Create a new agent (Agent 1), adding each element of the vector parameters of the current best agent to each element of the bias vector and each element of the offset vector. Values are repaired to this new agent so that it meets restrictions of range and data type. The fitness measure is then calculated for this new agent, based on RMSNE

Step 3: If Agent 1 has a lower fitness measure than the best current agent (the best solution to the problem), Agent 1 replaces the best agent. Each element of the bias vector is updated using the equation: (0.2 * bias vector + 0.4 * (offset vector + bias vector)), the number of successes is increased by one and the number of failures initialized at zero. The constants in the previous equation were taken from Molina et al. [20]

Step 4: If Agent 1 is not better than the best current agent, then a new agent (Agent 2) is created by subtracting each element of the parameter vector of the current best agent with each element of the bias vector and each element of the offset vector. This new agent is given its values to comply with the restrictions of range and type. The fitness measure is then calculated for this new agent using RMSNE

Step 5: If Agent 2 has a better (lower) fitness measurement than the current best agent, then the agent is replaced by Agent 2, each element of the bias vector is updated using the equation: (bias vector−0.4 * (offset vector + bias vector)), the number of successes is increased by one and the number of failures is reset to zero

Step 6: If Agent 2 is not better than the current best agent, then the number of failures is increased by one and the number of successes is reset to zero

Step 7: If the number of successes is greater than two, then the value of the standard deviation (rho) is duplicated and the number of successes reset to zero

Step 8:   If not, i.e. the number of failures is greater than one, then the value of the standard deviation (rho) is reduced to half and the number of failures is reset to zero

Step 9:   If the value of the standard deviation (rho) is less than one, then the value of the standard deviation is made to be equal to one (this is because there are calibration parameters that vary within a small range of integer values)

Step 10:   If the stop condition is not met, go to step 1 taking the new value of the standard deviation (rho) modified in steps 7, 8 or 9

Finally, the execution process of the calibration is performed with its respective validation and analysis of results. The results of this step are presented in the following sections.

3   Execution of Calibration and Results

The experimentation was performed using two CORSIM models previously studied [13]. The first model is considered low complexity due to the number of links that it contains (20). It is a hypothetical network provided by McTrans. The second experiment was performed using a real network, which corresponds to the Pyramid Highway in Reno, Nevada, United States. This model is considered medium complexity because it contains 126 links.

To perform the experiments, the MA-SW-Chains algorithm was implemented and included in the software tool, "Calibration Tool" [13], which provides four alternative algorithms: a memetic algorithm, which combines a genetic with simulated annealing (GASA), a memetic algorithm, which combines a genetic with Tabu search (GATS), a genetic algorithm without a local optimizer (GA), and the SPSA. In this research, the results obtained by the MA-SW-Chains algorithm were compared with the results from GASA and SPSA, which provided best results in the past.

3.1   Algorithm Parameters

Parameters used in MA-SW-Chains for the experimentation were as follows: Initial population 4, Percent of selection 60, Percent of crossover 75 and percent of mutation 7. The local search intensity parameter (Istr) of Solis and Wets was set to 30.

3.2   Experiments

The two models used in the experiments, volume and speed were considered simultaneously for the calibration. The Reno model included actual field measurements taken on 45 links of the model. For the field data of the McTrans model, default values were used. For each of the models, 50 independent runs of the algorithms were performed and the results averaged.

As a result, 50 independent runs on the McTrans model (Fig. 2 (a)) with an average of 0.094 was achieved in the objective function (RMSNE) with a standard deviation of 0.0545, which indicates an adequate quality in the calibration of the model. However, the behavior of the algorithm in the Reno model is much better (Fig. 2 (b)), with an average of 0.0955 was obtained in the objective function with a standard deviation of 0.0144. This indicates a greater stability in the behavior of the algorithm, obtaining in 78 % of the 50 runs a RMSNE < 0.10, which is promising for this model (this model is more complex).

The best RMSNE obtained in each model after running all experiments was 0.0163 for McTrans and 0.0813 for the Reno network. In Figs. 3 to 5 are the results obtained in the best experiment of each model.

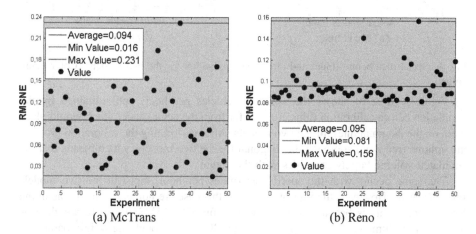

Fig. 2. The objective function for the two CORSIM models.

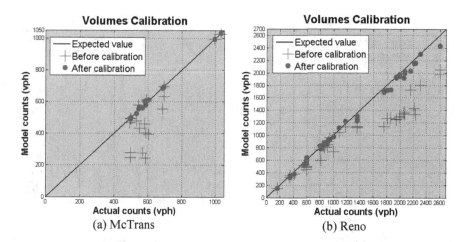

Fig. 3. Volume data before (blue) and after (red) calibration in the two models. (Color figure online)

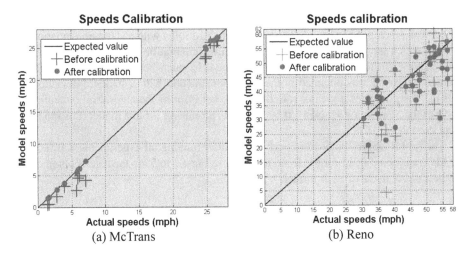

(a) McTrans   (b) Reno

Fig. 4. Speed data before (blue) and after (red) calibration for the two models. (Color figure online)

Figure 3 (a & b) shows the volume (vehicles per hour, vph) obtained by the simulation before calibration (blue crosses) far from the actual data (which would place them on the diagonal line) for the two models. After performing the calibration process, the volume (red dots) approach the diagonal for the two models, which means that the simulated volume largely coincides with the actual.

The same occurs with the speed data (miles per hour, mph) in the two models Fig. 4 (a & b). However, the improvements in this case are not as significant as in the Reno model. This is because in the objective function (RMSNE) more weight was assigned to volume (70 %), compare to speed (30 %). Although the weight parameter can be modified, in these experiments greater weight was assigned to volume because there was a greater confidence in the volume data (real data).

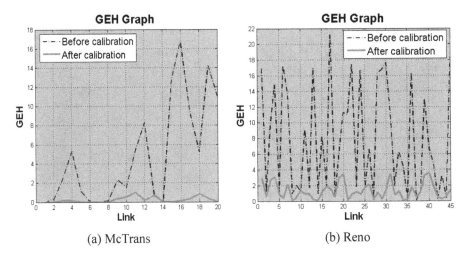

(a) McTrans   (b) Reno

Fig. 5. GEH statistic for the two models. (Color figure online)

Figure 5 (a & b) shows the results of statistical GEH, in which the dotted blue line indicates the value before the calibration process and clearly shows many links above 5, indicating that the model is not yet calibrated. The thick red line indicates its value after calibration. Following calibration, a marked improvement was obtained in both models, in which 100 % of the links had a GEH < 5; thus, fulfilling the defined calibration criteria.

4   Comparative Analysis of Results (MA-SW-Chains Vs GASA and SPSA)

To demonstrate the benefits of MA-SW-Chains for the calibration of microscopic traffic flow simulation models, a comparison was made with the state-of-the-art algorithms GASA and SPSA [13]. Table 1 shows the results obtained by the three algorithms.

Table 1. Comparison of MA-SW-Chains, GASA and SPSA

Experiments/ Criteria	McTrans			Reno		
	MA-SW Chains	GASA	SPSA	MA-SW Chains	GASA	SPSA
Run time	9,56 min	9,53 min	9,55 min	59,45 min	43,55 min	29,30 min
RMSNE before	0.2911	0.2911	0.2911	0.2217	0.2217	0.2217
RMSNE after	0.0163	0.0192	0.0404	0.0813	0.0857	0.0957
GEH (% of links < 5) before	55 %	55 %	55 %	47 %	47 %	47 %
GEH after	100 %	100 %	100 %	100 %	93 %	100 %
Time for parameter tuning	1 h	1 h	20 h	1 h	1 h	20 h

SPSA requires less time to run but more time for parameter tuning (20 h), which is not the case with the GASA and MA-SW-Chains algorithms. Time for parameter tuning is the time required for determining appropriate parameter values for the optimization algorithm (setting up the optimization algorithm). While the results showed in Table 1 are very similar among the algorithms, the potential of the MA-SW-Chains algorithm is shown in Fig. 6 using the speed of convergence in contrast to GASA and SPSA. In the McTrans model, the MA-SW-Chains algorithm reaches RMSNE values below 0.02 faster, approximately at four minutes while GASA requires eight minutes. SPSA did not reach this value and its convergence curve varies over time. On the other hand, in the Reno model, the MA-SW-Chains algorithm converges faster to values close to 0.09, at approximately 10 min, GASA at 14 min, and SPSA does not reach these values.

The proposed algorithm obtained best solution in the objective function (RMSNE) and met the calibration criteria for 100 % of the links in the two models, indicating that the memetic algorithm based on Solis and Wet local search chains obtains better and faster solutions.

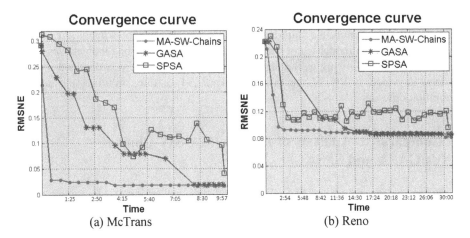

Fig. 6. Convergence curve of the objective function for the three algorithms in the two CORSIM models.

5   Conclusions and Future Work

This paper proposes the adaptation of the MA-SW-Chains algorithm, a memetic algorithm with Solis and Wet local search chains, to support the calibration of microscopic traffic flow simulation models. The use of this algorithm with the methodology proposed in [13] makes possible the calibration for any type of implementation because no information about the characteristics of traffic flow model is required. The MA-SW-Chains algorithm successfully calibrated two CORSIM vehicular traffic models, MacTrans and Reno, with the GEH statistical and RMSNE measure within the required limits.

Comparison of the MA-SW-Chains algorithm with the GASA and SPSA algorithms, previously discussed in the literature [11, 13] showed, indicated that MA-SW-Chains is an alternative that provides better results, measured by the objective function and statistical GEH.

As future work, the research group intends to evaluate other single-objective optimization approaches, such as the use of multiple offspring sampling (MOS) [21] and differential evolution using cooperative coevolution (DECC-G) [22]. In addition, queue lengths are to be included as an additional factor in the objective function.

Acknowledgements. The work in this research study was supported by the University of Cauca (Popayan, Colombia) and the University of Nevada Las Vegas, United States. We are grateful to Mr. Colin McLachlan for his help translating the first version of this document.

References

1. Hale, D.K., et al.: Optimization-based assisted calibration of traffic simulation models. Transp. Res. Part C Emerg. Technol. **55**, 100–115 (2015)
2. Hollander, Y., Liu, R.: The principles of calibrating traffic microsimulation models. Transportation **35**(3), 347–362 (2008)

3. Abdalhaq, B.K., Baker, M.I.A.: Using meta heuristic algorithms to improve traffic simulation. J. Algorithms **2**(4), 110–128 (2014)
4. Kim, K.-O., Rilett, L.: Simplex-based calibration of traffic microsimulation models with intelligent transportation systems data. Transp. Res. Rec. J. Transp. Res. Board **1855**, 80–89 (2003)
5. Park, B., Kamarajugadda, A.: Development and evaluation of a stochastic traffic signal optimization method. Int. J. Sustain. Transp. **1**(3), 193–207 (2007)
6. "Brian" Park, B., Yun, I., Ahn, K.: Stochastic optimization for sustainable traffic signal control. Int. J. Sustain. Transp. **3**(4), 263–284 (2009)
7. Chiappone, S., et al.: Traffic simulation models calibration using speed–density relationship: an automated procedure based on genetic algorithm. Expert Syst. Appl. **44**, 147–155 (2016)
8. Korcek, P., Sekanina, L., Fucik, O.: Calibration of traffic simulation models using vehicle travel times. In: Sirakoulis, Georgios Ch., Bandini, S. (eds.) ACRI 2012. LNCS, vol. 7495, pp. 807–816. Springer, Heidelberg (2012)
9. Lee, J.-B., Ozbay, K.: Calibration of a macroscopic traffic simulation model using enhanced simultaneous perturbation stochastic approximation methodology. In: Transportation Research Board 87th Annual Meeting (2008)
10. Balakrishna, R., Antoniou, C., Ben-Akiva, M., Koutsopoulos, H., Wen, Y.: Calibration of microscopic traffic simulation models: methods and application. Transp. Res. Rec. J. Transp. Res. Board **1999**, 198–207 (2007). doi:10.3141/1999-21
11. Paz, A., Molano, V., Gaviria, C.: Calibration of corsim models considering all model parameters simultaneously. In: 2012 15th International IEEE Conference on Intelligent Transportation Systems (ITSC). IEEE (2012)
12. Yuan, J., Ng, S.H., Tsui, K.L.: Calibration of stochastic computer models using stochastic approximation methods. IEEE Trans. Autom. Sci. Eng. **10**(1), 171–186 (2013)
13. Paz, A., et al.: Calibration of traffic flow models using a memetic algorithm. Transp. Res. Part C Emerg. Technol. **55**, 432–443 (2015)
14. Molina, D., Lozano, M., Herrera, F.: MA-SW-Chains: memetic algorithm based on local search chains for large scale continuous global optimization. In: 2010 IEEE Congress on Evolutionary Computation (CEC). IEEE (2010)
15. Li, X., et al.: 2015 IEEE Congress on Evolutionary Computation Competition on: Large Scale Global Optimization, p. 19. RMIT University (2015)
16. Alabert, A., et al.: No-Free-Lunch theorems in the continuum. Theor. Comput. Sci. **600**, 98–106 (2015)
17. Solis, F.J., Wets, R.J.-B.: Minimization by random search techniques. Math. Oper. Res. **6**(1), 19–30 (1981)
18. McTrans, CORSIM User's Guide (2011)
19. Altiparmak, F., et al.: A steady-state genetic algorithm for multi-product supply chain network design. Comput. Ind. Eng. **56**(2), 521–537 (2009)
20. Molina, D., et al.: Memetic algorithms based on local search chains for large scale continuous optimisation problems: MA-SSW-Chains. Soft. Comput. **15**(11), 2201–2220 (2011)
21. LaTorre, A., et al: Multiple offspring sampling in large scale global optimization. In: 2012 IEEE Congress on Evolutionary Computation (2012)
22. Yang, Z., Tang, K., Yao, X.: Large scale evolutionary optimization using cooperative coevolution. Inf. Sci. **178**(15), 2985–2999 (2008)

A New Strategy Based on Feature Selection for Fault Classification in Transmission Lines

Márcia Homci$^{(\boxtimes)}$, Paulo Chagas, Brunelli Miranda, Jean Freire,
Raimundo Viégas Jr., Yomara Pires, Bianchi Meiguins, and Jefferson Morais

Federal University of Pára, Rua Augusto Correa, 01,
Belém, Pará 66075-110, Brazil
{marciahomci,paulo.chagas,brunelli.miranda,jeanarouche,
rviegas,yomara,bianchism,jmorais}@ufpa.br
http://www.ufpa.br/ppgcc

Abstract. The transmission lines are the element most susceptible to faults on power systems, and the short circuit faults are the worst type of faults than can happen on this element. In order to avoid further problems due to these faults, a fault diagnostic is necessary, and the use of front ends is required. However, the selection process for choosing the front ends is not a simple one because it behaves differently for each. Therefore, this paper presents a new front end, called Concat front end, which integrates other front ends, such as wavelet, raw and Root Mean Square. Furthermore, we have applied feature selection techniques based on filter in order to decrease the dimension of the input data. Thus, we used the following classifiers: neural network, K-nearest neighbor, Random Forest and support vector machine. We used a public dataset called UFPAFaults to train and test the classifiers. As a result, the concatenation of front ends, on most cases, had achieved the lowest error rates. In addition, the feature selection techniques applied showed that it is possible to get higher accuracy using less features on the process.

Keywords: Short-circuit fault · Transmission lines · Front ends · Feature selection · Machine learning algorithms

1 Introduction

Electric energy is an essential condition to the economic development of a country, promoting the satisfaction and well fare of society. This way, the electric power systems must ensure the availability of electric energy for as long as possible and with an acceptable degree of quality. However, this is not always possible due to faults (disturbances) that happens in the operation of these systems.

Among the components of a typical electrical system, the transmission line is the element more prone to faults, especially if we consider its physical dimensions, functional complexity and the environment where it is. Short circuit faults are the worst type of faults that can happen in a transmission line [1]. When this happens, automatic devices, called protection relays, work in the electrical

© Springer International Publishing AG 2016
M. Montes-y-Gómez et al. (Eds.): IBERAMIA 2016, LNAI 10022, pp. 376–387, 2016.
DOI: 10.1007/978-3-319-47955-2_31

system interrupting the electric energy supply to consumers. Hence, it is evident the need of electric power systems to use mechanisms for short circuit faults diagnostic in transmission lines to restore the operation of the electrical system in the shortest possible time.

In recent years, applications of data mining techniques (preprocessing and machine learning algorithms) for short-circuit faults classification in transmission lines have attracted many attentions [1–5]. Discrete wavelet transform as a preprocessing tool or front end has been used in different applications [2,4–7]. In some researchers, wavelet transform has been combined with other topics such as artificial neural network, SVM and fuzzy systems to form hybrid frameworks [2,6,8]. Others front ends such as Fast Fourier Transform, Root Mean Square (RMS) and Raw are widely used for fault classification in transmission lines [5,9,10]. In [5], for example, a comparison between different front ends (raw, rms and wavelet) and machine learning algorithms (decision trees and neural network) are performed.

Despite the evolution in the use of front ends, it is not a trivial task to determine which one is better suited for the process of fault classification. Therefore, this work aims to present a new front end, called Concat front end, which integrate the front end samples, such as wavelet, raw and rms, into a single set of sequences. However, the concatenation process creates a data set of high dimensionality that can cause an increasing in the classifier's computational cost. In this case, another contribution of this work is to apply techniques of feature selection to decrease the dimensionality of data that will be used in the subsequent stage of fault classification.

This paper is organized as follows. Section 2 provides classification of time series representing faults. Section 3 discusses the front ends, presenting definitions and notation. The feature selection process adopted is discussed in Sect. 4. The scenarios used for the experiments and their respective results are shown in the Sect. 5, respectively. The Sect. 6 provides conclusions.

2   Classification of Time Series Representing Faults

In this work, the time series represent faults, which are basically short-circuits in transmission lines. There are many ways of representing and classifying time series. Then, this section presents a precise notation to avoid obscure points.

Most transmission systems use three phases: A, B and C. Hence, a short circuit between phases A and B will be identified as "AB". Considering the possibility of a short circuit to "ground" (G), the task is to classify a time series into one among ten possibilities: AG, BG, CG, AB, AC, BC, ABC, ABG, ACG and BCG. Each fault is considered a multivariate time series of variable duration. The n-th fault $\mathbf{X}_n$ in a database is represented by a $Q \times T_n$ matrix. The column $\mathbf{x}_t$ of $\mathbf{X}_n$, $t = 1, \ldots, T_n$, is a multidimensional sample represented by a vector of Q elements. For example, this work adopts $Q = 6$ (voltage and current of phases A, B and C) in the experiments. We highlight that as the number of multivariate samples T_n depends on n, the use of a conventional classifier (e.g., neural network) is not feasible, as it needs to deal with sequences of variable dimension.

In this case, the samples are frequently concatenated to create a frame $\mathbf{F}$. The frames have $Q \times L$ dimensions, where L is the frame length and their concatenation $\hat{\mathbf{Z}} = [\mathbf{F}_1, \ldots, \mathbf{F}_N]$ is a matrix of dimension $Q \times L_N$, and N is the number of frames, which can overlap themselves making their displacement (frame shift S) (number of sample between the beginning of two consecutive frames), be smaller than the window size. The representation of the frame number of a $\mathbf{X}_n$ fault is equal to

$$N_n = 1 + \lfloor (T_n - L)/S \rfloor, \tag{1}$$

where $\lfloor \rfloor$ is a floor function. When $S = L$ (there is no overlap) and a frame is a concatenation of samples $\mathbf{F}_n = [\mathbf{x}_{(1+L(n-1))}, \mathbf{x}_{(2+L(n-1))}, \ldots, \mathbf{x}_{(L+L(n-1))}]$, the $\mathbf{X} = \hat{\mathbf{Z}}$ matrices coincide. The frames (matrices) can be conveniently organized as a vectors of dimension $K = Q \times L$, and $\hat{\mathbf{Z}}$ is resized to create $\mathbf{Z} = [\mathbf{z}_1, \ldots, \mathbf{z}_N]$ of dimension $K \times N$. Henceforth, we assume that the processing is performed in $\mathbf{Z}$ (not $\mathbf{X}$).

The fault classification systems can be of two types: on-line classification systems and post fault systems (or off-line). Both of them will be briefly discussed in the next sections, highlighting their relations with the conventional and sequence classifications.

The on-line faults classification aims the frame-by-frame classification and it is usually performed, for example, in the level of protection relays [11]. The on-line fault classification systems deal with $\mathbf{z}$ vectors of fixed dimension K and must provide an output in a relatively short time. This time is often based on a frame corresponding to half of a cycle or a cycle with sine wave signal of 50 or 60 Hz. Assuming, for example, a signal of 60 Hz and a sampling frequency of $f_s = 2 \, \text{kHz}$, a corresponding cycle of $L = 2000/60$ which is approximately 33 samples. Thus, the on-line systems try to solve problems that can be treated as problems of conventional classification [12], where a decision must be taken for each 33 signal samples, for example.

Alternatively to the on-line classification, the post-fault classification deals directly with $\mathbf{Z}_n$ matrices of variable dimension $K \times N_n$ and can be performed in a supervisory center in a post-fault stage [4]. The post-fault classification systems try to solve problems that can be treated as problems of sequence classification [13]. There are techniques such as Hidden Markov Models (HMM) [14] and Dynamic Time Warping (DTW) [15] which deal directly with the sequences. Another alternative is Frame Based Sequence Classification (FBSC) discussed in [5], which uses conventional classifiers (eg. neural network) in the faults classification.

3   Front Ends

Regardless of the parametric representation used, a single sample usually don't have enough information to allow the accomplishment of reasonable decisions. For this reason, a front end transforms the samples into features generating the sequences that will be passed to the classification algorithms. As the choice of a front end is not a simple task, in this work we used a strategy of front ends concatenation, building a single set of sequences to be evaluated by the classifiers. The selected front ends are the same used in [5] and are described in the next sections for convenience.

3.1   Raw Front End

In the raw front end, the output parameters correspond to the sample values of the original signals, without any other processing to organize the samples for the $\mathbf{Z}$ matrix according to the selected values of L and S. For example, in [11] the frames are formed by concatenating samples of voltage and current waveforms and the $\mathbf{Z}$ vectors have dimension $K = 198$. More specifically, if $Q = 6$ (voltage and current waveforms), a raw front end could get frames $\mathbf{F}_n$ of dimension 6×5 taking for each central sample, its two neighbors at left and the two at its right. In this case, assuming a fault with $T_n = 10$ samples and $L = S = 5$, the values of K and N are 30 and 2, respectively, such that $\mathbf{X} = \hat{\mathbf{Z}}$. Thus, $\hat{\mathbf{Z}}$ and $\mathbf{Z}$ have dimensions of 6×10 e 30×2, respectively.

3.2   RMS Front End

This front end consists in the calculating of the RMS value windowed for each kind of Q waveform. Assuming a window length L and shift S, the n-th RMS value $z[n], n = 1, \ldots, N$ of a waveform $x[t], t = 1, \ldots, T$ is

$$z[n] = \sqrt{\frac{1}{L} \sum_{l=1}^{L} (x[l + (n-1)S])^2}, \tag{2}$$

where N is calculated with Eq. (1) and T is the sample number.

3.3   Wavelet Front End

The wavelet front end uses two filters to decompose the obtained signal through its convolution repeated times. The first is a low-pass filter that represents the low-frequency content, here called coefficient of approximation $\mathbf{a}$, and the second is a high-pass filter with the high-frequency content with the details represented by the coefficient of detail $\mathbf{d}$. Then a dyadic wavelet decomposition takes place, which has γ stages of filtering and decimation by two, i.e., for every two output samples of filter, one of them is discarded.

In this paper, we will use the two techniques approached in [5], where the first uses a concatenation of all coefficients and organize them in a $\mathbf{Z}$ matrix, called WaveletConcat, and the second calculates the coefficient energy, known as WaveletEnergy. The WaveletConcat takes into consideration that for $\gamma > 1$ the coefficients have different sampling frequencies. For this, instead of using a single L, the user specifies a L_{min} value for the waveforms with lower sampling frequency f_s ($\mathbf{a}$ and $\mathbf{d}$) and a higher value $L_i = 2^{\gamma-i} L_{\mathsf{min}}$ is automatically adopted for other details $i = 1, \ldots, \gamma - 1$. For example, assuming a wavelet application with 3 levels of decomposition ($\gamma = 3$), the frame length for $\mathbf{d}_1$ and $\mathbf{d}_2$ are $4L_{min}$ e $2L_{min}$, respectively. A similar approach is used for the frame shift $S_i = 2^{\gamma-i}S_{min}$, where S_{min} is other parameter defined by user. As a result of this process, we have a frame $\mathbf{F}$ with the coefficients organized, with dimension $Q \times L$, where $L = 2^\gamma L_{min}$ and his number of features will be $K = 2^\gamma L_{min}Q$, providing a number of frames

$$N = 1 + \lfloor (T_{\mathbf{a}} - L_{min})/S_{min} \rfloor \tag{3}$$

where $T_{\mathbf{a}}$ is the number of elements in $\mathbf{a}$.

The WaveletEnergy differs from the first because, instead of concatenate all the coefficients, the same represents $\mathbf{X}$ by the energy E in each frequency band obtained in the wavelet decomposition, which is calculated for short intervals. For example, if we have $L_{min} = S_{min} = 1$ and an $\mathbf{X}$ matrix of dimension 6×2000, it results in a 6×4 matrix, where the first line contains the medium energy of coefficient a calculated from the voltage of a phase, resulting in $K = 24$ and $N = 125$.

3.4   Concatenation of Fronts Ends

In this paper, a new strategy is proposed for use of front ends described above. The idea is to integrate the samples of front ends raw, WaveletConcat, WaveletEnergy and RMS, respectively, in a single set of sequences. This process is here called Concat front end.

First, it is performed a process of concatenation by signal, where the front ends are calculated for each signal (voltage and current, phases A, B and C) and grouped in the previously established order. For such, the frame length L and the frame shift S must be calculated, where $L = L_{min}2^\gamma$ and $S = S_{min}2^\gamma$. It is possible to generate all the front ends for each signal and to concatenate them into a matrix with the parameters of front ends defined. All that remains is to concatenate the resulting matrices in the following order: first the voltages of phase A, B and C, then next the currents of phases A, B and C. The Table 1 shows the number of features (columns) of each front end in the final matrix assuming $L_{min} = 9$, $S_{min} = 4$, $\gamma = 3$ e $Q = 6$.

It can be see that, depending of the front end, the number of features K increase too fast, generating high-dimensional matrices. The matrix generated by Concat front end will be the input data in the classifier, so a great number of features generates a high computational cost. As consequence of the process, we have the appearance of a problem well known to many applications, which is the generation of a sequence with a high level of dimensionality [16]. Therefore, this work adopts, before the stage of pattern recognition, techniques of feature selection to reduce the input data dimension, selecting the more relevant features. Such techniques of feature selection are discussed in the next section.

Table 1. Number of features per front end, considering $L_{min} = 9$, $L_{min} = 4$, $\gamma = 3$ e $Q = 6$. Note that $L = L_{min}2^{\gamma}$

Front end	Equation	Number of features
Raw	$K = L \times Q$	432
Waveletconcat	$K = L_{min}2^{\gamma} \times Q$	432
WavaletEnergy	$K = (\gamma + 1) \times Q$	24
RMS	$K = Q$	6
Concat front end	$K = (L_{min}2^{\gamma+1}) + \gamma + 2 \times Q$	894

4 Feature Selection

It is intuitive to think that a greater number of features will make more information available to the classifier, consequently the better would be its performance. However, with an increasing number of features, the data tend to become exponentially sparse resulting in samples equally sparse in distance, thus, reducing the classifiers capability of generalization. The treatment of the "curse of dimensionality", as the problem is known, can be performed in a general mode, through two distinct approaches [16]: feature extraction and feature selection. The feature extraction performs a mapping of the original space to one of small dimension, generating new features that will be used in the next stage. The feature selection consists of identifying and retaining only the features, which contributes the most for the execution of given task. In this work, we adopt only the approach of feature selection.

There are two ways to treat the problem of feature selection [17]: as a simple ordination of features based on some measure of importance, a technique called ranking, or as a search for an optimum subset. When the first approach is chosen, the features are ordered and only the K^* first features that had better represent the dataset are selected. On the other hand, when the second approach is chosen, the figure of merit becomes the subset of features that are evaluated according to some measure of importance. The problem of selecting a subset of features can be treated as a problem of heuristic search. In this process of heuristic search for the best subset features, we must first define the starting point that directly affects the direction that the search will take [17]. The main approaches proposed are [12]: Forward Selection, Backward Selection and Bidirectional Selection.

Another important point in the search process is the definition of the strategy to go through the search space. Some of the most used strategies are [17]: greedy hill climbing, best first and generic algorithm. All of these search strategies need a criteria for rating of the features subset. This way, the methods of feature selection can be classified basically into two groups: filters and wrappers. The last one uses a induction algorithm **L** that later will be adopted in the classifier project. The methods of filter type evaluate a given parameter through the use of heuristics based on the data statistics properties and are independent of **L**. In this paper, we use methods of filter type based on info gain (mutual information), gain ratio and Correlation-based Feature Selection (CFS). We adopted Forward Selection and Best first search heuristics for CFS [17].

5   Experimental Results

5.1   Simulation Setup

The dataset UFPAFaults was used in the experiments. UFPAFaults is a public labeled dataset with short-circuit events, which allows the user to compare algorithms and reproduce results over different sites. This dataset is composed by 27500 short-circuit faults, organized into five sets of $100, 200, \ldots, 1000$ faults each. Each fault corresponds to three voltage and three current waveforms stored as binary files with the associated label file. It is currently in its fifth version, called UFPAFaults5, and is available at [18].

In the experiments we used the front ends: raw, rms, WaveletConcat, WaveletEnergy and concat front end. The mother wavelet used was Daubechies 4 with $\gamma = 3$ levels of decomposition. The values of $L = L_{min}$ and $S = S_{min}$ were among the folowing pairs: (i) $L = L_{min} = 4$ and $S = S_{min} = 2$; (ii) $L = L_{min} = 9$ and $S = S_{min} = 4$; (iii) $L = L_{min} = 5$ and $S = S_{min} = 5$.

All the classifiers used belong to the package of data mining WEKA[1]: Artificial Neural Network (ANN) trained with the Backpropagation algorithm, K-nearest neighbor (KNN), Random Forest and Support Vector Machine (SVM). The choice of these classifiers was based on the fact the they are common representations of different paradigms of learning [12].

Often, the best performance of a classifier related to a specific database can only be reached by tedious parameter tuning. This task is called automatic model selection and corresponds, for example, to choosing parameters such as the number of neurons in the hidden layer for a neural network [5]. This task should not be confused with the feature selection presented in Sect. 4. Table 2 shows the grid adopted for model selection procedure.

Table 2. *Grid* of model selection procedure

Classifier	Parameter	*Grid* values	Best value
ANN	H	5, 10, 20, 40, 80 and 160	160
	L	0.1, 0.5 and 0.9	0.9
	M	0.1, 0.2 and 0.4	0.4
KNN	K	1, 3, 5, 7, 9, 11, 13 and 15	1
SVM	G	0.01, 1 and 100	100
	C	0.01, 1 and 100	0.01

---

[1] *Waikato Environment Knowledge Analysis.* Available at http://www.cs.waikato.ac. nz/ml/weka.

After many tests the most robust architecture for ANN had only one hidden layer with 160 neurons fixed in H, learning rate L of 0.9 and momentum rate M of 0.4, with the number of epochs fixed in 1000. For the SVM classifier, we used the polynomial kernel and the optimized parameters were gamma G with best value equal to 100 and error penalty C with best value equal to 0.01. For the KNN classifier, the parameter K (number of nearest neighbors) was varied. For the Random Forest the varied parameter was I, which represents the number of trees that will be generated in the model. In general, the higher the I, the higher the accuracy and processing. In [19] it is showed that although the accuracy grows with I, the error rate is likely to converge. Besides, it was showed that error rates for I with values starting in 64 are too close. Hence, to avoid unnecessary processing it was adopted a I value of 64, and for this reason the process of model selection was not applied for this classifier.

5.2   Performance of the Front Ends by Classifier

The ANN classifier was clearly the best classifier for wavelets front ends, while the other classifiers are competitive for the raw and RMS front ends, with SVM being slightly better. Overall, the Concat front end is the one that achieves the best results. The exception is the SVM classifier, which presented the worst performance when using features concatenation, but had a good behavior with the WaveletConcat front end, for example, the Fig. 1 presents the obtained results in terms of error rate for each front end.

Figure 2 shows a better visualization for performance comparison between the front ends, considering the best obtained values of L and S, where $L_{min} = L = 9$ and $S_{min} = S = 4$, and the concatenation of them relative to the error rate of classifiers. Analyzing the figure, we observe that the obtained results by classifiers (except for the SVM), after front ends concatenation, are better than the individual evaluation of each front end. As one of the main disadvantages of SVM is the choice of kernel for a specific dataset [20], the variation in the results of this classifier for each front end shows that a deeper analysis must be done for selection of kernel and parameters for specific front ends. Regarding the other classifiers, both neural network and random forest had an excellent performance with error rates below 1 %. The KNN classifier, although having presented worse results than ANN and RF, lessened its error rate when compared to the other front ends.

5.3   Performance of the Front Ends with Feature Selection

The results for classification applying feature selection are presented in Fig. 3, using $L_{min} = 9$ e $S_{min} = 4$. It was used a filter based on a search method and heuristic evaluation for the process of feature selection. In this work we used the ranker search method together with the heuristic evaluations of information gain and gain ratio, and the FowardSelection method with evaluation starting from correlated subsets. The number of features varied in $10, 20, 30, \ldots, 100, 200, 300$ and 400, and the respective errors were calculated for the previously mentioned

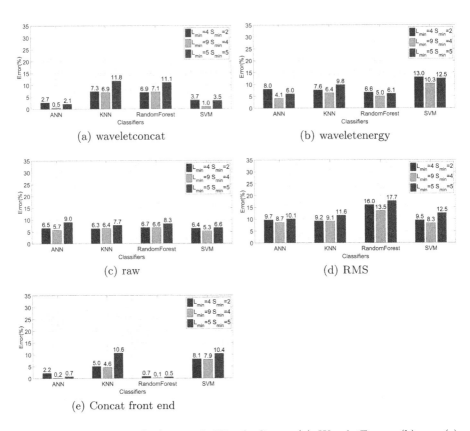

(a) waveletconcat

(b) waveletenergy

(c) raw

(d) RMS

(e) Concat front end

Fig. 1. Error rate using the front ends: WaveletConcat (a), WaveletEnergy (b), raw (c), RMS (d) and Concat (e).

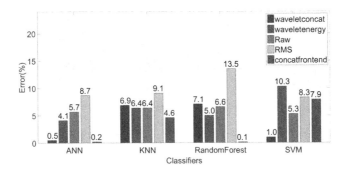

Fig. 2. Comparison of error rate between the adopted front ends, considering $L = L_{min} = 9$ and $S = S_{min} = 4$ and their concatenation.

classifiers, enabling them to be compared to the error rate of the Concat front end, which used all features ($K = 894$). It can be seen that the error rate varies greatly depending on the classifier used and the number of selected features. However, in general, the classifiers that used training files and tests with a lower number of features achieved results at least equivalent to results obtained when using all the features.

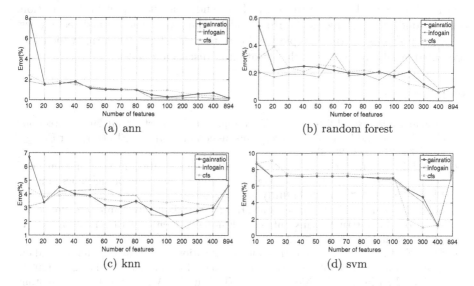

(a) ann

(b) random forest

(c) knn

(d) svm

Fig. 3. Misclassification error applying feature selection.

The classification heuristics had a behavior similar to all classifiers, displaying a greater difference only in the higher case of each classifier, where info gain presented the best case in 2 of 4 classifiers. Besides, we observed that improvements in error rates have a number of features in the range of 90 to 400, displaying that it is possible to reproduce n error rate with at least half the features totality.

6 Conclusions

This paper presented a new strategy for fault classification in transmission lines based on the concatenation of well-known front ends. The experiments showed that most times, the Concat front end provided better error rates than the other front ends. The only classifier that did not have better results with Concat front end was SVM, which needs more research regarding the choice of kernel and the classifiers parameters for the specific only for faults classification.

Another contribution of this paper was to adopt feature selection to reduce the dimensionality of input data after the front ends concatenation. We applied simple techniques of filter type and it was possible to obtain lower error rates using a fewer number of features. It is necessary to highlight the performance of ANN and Random Forest classifiers in the performed tests. Both classifiers displayed a regular behavior and had a very low error rates in relation to the other classifiers, thus obtaining good results with the totality of parameters and with feature selection.

For future works, the filters can be replaced by wrapper methods, in a way that it is possible to analyze the interactions between the features. Furthermore, new front ends can be added to Concat front end, since it is open to the addition of any front end, as long as the front end is adapted to the framed techniques specified previously. An example would be the use of S-Transform, which has presented good results in the classification of disturbances [21].

References

1. Fathabadi, H.: Novel filter based ANN approach for short-circuit faults detection, classification and location in power transmission lines. Int. J. Electr. Power Energy Syst. **74**, 374–383 (2016)
2. Hosseini, K.: Short circuit fault classification and location in transmission lines using a combination of wavelet transform and support vector machines. Int. J. Electr. Eng. Inf. **7**, 353–365 (2015)
3. Singh, M., Panigrahi, B., Maheshwari, R.P.: Transmission line fault detection, classification. In: International Conference Emerging Trends in Electrical and Computer Technology (ICETECT), Tamil Nadu, pp. 15–22 (2011)
4. Silva, K.M., et al.: Fault detection and classification in transmission lines based on wavelet transform and ANN. IEEE Trans. Power Delivery **21**(4), 2058–2063 (2006)
5. Morais, J., Pires, Y., Cardoso Jr., C., Klautau, A.: A framework for evaluating automatic classification of underlying causes of disturbances and its application to short-circuit faults. IEEE Trans. Power Delivery **25**(4), 2083–2094 (2010)
6. Gowrishankar, M., Nagaveni, P., Balakrishnan, P.: Transmission line fault detection and classification using discrete wavelet transform and artificial neural network. Middle-East J. Sci. Res. **24**(4), 1112–1121 (2016)
7. Livani, H., Evrenosoglu, Y.: A machine learning and wavelet-based fault location method for hybrid transmission lines. IEEE Trans. Smart Grid **5**(1), 51–59 (2013)
8. Nayeripour, M., et al.: Fault detection and classification in transmission lines based on a combination of wavelet singular values and fuzzy logic. Cumhuriyet Sci. J. **36**(3), 69–82 (2015)
9. Cardoso, C., et al.: Hierarchical Agglomerative clustering of short-circuit faults in transmission lines. In: 10th Brazilian Symposium on Neural Networks, Salvador, Brazil, pp. 87–92 (2008)
10. Das, D., Singh, N.K., Sinha, A.K.: A comparison of Fourier transform, wavelet transform methods for detection, classification of faults on transmission lines. In: IEEE Power India Conference (2006)
11. Zhang, N., Kezunovic, M.: A real time fault analysis tool for monitoring operation of transmission line protective relay. Electr. Power Syst. Res. **77**(3–4), 361–370 (2007)
12. Han, J., Kamber, M., Pei, J.: Data Mining: Concepts and Techniques. Morgan Kaufmann, Burlington (2012)
13. Li, M., Sleep, R.: A robust approach to sequence classification. In: International Conference on Tools with Artificial Intelligence (2005)

14. Telford, R., Galloway, S.: Fault classification and diagnostic system for unmanned aerial vehicle electrical networks based on hidden Markov models. IET Electr. Syst. Transp. **5**(3), 103–111 (2015)

15. Petitjean, F., et al.: Dynamic Time Warping averaging of time series allows faster, more accurate classification. In: 2014 IEEE International Conference on Data Mining, Shenzhen (2014)

16. Priddy, K.L., Keller, P.E.: The curse of dimensionality. In: Artificial Neural Networks: an Introduction, pp. 26–30. SPIE - The International Society for Optical Engineering, Washington (2005)

17. Chandrashekar, G., Sahin, F.: A survey on feature selection methods. Comput. Electr. Eng. **40**(1), 16–28 (2014)

18. UFPAFaults. http://www.laps.ufpa.br/freedatasets/UfpaFaults

19. Oshiro, T.M., Perez, P.S., Baranauskas, J.A.: How many trees in a random forest? In: Perner, P. (ed.) MLDM 2012. LNCS (LNAI), vol. 7376, pp. 154–168. Springer, Heidelberg (2012). doi:10.1007/978-3-642-31537-4_13

20. Burges, C.J.C.: A tutorial on support vector machines for pattern recognition. Data Min. Knowl. Disc. **2**(2), 121–167 (1998)

21. Reddy, M.J.B., Mohanta, D.K.: Detection, classification, localization of power system impulsive transients using S-transform. In: 9th International Conference on Environment and Electrical Engineering (EEEIC), Prague, Czech Republic (2010)

AI in Education, Affective Computing, and Human-Computer Interaction

Context Ontologies in Ubiquitous Learning Environments

Gabriela González[1]([✉]), Elena Durán[2], and Analía Amandi[3]

[1] CONICET, Facultad de Ciencias Exactas Y Tecnologías,
Universidad Nacional de Santiago del Estero, Santiago del Estero, Argentina
ggonzalez@unse.edu.ar
[2] Facultad de Ciencias Exactas y Tecnologías,
Universidad Nacional de Santiago del Estero, Santiago del Estero, Argentina
eduran@unse.edu.ar
[3] Facultad de Ciencias Exactas, ISISTAN (CONICET/UNICEN),
Tandil, Argentina
amandi@exa.unicen.edu.ar

Abstract. Modeling ubiquitous context in u-learning systems is a challenging task. The model must include support for a variety of context information that differs in nature and granularity level. As such, a representation mechanism that allows an adequate manipulation of this kind of information is needed. Many researchers have proposed using ontologies to embody ubiquitous learning context information. In this paper, an overview of such models is presented, followed by a discussion of their main characteristics regarding the ontology model itself and the support the model provides to the u-learning system.

Keywords: Context · Ontologies · Ubiquitous learning

1 Introduction

Ubiquitous learning (u-learning) can be seen as a new paradigm regarding delivery and adaptation of learning services for students. This new form of learning uses technology that is already integrated in their everyday lives, such as mobile devices and wireless communication networks. The main objective of u-learning is to provide personalized content and learning services to students as they move with their mobile device through an open or closed environment.

One of the most challenging areas in the implementation of systems that support u-learning is modeling the ubiquitous context of a student. There has been a growing body of research around this topic, and several approaches and structures to model context have been proposed, out of which ontologies seem to be the most promising one. This is due to the fact that they promote interoperability and allow reasoning over the represented data.

Existing ubiquitous context ontologies for u-learning scenarios found in the literature, differ in structure and composition, expressivity level and language used. The goal of this paper is to provide some background on current ontology-based models for u-learning found in the literature. The main characteristics of these models are compared and analyzed aiming to provide a starting point for works that deal with modeling the ubiquitous context required for personalization in ubiquitous learning environments.

© Springer International Publishing AG 2016
M. Montes-y-Gómez et al. (Eds.): IBERAMIA 2016, LNAI 10022, pp. 391–403, 2016.
DOI: 10.1007/978-3-319-47955-2_32

The paper structure is as follows. Section 2 presents some related work. Section 3 provides an overview of the most relevant aspects of ubiquitous learning and ontologies, in relation to this review. Section 4 briefly describes existing ontology-based context models for u-learning, and Sect. 5 analyzes and compares these models. Finally, Sect. 6 draws some concluding remarks.

2 Related Work

Some related reviews are presented below.

Strang and Linnhoff-Popien [1] review different approaches to model context for ubiquitous computing, including ontologies. They propose a desired characterization for an ubiquitous system and then evaluate the ontology models accordingly.

Krummenacher *et al.* [2] analyze the modeling of context with ontologies. They evaluate ontology models with regards to a set of criteria related to ontology engineering and context modeling, without considering a specific domain.

Bolchini *et al.* [3] present a survey of context models from a data-oriented perspective. They provide an analysis framework for context models based on the notion that, in context-aware systems, context data is used to customize the way inputs are processed. In the evaluation, they include ontology-based models as well.

Bettini *et al.* [4] provide a survey of context modeling and reasoning techniques, with a strong focus on ontology-based models. They also address additional issues, such as high-level context abstractions and uncertainty of context information.

Verbert *et al.* [5] survey context-aware recommender systems in the domain of learning, i.e. systems that try to filter learning material or learning activities with regards to the context of a student. They analyze context categories used by the systems as well as the level of support that these systems are able to provide to students.

Al-Yahya *et al.* [6] survey ontologies used in e-learning domain, including ontologies to model student data. The student data also includes context-aware features related to the learning environment of the learners.

Unlike works previously mentioned, this review focuses on u-learning models for personalization, concentrating on characteristics that are specific to ontology-based models. It attempts to identify the suitability of using an ontology model to represent the information required for u-learning personalization.

3 Background

3.1 Ubiquitous Learning

Ubiquitous learning [7] aims to provide learners with the right information at the right time in the right way, using mobile devices, wireless communications and sensor technologies. This means that students can move through a physical environment and the u-learning system will dynamically support their learning, because it can communicate with the embedded computers in the surroundings to obtain information of the learning context and adapt accordingly.

Support that the system can offer to students is of two types [8]. One type is to try to enhance the learning process using ubiquitous technology. This will be useful when the physical environment is semantically related to learning objectives and activities, and it is often associated with pedagogical theories of situated, and experimental learning [7]. The other type deals with supporting the learning process while a learner makes use of ubiquitous computing technology. This will be used when the environment conditions might have an impact on the learning experience.

In any case, one key feature to realize this type of learning is context awareness. A context-aware system can perceive characteristics of the environment that surround him, i.e. his context, and react to changes in this environment [9]. This means that the system is able to acquire context information, store it in a particular internal structure and later process it to provide the required adaptation functions.

Context [10] is any information that can be used to characterize the situation of an entity. An entity is a person, place, or object that is considered relevant to the inter-action between a user and an application, including the user and applications them-selves. In u-learning, context is of interest because it can help to integrate educational activities with real world objects and situations, or because it can aid in providing adapted material based on context characteristics.

According to [11] there are five types of situation parameters, i.e. context infor-mation, that are useful for activities conducted in a u-learning environment.

Personal contexts sensed by the system: learner's location and time of arrival, temperature, perspiration level, heartbeat, blood pressure, etc.

Environmental contexts sensed by the system: sensor's ID and location, tempera-ture, humidity, air ingredients, and other environment parameters around the sensor, and objects that are approaching the sensor.

Feedback from the learner via the mobile learning device: observed or sensed data of target items and acquired photos or interactions with the learning system.

Personal data retrieved from databases: learner's profile and learning portfolio.

Environmental data retrieved from databases: detailed information of the learning site, equipment located in it, persons who use it or manage it, etc.

Since this work is mainly focused on new features of context required for ubiq-uitous environments, it will not consider the category *Personal data retrieved from database* in the analysis, as this is really part of a user model, as opposed to a context model [12].

As stated before, the context information will usually be obtained from sensors or databases. However, it is not always possible to do this, for example, because there is no sensor to capture a specific piece of data or because the data obtained is not reliable or complete. In this cases, it is desirable for a context-aware system to have context composition capabilities [13], i.e. the ability to combine several simpler contexts to derive a new one. There are four types of context composition: combining homoge-neous or heterogeneous contexts, deriving high level context from lower level ones and vice versa.

3.2   Ontologies

In Computer Science, "An ontology is a formal, explicit specification of a shared conceptualization. A 'conceptualization' refers to an abstract model of some phenomenon in the world by having identified the relevant concepts of that phenomenon. 'Explicit' means that the type of concepts used, and the constraints on their use are explicitly defined. 'Formal' refers to the fact that the ontology should be machine readable, which excludes natural language. 'Shared' reflects the notion that an ontology captures consensual knowledge, that is, it is not private to some individual, but accepted by a group" [14].

The main purpose of an ontology is to provide a common understanding of a particular domain. This common understanding reflects a consensus about a body of knowledge, which acts as a unifying framework that allows people and/or systems to work together seamlessly. The group of people, either large or small, that agree on this consensus are said to commit themselves to that ontology [14].

Since ontologies provide a shared understanding, one way to favor compatibility when designing a new one is to reuse a Top Level or Foundational ontology. According to Guarino [15], a Foundational ontology describes very general concepts like space, time, matter, object, event, action, etc., which are independent of a particular problem or domain. Then, more specific ontologies, such as domain, task or application ontologies, will specialize the terms introduced in the top-level ontology.

Another important aspect when designing an ontology, is the level of expressivity to be used. Gómez-Pérez et al. [16] differentiate Lightweight ontologies from Heavyweight ontologies. The first ones can be controlled vocabularies, glossaries, thesauri, or informal is-a hierarchies, while the heavyweight ones can be formal is-a hierarchies, formal is-a hierarchies including instances, including relations, including value restrictions, or expressing general logical constraints. It is worth noting that according to some works [16, 17], lightweight ontologies are not considered to be ontologies.

The choice of a level of expressivity for an ontology can impact the available languages that could be used to codify it, i.e. not every language allows to represent complex expressions. When choosing the language it is also useful to take into account the compatibility and current maintenance status of such language [16].

4   Context Ontologies in U-Learning

The following summarizes the reviewed works that use ontologies to model context information, to personalize learning services and content in u-learning systems.

Hsu et al. [18] propose a mobile learning framework that supports context-aware personalized learning in museums. The framework aims at providing relevant educational content and services to students according to his preferences and location within the place. The framework has an ontological layer which contains a Context Ontology with environment information. This context information is then used in an adaptive layer to trigger relevant content and services to students while they explore the physical place.

Hu *et al.* [19] present a semantic ontology-based context model called SmartContext, that provides personalized services to tutors and students. The Context Ontology is built based on a context template that holds relevant context information, and a context middleware component supports context merging and reasoning.

Yang [20] presents a context model as part of a context aware ubiquitous learning environment that emphasizes collaborative learning through the provision of several learning services. Required context information is defined from the learners' point of view, i.e. the surrounding environment affecting services discovery and access; and from the services perspective, i.e., the surrounding environment affecting learning services delivery and execution. This context information is modeled as part of the Learner and Service ontologies that comprise the system.

The Web-Based Educational System SIMBAD is presented by Bouzeghoub *et al.* in [21]. SIMBAD delivers adapted learning resources depending on the context of the learner, i.e. the particular contextual information during a time interval.

In Schmidt [22] a context-aware architecture for a learning support system is presented. The context information is modeled in a Context Ontology, which is used to store student related data. This data is obtained from different sources and pre-processed outside the ontology which receives it directly as high level context. This context data, along with other, is used by a matching service to handle recommendation and adaptation functions.

In [23], Hong and Cho present CALA, an architecture for context-aware learning services in an u-learning environment. Context information is represented in CALA-ONT ontology, and it is obtained from a context provision module that feeds on various sensors and devices. Execution of a particular learning service is triggered by a set of predefined rules contained in the Learning Service Coordination Module.

Siadaty *et al.* [24] present m-LOCO, an ontology-based framework for context-aware mobile learning. The main purpose of this framework is to aid in the provision of learning contents in mobile learning environments, and as such, context is defined in relation to its delivery medium, i.e., mobile devices. The context ontology, which is really a mobile device ontology, is based on the W3C's CC/PP (Composite Capability/Preference Profiles) specification.

Yu *et al.* [25] present a context-aware e-learning infrastructure that provides adaptive content recommendation. The Context ontology groups information about student context and interacts primarily with two other components. The context aggregator, which aggregates context data from various sensors and asserts it into the context knowledge base, and the context reasoner, which infers high-level context information and checks for knowledge consistency in the context KB.

Pernás *et al.* [26] present an ontology network model for a context-aware learning system that allows to identify the situation of the student. The ontology network allows to relate existing ontologies, such as WGS84 geo positioning, with new ones.

In [27], Soualah-Alil *et al.* present an adaptive context-aware m-learning system, that offers an industrial trainee the most suitable learning activity to be carried out in a specific situation. This situation refers to the context of mobility related to the industrial training environment.

González [28] presents a student model aimed at providing personalized services in u-learning environments. Context information is modeled in the student ontology and the classes presented are tailored to a university context. Adaptation is triggered by predefined rules defined within the ontology.

Salazar Ospina [29] introduces a model of a multi-agent context-aware system that offers personalized services and recommendations. Context information is contained in the Student ontology, and it groups data related to technological and spatial contexts of a student. This information is then used by an agent to provide recommendations of tutors for each learner.

Zhang *et al.* [30] propose a layered context model aimed at providing adapted learning support to professionals in work-based learning domain. The context model consists of a common layer, with a common ontology, and a particular domain layer, with a work-based learning generalized ontology and a specific work ontology.

Abech *et al.* [31] propose OntoAdapt, an ontology with student, learning object and context information, that allows the adaptation of learning objects based on context data. Students must use a mobile LMS and their access device is used both as the source of context data and as the presentation device of the adapted object.

5   Analysis and Comparison of Context Ontologies

To compare the u-learning context ontologies described in the previous section, a comparison framework is proposed. This framework is divided in two parts.

One part relates to the ontology model itself, and it is based on the main foundational design decisions to take when designing an ontology, as described in section two. The proposed questions used to assess ontology models are: *1. Does the ontology rely on any foundational ontologies? 2. Which is the level of expressivity of the ontology? 3. What language does the ontology use?*.

The other part relates to the support provided by the ontology to the ubiquitous learning model. Here the focus is on identifying if the most relevant characteristics of context models for u-learning, as described in section two, are present on the reviewed ontology-based models. The proposed questions to assess this, are: *1. What is the type of u-learning environment involved? 2. What u-learning situation parameters are supported by the ontology? 3. Does the ontology provide a mechanism for context composition? 4. In which model is context information contained? (e.g. Context model, Student model, etc.).*

5.1   Analysis of Ontology Models

Considering the first question, it can be seen that only [21, 26] make use of a foundational ontology. In [21], it is used only for one concept, Time, out of four main concepts of context information represented, while [26] uses a complete vocabulary as part of an ontology network. The main advantage of using a foundational ontology is that new terms defined will be backed up by an already accepted taxonomy of higher level concepts. This adds value to the ontology as a shareable artifact.

With regards to the second question, it is noted that all ontologies are heavyweight ontologies, i.e. they provide at least a formal is-a hierarchy. Most of them, [18, 19, 21, 22, 24–27, 29–31], provide definitions for binary relations; while some, [20, 23], also define disjoint class axioms. The works [23] and [28] present more expressive models adding properties of relations and properties of relations plus complex class definitions, respectively. While having a more expressive model allows representing context in a more detailed way, it also comes at the price of a higher complexity of reasoning.

Considering the third question, most ontologies reviewed, [20, 21, 23, 25, 27–31] use some form of OWL as a representation language, while only [19] relies on RDF. Both OWL and RDF languages are W3C recommendations and as such, they are a reliable option when it comes to specify an ontology in a standard formalism.

Table 1 synthetizes the questions and answers previously discussed.

Table 1. Comparison of ontology characteristics.

Ref.	Question 1	Question 2	Question 3
[18]	No	Heavyweight: relations	Not spec.
[19]	No	Heavyweight: relations	RDF
[20]	No	Heavyweight: relations, disjoint classes	OWL-S
[21]	OWL-Time	Heavyweight: relations	OWL
[22]	No	Heavyweight: relations	Not spec.
[23]	No	Heavyweight: relations, properties of relations, disjoint classes	OWL-DL
[24]	No	Heavyweight: relations	Not spec.
[25]	No	Heavyweight: relations	OWL
[26]	WGS84 geo positioning	Heavyweight: relations	Not spec.
[27]	No	Heavyweight: relations	OWL-DL
[28]	No	Heavyweight: relations, properties of relations, disjoint classes, complex class definitions	OWL-DL
[29]	No	Heavyweight: relations	OWL
[30]	No	Heavyweight: relations	OWL-DL
[31]	No	Heavyweight: relations	OWL-DL

5.2   Analysis of the Support Provided by the Ontology to the U-learning Model

Considering the first question, it can be noted that most works, [19–22, 24–29, 31] deal with context ontologies that will be used in an ubiquitous learning environment where content and learning objectives are not related to real world objects that surround a student. Only [18, 23, 30] work with learning environments that involve learning in the real world. There is no significant difference between the information used in the model for each type of setting, i.e. both types of environments make use of the same context information categories.

With regards to the second question, all models except [24] include location information. This is clearly the most significant piece of context information when it comes to u-learning, not only because it allows providing personalization based on it, but also because it can be a pointer to other context data, such as people and objects nearby. Also, almost all ontologies, [20–27, 29–31], contain device information in order to be able to adapt content to the device of a student. For example, to change font size according to screen size or to select the format of learning material to be delivered based on network speed. When it comes to environmental data, almost half of the works use it [21, 23, 26–28, 30] and only [30] includes feedback provided from a learner through the mobile device, in this case, sensing QR tags and input of desired activity (e.g. just-in-time learning, review or progress). It is worth mentioning that [20] considers information about the situation of a student, i.e. the kind of task that he/she is doing, such as, in a meeting or reading, although it is not clear how this information is obtained. Also, [27] attempts to infer the situation of the learner using the context information contained in the ontology. On the other hand, [22] registers the stress level of a learner as part of the context model. This kind of information can have a significant impact in the adaptation process in an educational setting and belongs to a new trend of research called affective computing [12].

Considering the third question, most ontology models, [19–21, 28, 29] support combining heterogeneous user contexts, for example, if student S is using device D and device D is located in place P then student S is in place P. This can be useful to derive unavailable context information, but it can also be used to obtain information that is available by other means, in a more reliable or efficient way. On the other hand, only [21, 25, 27] deal with inferring high level context from lower level context data. In all cases they employ rules defined within the ontology.

With respect to the fourth question, [18, 20, 25, 28, 29], model context information in the student model, while [19, 21–24, 26, 27, 30, 31], use specific models. More specifically, in [21, 26, 31], context information is separated into different models, grouping different kind of data, such as device, location and time. In [30], there are layered ontologies of context information, from general to specific.

Table 2 synthetizes the questions and answers previously discussed. It uses the following abbreviations: To describe the type of u-learning environments, T1 for u-learning environments that deal with learning in connection with the real world and T2 for environments where real world objects are not semantically related to learning objects; To describe contextual situation parameters, PC for *Personal contexts sensed by the system*, EC for *Environmental contexts sensed by the system*, FL for *Feedback from a learner via the mobile learning device* and ED for *Environmental data retrieved from databases*.

Table 2. Comparison of u-learning support provided by the ontology models.

Ref.	Question 1	Question 2	Question 3	Question 4
[18]	T1	PC: location, time	No	User Model
[19]	T2	PC: location, time	Combining heterogeneous contexts	Context model

(Continued)

Table 2. (*Continued*)

Ref.	Question 1	Question 2	Question 3	Question 4
[20]	T2	PC: location, situation. ED: device network channel	Combining heterogeneous contexts	Student model, Service model
[21]	T2	PC: location, time EC: noise levels, temperature, traffic condition, network connectivity ED: Device type, CPU speed, display size, input method, support wired network	Combining heterogeneous contexts and deriving high level context from lower level context	Environment model, Device model, Location model, Time model
[22]	T2	PC: stress level ED: device bandwidth, platform, resolution, equipment	No	Context model
[23]	T1	PC: location EC: lighting, humidity, temperature, weather condition, persons, activities, computational entities Room curtain status, door status. Device status ED: place latitude and longitude	Combining heterogeneous contexts	Context model
[24]	T2	ED: device screen size, color/image capabilities, operating system, list of audio and video encoders. Network latency and reliability	No	Device model
[25]	T2	PC: location, time ED: device hardware and software characteristics, and network conditions	Deriving high level context from lower level context	Learner model
[26]	T2	PC: location, situation, time EC: network connectivity ED: device screen size, type	Deriving high level context from lower level context	Location model, Device model, Situation model

(*Continued*)

Table 2. (*Continued*)

Ref.	Question 1	Question 2	Question 3	Question 4
[27]	T2	PC: time, laps, location. EC: noise, level, comfort ED: location type, coordinates, device type, software capabilities, hardware capabilities	No	Context model
[28]	T2	PC: location EC: noise level, lighting level, students, activities, computational entities ED: Place GPS coordinates	Combining heterogeneous contexts	Student model
[29]	T2	PC: location, time ED: device type, SO, SO version	Combining heterogeneous contexts	Student model
[30]	T1	PC: work environment, learning device, support device EC: noise level, lighting level, learner, activities, network connectivity ED: device screen size, mobility, operating system, software capabilities, work environment GPS coordinates	No	Context model
[31]	T2	PC: device, location EC: network connectivity, battery level ED: device model, battery level, available storage capacity, screen size, operating system version	No	Context model, Student model, Device Model

6   Conclusions

The use of ontologies for ubiquitous context modeling in u-learning environments is a growing research area. The u-learning context ontologies found in the literature provide expressive models that can handle the required context information and also offer some basic reasoning mechanisms, either by the ontology structure itself or by the addition of rules. These mechanisms are important because they help to infer context information that may otherwise not be explicitly available.

Another positive feature found in the reviewed ontologies is that they use standard languages to codify the information. Doing so increases interoperability and maintainability of the constructed models.

On the other hand, some open issues can be pointed out. First, all models deal only with a subset of context information that is of interest in a u-learning setting. As such, there is no proposal of a generic context model for this area. This is probably due to the fact that specific context information required is highly dependent on the system that will use it and to the purpose it will be used for. However, it can also be argued that the main goal of an ontology is to provide a common understanding and vocabulary for a particular domain. Therefore, it might be beneficial to have a generic context ontology to unify concepts in this area. In line with this, it should be studied the convenience of reusing a foundational ontology to formulate a context model, as doing so would increase its reuse potential and conform to already accepted agreements.

Second, while u-learning systems can be considered a subset of ubiquitous or context-aware systems, they pose particular challenges related to educational settings. The type of adaptation that these systems provide is backed by pedagogical theories that try to enhance the learning experience of students. Inferring high level context relevant to such experiences, e.g., level of concentration based on environmental and personal contexts, is explored only superficially in some of the reviewed works.

Finally, context information has variable granularity levels and can be more or less dynamic in nature. Attempting to group context information according to these criteria has two potential benefits. On one hand, it can result in a more expressive model, formalizing both low and high context information. This can, in turn provide an enhanced inference capability. On the other hand, placing dynamic and static information separately can help manipulation of the model at runtime, because data that needs to be updated more frequently will be grouped together.

By highlighting some relevant characteristics of ontology-based context modeling in the area of ubiquitous learning, we intend to aid future researchers in choosing, or building their own, ontology-based context model. Well-designed context models should be capable of integrating heterogeneous context data, a compulsory prerequisite to build efficient and maintainable ubiquitous learning applications.

References

1. Strang, T., Linnhoff-Popien, C.: A context modeling survey. In: Graphical Models, pp. 1–8 (2004)
2. Krummenacher, R., Lausen, H., Strang, T., Kopecky, J.: Analyzing the modeling of context with ontologies. In: International Workshop on Context-Awareness for Self-Managing Systems (Devices, Applications and Networks) - CASEMANS 2007, pp. 11–22 (2007)
3. Bolchini, C., Curino, C.A., Quintarelli, E., Schreiber, F.A., Tanca, L.: A data-oriented survey of context models. ACM SIGMOD Rec. **36**, 19 (2007)
4. Bettini, C., Brdiczka, O., Henricksen, K., Indulska, J., Nicklas, D., Ranganathan, A., Riboni, D.: A survey of context modelling and reasoning techniques. Pervasive Mob. Comput. **6**, 161–180 (2010)
5. Verbert, K., Manouselis, N., Ochoa, X., Wolpers, M., Drachsler, H., Bosnic, I., Duval, E.: Context-aware recommender systems for learning: a survey and future challenges. Learn. Technol. IEEE Trans. **5**, 318–335 (2012)
6. Al-Yahya, M., George, R., Alfaries, A.: Ontologies in E-learning: review of the literature. Int. J. Softw. Eng. its Appl. **9**, 67–84 (2015)
7. Ogata, H., Yano, Y.: Context-aware support for computer-supported ubiquitous learning. In: Proceedings - 2nd IEEE International Workshop on Wireless and Mobile Technologies in Education, pp. 27–34 (2004)
8. Bomsdorf, B.: Adaptation of learning spaces: supporting ubiquitous learning in higher distance education. In: Mobile Computing and Ambient Intelligence: The Challenge of Multimedia (2005)
9. Schilit, B., Adams, N., Want, R.: Context-aware computing applications. In: WMCSA 1994 Proceedings of the 1994 First Workshop on Mobile Computing Systems and Applications, pp. 85–90. IEEE Computer Society, Washington (1994)
10. Dey, A.K., Abowd, G.D.: Towards a better understanding of context and context-awareness. Comput. Syst. **40**, 304–307 (1999)
11. Hwang, G.J., Tsai, C.C., Yang, S.J.H.: Criteria, strategies and research issues of context aware ubiquitous learning. Educ. Technol. Soc. **11**, 81–91 (2008)
12. Brusilovsky, P., Millán, E.: User Models for Adaptive Hypermedia and Adaptive Educational Systems. In: Brusilovsky, P., Kobsa, A., Nejdl, W. (eds.) Adaptive Web 2007. LNCS, vol. 4321, pp. 3–53. Springer, Heidelberg (2007)
13. Poslad, S.: Ubiquitous Computing: Smart Devices, Environments and Interactions. Wiley, Hoboken (2009)
14. Studer, R., Benjamins, V., Fensel, D.: Knowledge engineering: principles and methods. Data Knowl. Eng. **25**, 161–197 (1998)
15. Guarino, N.: Formal Ontology and Information Systems. Proceedings of the 1st International Conference, vol. 46, pp. 3–15. IOS Press, Amsterdam (1998)
16. Gómez-Pérez, A., Fernández-López, M., Corcho, O.: Ontological Engineering. Springer, London (2004)
17. Breitman, K.K., Casanova, M.A., Truszkowski, W.: Semantic Web: Concepts, Technologies and Applications. Springer, London (2007)
18. Hsu, T.-Y., Ke, H.-R., Yang, W.-P.: Knowledge-based mobile learning framework for museums. Electron. Libr. **24**, 635–648 (2006)
19. Hu, B., Moore, P.: SmartContext: an ontology based context model for cooperative mobile learning. In: Shen, W., Luo, J., Lin, Z., Barthès, J.-P.A., Hao, Q. (eds.) CSCWD 2006. LNCS, vol. 4402, pp. 717–726. Springer, Heidelberg (2007)

20. Yang, S.J.H.: Context aware ubiquitous learning environments for peer-to-peer collaborative learning. Educ. Technol. Soc. **9**, 188–201 (2006)
21. Bouzeghoub, A., Do, K.N., Lecocq, C.: Contextual adaptation of learning. In: IADIS International Conference Mobile Learning 2007, pp. 41–48 (2007)
22. Schmidt, A.: Impact of context-awareness on the architecture of E-learning solutions. In: Pahl, C. (ed.) Architecture Solutions for E-Learning Systems, pp. 306–319. IGI Global, Hershey (2007)
23. Hong, M., Cho, D.: Ontology context model for context-aware learning service in ubiquitous learning environments. Int. J. Comput. **2**, 193–200 (2008)
24. Siadaty, M., Torniai, C., Gašević, D., Jovanović, J., Eap, T.M., Hatala, M.: m-LOCO: an ontology-based framework for context-aware mobile learning. In: Proceedings of the 6th International Workshop on Ontologies and Semantic Web for Intelligent Educational Systems at 9th International Conference on Intelligent Tutoring Systems (ITS 2008), Montreal, Canada, pp. 21–35 (2008)
25. Yu, Z., Zhou, X., Shu, L.: Towards a semantic infrastructure for context-aware e-learning. Multimed. Tools Appl. **47**, 71–86 (2010)
26. Pernas, A.M., Diaz, A., Motz, R., Oliveira, J.P.M.De: Enriching adaptation in e-learning systems through a situation-aware ontology network. Interact. Technol. Smart Educ. **9**, 60–73 (2012)
27. Soualah-Alila, F., Mendes, F., Nicolle, C.: A context-based adaptation in mobile learning. Bull. Tech. Comm. Learn. Technol. **15**, 14–18 (2013)
28. González, G.: Ontología del perfil de usuario para personalización de sistemas de u-learning universitarios. In: 43° Jornadas Argentinas de Informática- JAIIO 2014 - 17° Concurso de Trabajos Estudiantiles, EST 2014, pp. 90–114 (2014)
29. Salazar Ospina, O.M.: Modelo de Sistema Multi-Agente ubicuo, adaptativo y sensible al contexto para ofrecer recomendaciones personalizadas de recursos educativos basado en ontologías (2014)
30. Yin, C., Zhang, B., David, B., Xiong, Z.: A hierarchical ontology context model for work-based learning. Front. Comput. Sci. **9**, 466–473 (2015)
31. Abech, M., Da Costa, C.A., Barbosa, J.L.V., Rigo, S.J., da Rosa Righi, R.: A model for learning objects adaptation in light of mobile and context-aware computing. Pers. Ubiquitous Comput. **20**, 167–184 (2016)

An Interactive Tool to Support Student Assessment in Programming Assignments

Lina F. Rosales-Castro[1]([✉]), Laura A. Chaparro-Gutiérrez[1],
Andrés F. Cruz-Salinas[1], Felipe Restrepo-Calle[1],
Jorge Camargo[2], and Fabio A. González[1]

[1] MindLab Research Group, Universidad Nacional de Colombia, Bogotá, Colombia
{lfrosalesc,lachaparrog,afcruzs,ferestrepoca,
fagonzalezo}@unal.edu.co
[2] Laboratory for Advanced Computational Science and Engineering Research,
Universidad Antonio Nariño, Bogotá, Colombia
jorgecamargo@uan.edu.co

Abstract. The paper presents an interactive tool for analysis of a set of source code submissions made by students when solving a programming assignment. The goal of the tool is to give a concise but informative overview of the different solutions submitted by the students, identifying groups of similar solutions and visualizing their relationships in a graph. Different strategies for calculating the solution groups as well as for visualizing the solution graphs were evaluated over a set of real codes submitted by students of an algorithms class.

Keywords: Programming education · Visualization · Source code analysis

1 Introduction

Tools for automatic program evaluation make it easier to automatically evaluate student performance in programming courses. These tools let the instructor to improve the time expended grading an assignment and gives her/him the opportunity to attend students' problems [6]. However, when the ratio between instructor and students is bigger than 1 to 10, it becomes almost impossible for the teacher to understand what the students are doing in the process of solving a problem [15]. In addition, tools for automatic programming assignment grading, such as DOMJudge [3], only give a limited feedback to the student, basically whether the program runs or not and if the produced output was correct or not given a known answer. Moreover, continuous evaluation generates hundreds of submissions even in a moderately small class. Reviewing them by hand would be cumbersome.

This paper presents a tool that helps the instructor to evaluate students, by providing a visual representation of the set of submissions of a programming assignment. In particular, the tool compares the source code of the different submissions and graphically displays the similarities and dissimilarities as well as

M. Montes-y-Gómez et al. (Eds.): IBERAMIA 2016, LNAI 10022, pp. 404–414, 2016.
DOI: 10.1007/978-3-319-47955-2_33

the clusters that group similar solutions. Different strategies for visualization and clustering of the submissions were evaluated over a set of codes submitted by students of an Algorithms class for a specific problem during two practical sessions.

The rest of the paper is structured as follows. The Sect. 2 presents a state of the art in automatic source code assessment and analysis. The Sect. 3 presents the tool we are proposing to improve the source code assessment and visualization. The Sect. 4 includes the tool evaluation, and finally, the last Section discusses the conclusions and future work.

2   Automatic Source Code Assessment and Analysis

There is a wide group of tools to automatically grade source code, which are originally intended for programming contests. These tools, denominated online judges, attempt to grade a programming solution in terms of how it performs on a set of test cases for a given problem. Uva Online Judge [18] is one of the oldest and most recognized among them. It was created to help programming contest competitors to improve their skills during competitions, but with time, people around the world started to use it for personal study. DOMjugde [3] another online judge, was developed by a community of developers around the world as an open source project. This online judge was developed to run programming contest such as ACM-ICPC, but it has been adapted to teaching environments in universities around the world including the Universidad Nacional de Colombia.

Regarding source code analysis tools intended to support the software engineering process, we can find a wide range of tools with different goals, such as: reverse engineering, authorship recognition, debugging, clone detection, visualization, among others [12]. For instance, in [19] a tool was proposed to improve the quality of software by measuring code metrics for C# using an Abstract Syntax tree (AST) representation of the source code. Related to clone detection, in [13], authors also use an AST representation and take advantage of suffix trees to propose an approach to detect syntactic clones in linear time. However, related to student's source code analysis and visualization there is not much related work [11].

The motivations to use source code analysis in this context are mainly for the improvement of the feedback given to both, students and teachers. In [17] authors put emphasis on feedback is a key ingredient for the development of quality in students learning. Also for the teacher, when he wants to give a proper and fast feedback to the students about the solutions tried, this is very difficult because even a student can solve a problem in different ways [10].

Among the tools that attempt to use source code analysis in an academic context is CodeWebs [15], which is a source code search engine that intends to identify similar solutions using clustering techniques, and identify groups that may need the same feedback from an instructor. Another approach is the one given by [15], which builts a solution space representing all the paths that a student could take to solve a problem. Authors propose to use a tool that gives hints to the students in the process of reaching a solution instead of just giving her/him the solution. OverCode [11] also creates clusters of code with similar performance, from which one is presented to the teacher as a representation of certain cluster. The tool aims to give to the teacher an idea of solution variation so she/he can provide feedback to the students.

3   Source Code Analysis and Visualization Tool

The proposed tool is aimed at building a visual representation of the solution space where the students are solving certain programming problem. With this tool we want to improve how the instructor does an assessment based on providing an overview of what the students are trying to do while they are solving an assignment. Also, teachers may have an idea of how similar the given solutions are and then give feedback with less effort to a higher number of students, i.e. feedback with more impact. In addition, it could be possible to identify plagiarism cases where solutions are very close, according to the distance calculated between solutions.

Figure 1 shows the tool's data flow. First, students submit their solution to the Online Judge DOMJudge and the solutions are judged and stored in a database separated by an identifier of the practical session and submission id. From this database we extract a set of solutions belonging to the same practical session and then we represent each solution as an Abstract Syntax Tree and then by a string (parenthesized representation of the AST). Then, the similarity is calculated between each pair of source codes obtaining a matrix representing all the solutions. After that, three steps are carried out: (1) clustering using Spectral Clustering [14] and Affinity propagation [8]; (2) graph calculation; and (3) dimensionality reduction using Multidimensional Scaling [20] and the Fruchterman-Reingold force-directed. Finally, a graph presenting the solution space is visualized.

The tool has two components, the first one implemented in Java and the second one in Python. The Java component implements the transformation of the source code into an Abstract Syntax Tree (AST) representation, calculates the distances between each AST, and presents the user interface. The Python component implements clustering and visualization functionalities.

3.1   Source Code Database

A set of problems is presented to the students (each one has multiple test cases), and the task for the student is to program her/his own solution. The solutions are submitted by the student to an Online Judge, DOMjugde, which gives its verdict based on known solutions for the test cases. The verdict of the Online Judge could be: Correct, Wrong Answer, Time Limit, and Run-Time error. In this context, there are multiple ways of solving each problem and we can track of all the attempts for each student. The source code is obtained from a database connected to the DOMJudge. The database has all the source code from all the submissions of several semesters for the Algorithms course. The source code is

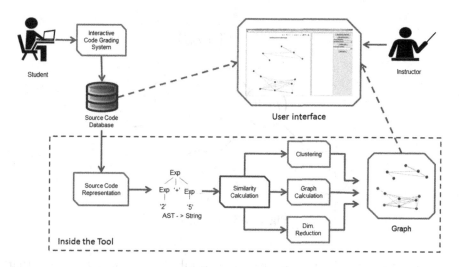

Fig. 1. Source code analysis and visualization tool architecture

obtained directly from the database by packages of codes from the same class. Internally, a division of sources code from each problem was done.

3.2 Source Code Representation

The source code representation is obtained using ANTLR, which is a parser generator for computer languages. It is able to parse source code of a given context-free grammar in a efficient way using its strategy "Adaptive LL(*)" [16]. ANTLR acts as a black box generating a parser that can build and walk parse trees in order to interpret the given grammar.

Our tool uses grammars previously designed and hosted in the ANTLR repository to parse each source code. Currently Java 7 [4] and Python 3 [5] grammars are supported. The parser is built and the AST is obtained in a string parenthesized representation provided by ANTLR. This representation encodes the tree structure in a top down fashion placing each node label in the string and representing sub trees as inner parentheses. This process continues in a recursive way until the leafs are reached. For example, the code fragment (m^e) mod n is first represented by its AST (Fig. 2) and then the AST is linearized as the string "(mod (pow m e) n)".

3.3 Similarity Calculation

ANTLR returns a string that contains a representation of the AST. With this representation of the code, the distance metrics are calculated between each pair of codes to be analyzed. The metric distance used is Jaro-winkler distance.

The Jaro-winkler distance is a variant of the Jaro distance metric. This method is based on the number and order of the common characters between

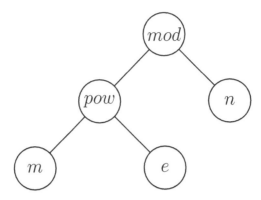

Fig. 2. AST example

two strings [7]. The idea of this metric is to take into account the errors that can transform an alphanumeric string into another.

A character is considered in common if the distance between the same character in both strings is no further apart than $m/2-1$ where $m = max(|s_1|, |s_2|)$. According to Winkler [21] half of a transposition occurs when the first assigned character in one string does not match the first assigned character in the second string. Then this is evaluated for the second assigned character in the other string, etc. The amount of mismatched characters is divided by two and it is the value of transpositions t.

The Jaro-Winkler distance is given by:

$$d = \begin{cases} 0 & \text{if } m = 0 \\ \frac{1}{3}\left(\frac{m}{|s_1|} + \frac{m}{|s_2|} + \frac{m-t}{m}\right) & \text{otherwise} \end{cases} \tag{1}$$

Then a distance matrix M is constructed, where each position d_{ij} is the distance between the source code i and j.

Using M, a similarity matrix S is obtained where its entries are calculated as follows

$$s_{ij} = \exp{-\gamma d_{ij}\sigma} \tag{2}$$

where d_{ij} are the entries from M.

3.4   Clustering

Two clustering techniques were used using the distance matrix obtained from the Java module. The first one was spectral clustering [14] and the second one was affinity propagation [8], both of them are implemented by the Scikit-Learn clustering module.

3.5 Graph Calculation and Dimensionality Reduction

The visualization was made using a graph, where each node is a student's code solution and the edges are built based on the similarity between solutions, as calculated from the distance matrix.

The nodes of the graph are marked by the order of submission, where a number 1 represents the first submission of the problem and the biggest number represent the last submission. Each node has a label that indicates the user of the student and the verdict of the submission from the online judge.

The position of each node over the space is determined in two ways: dimensionality reduction using Multidimensional Scaling (MDS) using the Scikit-Learn Maninfold learning module [2] and the Fruchterman-Reingold [9] force-directed (Spring) algorithm using the Plotly Python Library [1].

3.6 User Interface

The user interface has three main components: the graph visualizer, the options panel and the source code visualizer. The Fig. 3 shows the User interface

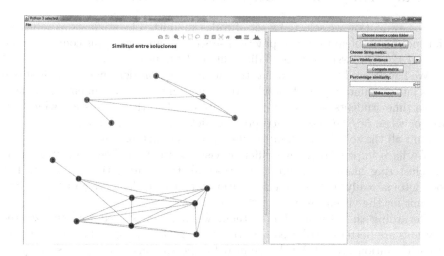

Fig. 3. User interface

The graph visualizer shows the graph when it has been calculated by the Python script. This component allows the interaction with the user as follows:

- Show the name of the student and the verdict of the judge when the mouse is over a node
- Zoom out and zoom in
- Save the graph as a .png image
- Autoscale

– Translate

The options panel has four components:

– The data selector button: this button allows to select the dataset or source code. A folder with the data set must be selected, the source codes could be in python or in Java.
– The distance drop down list selector: this drop down list shows the distance that could be selected.
– The Compute matrix button: this button starts the distance matrix calculation once the data is selected. The matrix is calculated using the distance metric selected in the distance drop down list selector.
– The script load button: once the data is selected and the distance matrix is computed, the interface allows to the user to load the Python script to calculate the visualization.

The source code visualizer allows the instructor to select a node and visualize the related source code.

4   Clustering, Visualization and Tool Evaluation

For the evaluation we selected a set of codes submitted by students as solutions of a specific problem from two practical sessions of an algorithms course, having in total 120 submissions from 30 different students in Python.

We have tested two approaches to visualize the solution space. The first one takes into account only the last submitted solution from each student, since most of the time students stop submitting solutions when the submission window is over or when he/she gets an accepted verdict. The second approach takes into account all the solutions submitted during the submission window.

We have implemented two different visualizations methods and two different clustering methods. Figure 2 describes all the combinations of them. The tool automatically selects the last solution for each student, which is taken into account for the visualization.

Regarding the visualization method, when MDS is used to visualize, the solutions are better distributed over the visualization space and the coordinates of each solution are totally determined by the similarity measure, so they do not variate from run to run as it is the case with the Fruchterman-Reingold force-directed algorithm. In this way, an instructor will obtain the same graph visualization each time the MDS algorithm is run.

Related to the clustering technique, in Figs. 4 (a) and (b) we show the result obtained with affinity propagation clustering. We can identify that the solutions belonging to the red cluster all have a correct verdict from the online judge. The ones belonging to the yellow cluster have a verdict time-limit and wrong-answer. The solutions belonging to the orange cluster have both, correct solutions and wrong-answer solutions, but not time-limit solutions.

Clustering the same solutions using spectral clustering and visualizing with MDS and Fruchterman-Reingold force-directed, generate the results depicted in

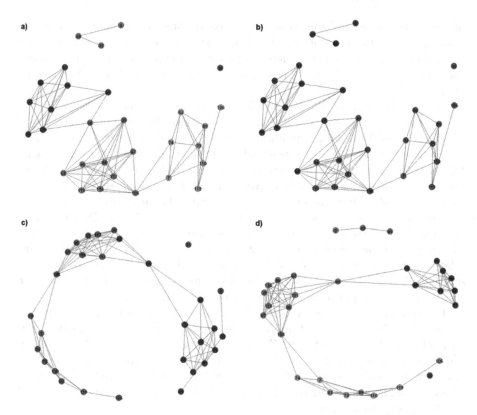

Fig. 4. Visualization of the solution graph for the last solution submitted by different students. Each node corresponds to a solution, its color indicates the cluster it belongs to and the arcs indicate that the similarity between the solutions is above a given threshold. Each subfigure shows a results corresponding to a different clustering-visualization combination: (a) MDS and affinity propagation, (b) MDS and spectral clustering, (c) Fruchterman-Reingold force-directed algorithm and spectral clustering and (d) Fruchterman-Reingold force-directed algorithm and affinity propagation

Fig. 4 (c) and (d) respectively. Note that the same distribution of the graph but grouped in two clusters is produced. The first one includes the light blue and the red one, and second one includes the orange one, the yellow one and the dark blue one. This approach also separates correct verdicts from time-limit and wrong-answer, but with minor details.

As a conclusion, we can say that the best result is obtained using MDS for visualization and affinity propagation for clustering.

The purpose of visualize all the solutions of a student is to identify what a student is doing to solve the problem and how her/his solutions are related to her/his own alternative solutions (or versions of them). In the online judge, DOMJudge, a student could obtain a wrong-answer as verdict when her/his logic is correct and she/he is only making a minor error. For example, in a loop or in a variable initialization, to correct a wrong version only requires a minimum change in the code, and sometimes, based on the instructors experience, these small errors are the hardest ones to identify since they require a conscientious revision of the code. If the instructor can identify a small error that most of the students are making, the instructor could give a faster and better feedback to each student.

We can also identify that if a solution is obtaining a *wrong-answer*, a *compilation* or a *run time error*, and it is very near to ones that are obtaining *correct*, the change needed in the *wrong-answer* solution to make it *correct* is not so big. An example can be seen in the small red box in the Fig. 5, where the submissions 103, 104, 105, and 106 are from the same student and, as can be seen, they are very close in the space according to MDS. In this particular case, the first three submissions got *wrong-answer* as verdict, but the 106 got a *correct answer*.

Also, in Fig. 5 we can identify that the solutions 75, 76, 77, 78, 79, and 81 are from the same student. The first four solutions got a verdict *run-time-error*, the 79 got a *wrong-answer*, and the 81 got a *correct* answer. All of them are relatively close each other according to the visualization. We can see that the solution 81 is near to the solution 90. Note that solution 90 is from another student and it also got a *correct* answer from the online judge.

If we compare the solutions using the source code visualizer, we can see that the solution 75 and the solution 76 are more similar between them since they only differ in three lines of code. When we compare the solution 77, 78 and 79, the student has introduced a new logic, adding different components as a new data structure to the solution. This can be reflected in the big red box in Fig. 5, where the solutions 75 and 76 belong to one cluster and the solution 77, 78, 79 and 81 belongs to another.

5   Conclusions and Future Work

The paper presented a tool for interactive visualization and analysis of a set of source code submissions made by students in a programming class. The tool automatically compares the submissions and calculates a graph that reflects the similarities between the different source codes. Groups of similar submissions are identified and visualized. In an initial exploratory evaluation of the tool different strategies for clustering and visualization were evaluated. The tool has an important potential to provide information that may help the instructor to understand how students address the different programming problems, the kind of solutions that they submit and the mistakes they make.

This tool clearly gives a better feedback than the one given by online judges, such as DOMJudge [3] and Uva Online Judge [18], in an academic contexts because it can show not only the evaluation of individual student solutions but also the similarity with other solutions implemented. Our tool do not uses the AST for individual code metrics as in [19], but takes advantage of the syntax in the AST to compare one solution with others. In contrast with the tool presented in [15], our tool, in addition of showing similar solutions using clustering techniques, also shows the source code of the solution, as future work, this functionality will show differences between selected solutions. Our tool also gives the

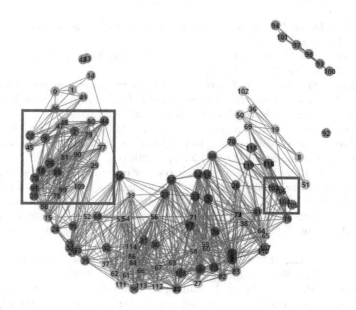

Fig. 5. Solutions visualized using MDS and affinity propagation. The red boxes indicate: the big one, solutions 75, 76, 77, 78, 79 and 81 in the solution space; the small one, solutions 103, 104, 105 and 106 all from the same student (Color figure online)

possibility of analyzing all the attempts of the students, this helps to understand the path that a student could take to solve a problem.

A more systematic evaluation of the tool as a support of a real course evaluation process is part of our future work. The evaluation will also include how the users interact with the tool and usability measures. Also, as part of our future work we plan to make an implementation of the tool based on web technologies. We also plan to explore strategies for automatic or semiautomatic students feedback.

References

1. Plotly python library (2014). https://plot.ly/python/. Accessed 14 April 2016
2. Scikit-learn clustering (2014). http://scikit-learn.org/stable/modules/manifold. Accessed 14 April 2016
3. Domjudge (2016). https://www.domjudge.org/. Accessed 15 May 2016
4. Java 7 grammar (2016). https://github.com/antlr/grammars-v4/tree/master/java. Accessed 5 April 2016
5. Python 3 grammar (2016). https://github.com/antlr/grammars-v4/tree/master/python3. Accessed 17 April 2016
6. Ala-Mutka, K.M.: A survey of automated assessment approaches for programming assignments. Comput. Sci. Educ. **15**(2), 83–102 (2005)
7. Cohen, W.W., Ravikumar, P.D., Fienberg, S.E., et al.: A comparison of string distance metrics for name-matching tasks. IIWeb. **2003**, 73–78 (2003)

8. Frey, B.J., Dueck, D.: Clustering by passing messages between data points. Science **315**(5814), 972–976 (2007)
9. Fruchterman, T.M., Reingold, E.M.: Graph drawing by force-directed placement. Softw. Pract. Exp. **21**(11), 1129–1164 (1991)
10. Gaudencio, M., Dantas, A., Guerrero, D.D.: Can computers compare student code solutions as well as teachers?. In: Proceedings of the 45th ACM technical symposium on Computer science education, pp. 21–26. ACM (2014)
11. Glassman, E.L., Scott, J., Singh, R., Guo, P.J., Miller, R.C.: Overcode: visualizing variation in student solutions to programming problems at scale. Acm Trans. Comput. Hum. Interact. (TOCHI) **22**(2), 7 (2015)
12. Jackson, D., Rinard, M.: Software analysis: a roadmap. In: Proceedings of the Conference on the Future of Software Engineering, pp. 133–145. ACM (2000)
13. Koschke, R., Falke, R., Frenzel, P.: Clone detection using abstract syntax suffix trees. In: Proceedings of the 13th Working Conference on Reverse Engineering, pp. 253–262. IEEE Computer Society, Washington, DC (2006) doi:10.1109/WCRE. 2006.18
14. Ng, A.Y., Jordan, M.I., Weiss, Y., et al.: On spectral clustering: analysis and an algorithm. Adv. Neural Inf. Process. Syst. **2**, 849–856 (2002)
15. Nguyen, A., Piech, C., Huang, J., Guibas, L.: Codewebs: scalable homework search for massive open online programming courses. In: Proceedings of the 23rd International Conference on World Wide Web, pp. 491–502. ACM
16. Parr, T., Harwell, S., Fisher, K.: Adaptive LL(*) parsing: the power of dynamic analysis. ACM SIGPLAN Not. **49**, 579–598 (2014)
17. Pieterse, V.: Automated assessment of programming assignments. In: Proceedings of the 3rd Computer Science Education Research Conference on Computer Science Education Research, pp. 45–56. Open Universiteit, Heerlen (2013)
18. Revilla, M.A., Manzoor, S., Liu, R.: Competitive learning in informatics: the UVA online judge experience. Olymp. Inform. **2**, 131–148 (2008)
19. Singh, P., Singh, S., Kaur, J.: Tool for generating code metrics for c# source code using abstract syntax tree technique. SIGSOFT Softw. Eng. Notes **38**(5), 1–6 (2013). doi:10.1145/2507288.2507312
20. Torgerson, W.S.: Multidimensional scaling: I. Theory and method. Psychometrika **17**(4), 401–419 (1952)
21. Winkler, W.E.: String comparator metrics and enhanced decision rules in the Fellegi-Sunter model of record linkage. In: Proceedings of the Section on Survey Research, American Statistical Association, pp. 354–359 (1990)

Hidden Markov Models for Artificial Voice Production and Accent Modification

Marvin Coto-Jiménez[1,2](✉) and John Goddard-Close[2]

[1] University of Costa Rica, San José, Costa Rica
marvin.coto@ucr.ac.cr
[2] Metropolitan Autonomous University, México, D.F., Mexico
jgc@xanum.uam.mx

Abstract. In this paper, we consider the problem of accent modification between Castilian Spanish and Mexican Spanish. This is an interesting application area for tasks such as the automatic dubbing of pictures and videos with different accents. We initially apply statistical parametric speech synthesis to produce two artificial voices, each with the required accent, using Hidden Markov Models (HMM). This type of speech synthesis technique is capable of learning and reproducing certain essential parameters of the voice in question. We then propose a way to adapt these parameters between the two accents. The prosodic differences in the voices are modeled and transformed directly using this adaptation method. In order to produce the voices initially, we use a speech database that was developed by professional actors from Spain and Mexico. The results obtained from subjective and objective tests are promising, and the method is essentially applicable to accent modification between other Spanish accents.

Keywords: HMM · Speech synthesis · Accents · Castilian Spanish · Mexican Spanish

1 Introduction

Speech is the most natural and primary form of human communication. In the past few decades, a great deal of research has been conducted to model this form of communication in the context of human-computer interaction. Two of the main fields considered are Automatic Speech Recognition (ASR) and Speech Synthesis. ASR tries to transcribe a speech signal into written words, analogous to a person listening, while Speech Synthesis does the opposite, by converting text into a speech signal, similar to a person speaking [1]. Whilst neither is completely solved, both have produced a number of successful applications.

ASR is primarily a task of pattern recognition, where a speech signal is converted into a set of relevant parameters, and these are then used to classify the signal into a set of categories representing some linguistic content, e.g. phonemes, syllables or words. Many of the concepts and techniques developed in ASR, have driven advances in Speech Synthesis. For small application domains e.g. speaking

© Springer International Publishing AG 2016
M. Montes-y-Gómez et al. (Eds.): IBERAMIA 2016, LNAI 10022, pp. 415–426, 2016.
DOI: 10.1007/978-3-319-47955-2_34

the time or a bank statement, the use of prerecorded words or phrases is usually enough. However, when a large domain is required, where it is not possible to prerecord every word, other methods must be developed.

One of the most recent, successful techniques for artificial speech production is based on concatenative speech synthesis, where a large database of prerecorded phonetic units, e.g. phonemes, are used, together with algorithms to splice them together to form words and utterances. This produces good quality but is not very flexible, and different databases have to be recorded for different languages and accents.

A more recent technique is based on HMMs, and known as statistical parametric speech synthesis. This method produces the acoustic parameters of a voice, using a time-series stochastic generative model [2] from a set of Gaussian probability density functions (pdf). The HMMs in the model are trained using a database of quality recorded utterances, and the voice of the speaker is reconstructed using the parameters found by the model.

Other challenges in speech sythesis, due to the complexity of human speech production, can also occur. For example, in human verbal communication there are many para-linguistic aspects which convey interesting and relevant information such as: age, emotion, gender, as well as the regional and social accents of the speaker [3]. Accent variations are defined by diversities in pronunciation, rhythm and pitch variation [4,5]. Statistical parametric speech synthesis captures a speaker's characteristics, including some of these para-linguistic aspects.

For some applications of speech synthesis, such as film and video dubbing, the accent of the target language is important. For example, it is often the case that a foreign language picture is dubbed using a Castilian Spanish version for European audiences, and a Latin American Spanish version for audiences in that part of the world. There are differences between both accents for some of the phonemes and pitch contours [6], and usually different actors who speak the particular native Spanish variation are employed to produce the desired accent. Ideally, a system that translates voice-to-voice will need to take into consideration the target accent of the synthetized voice.

In this paper, we use statistical parametric speech synthesis to produce two voices with Mexican Spanish and Castilian Spanish accents. We then propose an HMM adaptation technique to transform one of the voices into the other, and vice versa. We do this by analyzing the phonetic differences of both languages, and introducing the masking of phonemes to conveniently let the adaptation fulfill the inexistent sound of the target language.

The rest of this paper is organized as follows: Sect. 2 briefly describes the statistical parametric speech sythesis model used to produce the two voices, Sect. 3 gives the details of the HMM-adaptation technique, Sect. 4 describes Castilian and Mexican Spanish accents and our proposal to modify accents. Section 5 presents the experimental conditions and procedures, Sect. 6 shows the results and finally Sect. 7 presents some conclusions and ideas for future work.

2 Artificial Voice Production

We use statistical parametric speech synthesis to produce two voices, one with a Castilian Spanish accent and the other a Mexican Spanish accent. We shall use these voices in our proposal to convert one into the other, as described in the following sections. We should remark that the creation of Spanish voices using this technique is quite recent.

Statistical parametric speech synthesis produces the acoustic parameters of a voice from a set of HMMs, which have been trained on a speech database of quality recorded utterances of the target speaker. Figure 1 shows a typical HMM implementation for voice parameters generation, where each state has an output described by a set of pdfs for spectrum (mel frequency cepstral coefficients, mfcc), fundamental frequency $F0$ and aperiodicity streams.

Each state generates vectors from the pdfs, and a vocoder such as STRAIGHT [7] can be used to generate the speech waveform.

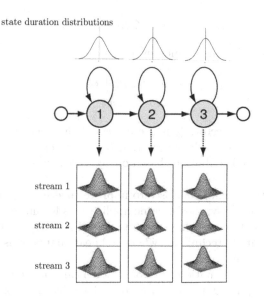

Fig. 1. Typical HMM implementation in speech synthesis with $F0$, mfcc and aperiodicity pdfs for each stream

Those pdfs are assumed to be simple multivariate Gaussian distributions, so each observation $o(t)$ from a HMM state can be described by [2]

$$b_i(o_t) = \mathcal{N}(o_t; \mu_i, \Sigma_i) \tag{1}$$

where μ_i is the mean vector and Σ_i the covariance matrix of the Gaussian.

3   HMM Adaptation

One of the prominent capabilities of HMM-speech synthesis is the ability to change the characteristics of the synthesized speech by modifying the HMM parameters, such as using an adaptation technique [8], obtaining quality voices from and average voice to a small amount of target speaker information. This technique has also been reported as working well in cross lingual voice conversion [9–13], where the algorithm is applied for an average voice in some language to a target voice in another language.

When the average voice model and target speaker use the same language, it can be refereed as intra-lingual adaptation. Usually, a few minutes of speech data for the target speaker is enough to produce quality voices from a good average model in intra-lingual adaptation. In speaker adaptation, the mean and covariance of pdfs from a well-trained average model can be transformed to match those of the target speaker. A transformation function, i.e. linear transformation, can be used to do so, such as MLLR for Gaussian means

$$\overline{\mu}_i = \zeta_k \mu_i + \epsilon_k \tag{2}$$

or CMLLR considering mean and covariance

$$\overline{\mu}_i = \zeta_k \mu_i + \epsilon_k \tag{3}$$
$$\overline{\Sigma}_i = \zeta_k \Sigma_i \epsilon_k \tag{4}$$

where μ_i is the mean vector, ζ_k is the matrix of the linear transform, Σ_i is the covariance matrix, and ϵ a bias vector. A regression class tree are generated to cluster the gaussian distributions based on acoustic similarity, to share the same transform.

This adaptation technique arises to improve speech recognition [14,15], and then applied to speech synthesis. Some applications for improving speech recognition with different accents have been proposed recently in [16]. Figure 2 illustrates the adaptation technique, where a linear mapping is applied between average and target models.

This kind of transformation in a voice model can change the target speaker's prosodic features [17]. Our aproach is to explore this transformation in HMM-speech synthesis adaptation as an accent modification of Spanish voices, as a special case of cross-lingual adaptation, where average voice model and target speaker are in different languages. It is necessary to explore the differences of the accents to be considered, and how to proceed with the distribution mapping of the models that doesn't match between them due to phonetic differences.

In cross-lingual adaptation, language mismatch between average voice language (L_{pdf}) and target language voice (L_{data}) may occur during training, in regression tree structure, or in synthesis. Four possible cases of mismatch arise, as described in [18].

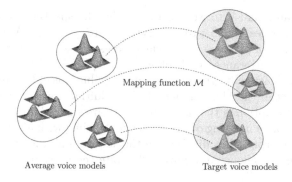

Fig. 2. Speaker adaptation technique with a linear mapping $\mathcal{M}$

4   HMM State Mapping of Spanish Dialects

For cross lingual speaker adaptation, HMM state mapping has been proposed in [8]. The techniques allows the average voice models to be altered by given adaptation data in different language [18]. In that case, there is adaptation data in L1 and an average voice model set for synthesis in L2 [10]. By establishing mapping rules in the form of linear transforms $\hat{\Lambda}$ between input and output languages, HMM state mapping is capable of relating the two different languages. In a similar fashion, for accent modification in both ways we need average accent models A1 and A2.

In Spanish accent conversion from Castilian Spanish to Mexican Spanish, beggining with the same database for both accents pronounced in its particular way, most of phonemes of A1 are shared with A2, it means that most of the HMM λ will have its equivalent in the other accent, but a difference should be introduced to achieve those unseen in the other language.

Castilian Spanish has a description of 29 phonetic symbols in SAMPA [19]. As exposed in some papers as [20], for Latin American spanish accents, including Mexican, graphemes [j] and [g] followed by either [e] or [i] are transcribed in Spain as /x/; in Latin American Spanish both are transcribed as /h/. The other main difference arises in graphemes [s] and [z], which are transcribed in Spain as /s/ and /T/ respectively. Grapheme [c] followed by [e] or [i] is transcribed in Spain as /T/. In Latin America, all that cases are transcribed as /s/. Mexican Spanish has some particular phonemes, specially after the /x/ grapheme, with no clear rules to the different pronunciations, appearing specially in proper names. That particular case will be considered in future experiments.

For accent modification with HMM-synthesis adaptation, let $\mathbb{S}_{in}$ and $\mathbb{S}_{out}$ be the state distributions of input and output accents, respectively. As mentioned above, most of the state distributions will be shared in phonetic content. Our proposal is to map A1 and A2 from each other with the following rules.

– From Castilian Spanish to Mexican Spanish: The speaker adaptation establishes mapping rules from $\mathbb{S}_{in}$ to $\mathbb{S}_{out}$, $\mathcal{M}_d : \mathcal{M}_d(\mathbb{S}_{in}) = \mathbb{S}_{out}$. As the former has a bigger number of phonemes, it became necessary to substitute the phonetic transcription of utterances with phonemes /T/ and /s/ to /s/ in all cases. The mapping practically ignores the state distributions of Castilian phonemes as non existent in the Mexican accent. For additional experiments where the phonetic transcription is the same, but there are slight differences in pronunciation, we apply a masking of the phoneme, labeling with a special prefix, 'x', so it will be interpreted as different on the adaptation process and will be the same as a unseen phoneme. Figure 3 illustrate the procedure of the mapping for this case.

– From Mexican Spanish to Castilian Spanish: The adaptation establish mapping rules from unseen distributions in $\mathbb{S}_{out}$. All the state distributions in $\mathbb{S}_{in}$ will be transformed into $\mathbb{S}_{out}$. In this case, the utterance to synthetize requires the Castilian phonetic transcription. The new distributions for the unseen phonemes will be created through mapping from the castillian average voice. Figure 4 illustrates this process of generating unseen distributions for the new phonemes.

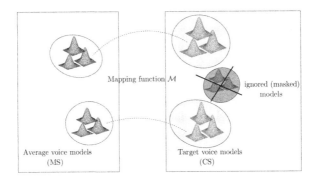

Fig. 3. State mapping from Mexican Spanish to Castilian Spanish

The number of transformations from adaptation data to target voice data are estimated from the amount of available data. If only a small size corpus is available, a global transformation for all the distributions is applied.

5   Experiments

5.1   Corpus Description

Two Mexican professional actors (male and female) recorded a set of 184 Mexican Spanish utterances with seven emotions: neutral, angry, joy, fear, surprise, sadness and disgust. These included isolated words as well as affirmative and interrogative sentences as shown in Table 1.

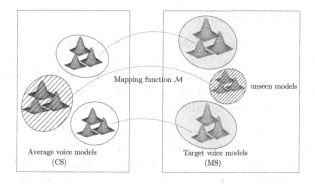

Fig. 4. State mapping from Castilian Spanish to Mexican Spanish

Table 1. Spanish Corpus Contents for each gender and emotion

Identifier	Corpus contents
1–100	Affirmative
101–134	Interrogative
135–150	Paragraphs
151–160	Digits
161–184	Isolated words

The selection of the words, sentences and paragraphs were the same as that of [21]. The records were carried out in a professional studio where all the technical aspects and the recording conditions were completely controlled. In similar conditions were recorded male and female Castilian Spanish databases in the ELRA Catalogue, which we used for the Castilian average voice or the target language. We also used an audio book with a Mexican voice with over 500 utterances for the Mexican average voice.

5.2   Experimental Conditions

Context-dependent HMM with five states and standard mixture-of-diagonal-Gaussian models were used. The number of HMM states is closely related to the phonetic units (phonemes) length, and have been successfuly applied in speech recognition. Speech features were 39th-order mel-cepstra, log $F0$, and their delta and delta-delta coefficients. The CMLLR algorithm and 85 adaptation utterances were used, adapted from the HTS-Straight adaptation demo in [22], with over 700 utterances for average voices. Decision clustering trees and regression class trees share the same questions for both methods of accent modification. Several experiments were carried out to perform different average voices and mapping for the target voice, including mixing gender, emotions, and the audiobook voices.

5.3 Subjective Evaluation

In order to conduct a subjective evaluation of the voices, we decided to put synthesized utterances on-line via a web browser interface (http://goo.gl/forms/O09clz4IYnzuPhC82), in order to try to have a diverse spanish speaking public. In the end, 35 subjects logged on and answered the evaluation.

The following results were evaluated:

- Castilian Spanish voice without speaker adaptation (CS)
- Mexican Spanish voice without speaker adaptation (MS)
- Accent modification from Mexican Spanish to Castilian Spanish (MS to CS)
- Accent modification from Castilian Spanish to Mexican Spanish (CS to MS)

Each synthetic utterance was judged using a three point response scale for accent identification, including one option for unidentifiable accent. Mexican and Castilian spanish native speakers evaluate four utterances for each case.

5.4 Objective Evaluation

It is known that accents are partly conveyed by the differences in vowel duration patterns and speaking rate [23]. For this accent modification, our proposal is to assess the quality of the accent modification by exploring how vowel duration, speech rate, and some pitch contours are affected.

6 Results and Discussion

The results of the subjective evaluation regarding the accent identification of the selected utterances by the listeners, are shown in Fig. 5. It can be seen how the majority of opinions correctly identified the Castilian accent voice, with a small percentage of misidentification. The Mexican accent voice that converted to a Castilian accent was viewed by most as having a Castilian accent, but some reviews indicated that listeners could not identify the accent.

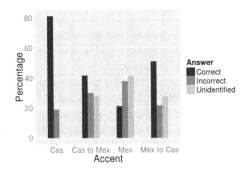

Fig. 5. Percentage of correct, incorrect and unidentifiable accent opinions for Castilian, Mexican, Mexican to Castilian and Castilian to Mexican voices

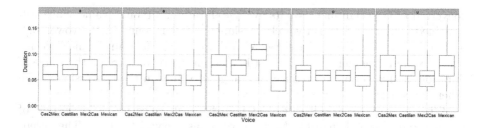

Fig. 6. Boxplots of the five vowel duration for Castilian, Mexican, Mexican to Castilian and Castilian to Mexican voices

Regarding the Mexican accent, the results are less conclusive, although the majority of listeners could identify the converted Castilian accent to the Mexican accent with the target accent. The low rate of correct identifications in the Mexican accent voice may be due to their inferior quality compared to that of the Castilian Spanish voice. The difference in quality could be explained by the more consistent speaking style of the Castilian speaker, in comparison with the expresive recordings of the Mexican speaker.

The results of objective evaluations based on duration of vowels can be seen in Fig. 6. Here, the most important differences are found in the vowel /i/, where the duration in Mexican Spanish tends to be lower. Mexican Spanish accent voice

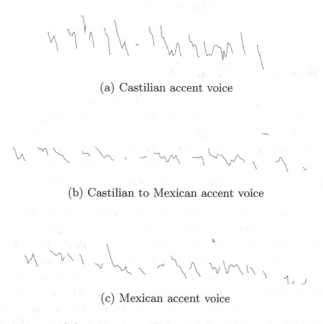

(a) Castilian accent voice

(b) Castilian to Mexican accent voice

(c) Mexican accent voice

Fig. 7. Pitch contours of the utterance "Con estoico respeto a la justificia adyacente guardó sus flechas"

converted to Castilian increases its average duration, although values have still remained above the Spanish vowels. A similar effect is found in the vowel /u/, with a tendency to have greater range in duration in Mexican Spanish accent, which explains the conversion to Castilian accent tends to increase the range of duration of this vowel and reduce the range of values in the conversion in the other direction.

With regards to speech rate, Castilian Spanish voice has a lower rate of phonemes per second than the Mexican voice, as shown in Table 2. The accent modification affects the speech rate in the direction of bringing it closer to the target accent.

Finally, Fig. 7 shows how a pitch contour of the Castilian voice is transformed, becoming more similar to the Mexican voice by accent adaptation. Those three pitch contours have the same vertical scale, and the same utterance is pronounced using each of the three voices.

Table 2. Speech rate of 100 utterances in each accent and accent converted voices. MS: Mexican Spanish. CS: Castilian Spanish

Voice	Speech rate
MS	13.44
CS	11.07
MS to CS	12.99
CS to MS	11.30

7   Conclusions and Future Work

In this paper, we have investigated how the HMM-based speach synthesis adaptation technique can be used to modify the accent of synthetized speech from Mexican Spanish to Castilian Spanish and vice versa. Subjective results have shown favorable accent identification, especially in the conversion of Mexican accent to Castilian accent. In the other direction, the results have not been as clear, which may be due to voice quality obtained from databases of Mexican accent.

Objective evaluations have shown how prosodic elements of the voices are affected, tending the voices to make it sound like the target accent. This coincides with the results of subjective evaluations and shows how the adaptation technique helps to change prosody in this accent modification.

For future work, we must continue testing larger databases and include other techniques besides adaptation, such as deep learning. We should also experiment with other Latin American accents in conversions, which can lead to greater personalization of synthesized voices for a wide variety of applications. Some additional objective measures for accent assessment should be taken into consideration, in addition to speaker similarity and intelligibility after conversion.

Acknowledgements. This work was supported by the SEP and CONACyT under the Program SEP-CONACyT, CB-2012-01, No.182432, in Mexico, as well as the University of Costa Rica in Costa Rica. We also want to thank ELRA for supplying the original Emotional speech synthesis database.

References

1. Hermansky, H.: Should recognizers have ears? Speech Commun. **25**(1), 3–27 (1998)
2. Tokuda, K., Nankaku, Y., Toda, T., Zen, H., Yamagishi, J., Oura, K.: Speech synthesis based on hidden markov models. Proc. IEEE **101**(5), 1234–1252 (2013)
3. Lazaridis, A., Khoury, E., Goldman, J.-P., Avanzi, M., Marcel, S., Garner, P.N.: Swiss french regional accent identification. In: Proceedings of Odyssey (2014)
4. Woehrling, C., de Mareüil, P.B.: Identification of regional accents in french: perception and categorization. In: INTERSPEECH (2006)
5. Leemann, A.: Comparative analysis of voice fundamental frequency behavior of four swiss german dialects: Elektronische daten, Ph.D. dissertation, Selbstverlag (2009)
6. Beckman, M., Daz-Campos, M., McGory, J.T., Morgan, T.A.: Intonation across spanish, in the tones and break indices framework. Probus **14**(1), 9–36 (2002)
7. Kawahara, H.: Straight, exploitation of the other aspect of vocoder: perceptually isomorphic decomposition of speech sounds. Acoust. Sci. Technol. **27**(6), 349–353 (2006)
8. Wu, Y.-J., Nankaku, Y., Tokuda, K.: State mapping based method for cross-lingual speaker adaptation in hmm-based speech synthesis. In: Interspeech, pp. 528–531 (2009)
9. Wu, Y.-J., King, S., Tokuda, K.: Cross-lingual speaker adaptation for HMM-based speech synthesis. In: 6th International Symposium on Chinese Spoken Language Processing, ISCSLP 2008, p. 14. IEEE (2008)
10. Liang, H., Dines, J., Saheer, L.: A comparison of supervised and unsupervised cross-lingual speaker adaptation approaches for HMM-based speech synthesis. In: 2010 IEEE International Conference on Acoustics Speech and Signal Processing (ICASSP), pp. 4598–4601. IEEE (2010)
11. Oura, K., Tokuda, K., Yamagishi, J., King, S., Wester, M.: Unsupervised cross-lingual speaker adaptation for HMM-based speech synthesis. In: 2010 IEEE International Conference on Acoustics Speech and Signal Processing (ICASSP), pp. 4594–4597. IEEE (2010)
12. Yoshimura, T., Hashimoto, K., Oura, K., Nankaku, Y., Tokuda, K.: Cross-lingual speaker adaptation based on factor analysis using bilingual speech data for HMM-based speech synthesis. In: 8th ISCA Speech Synthesis Workshop, pp. 317–322 (2013)
13. Nagahama, D., Nose, T., Koriyama, T., Kobayashi, T.: Transform mapping using shared decision tree context clustering for HMM-based cross-lingual speech synthesis. In: Fifteenth Annual Conference of the International Speech Communication Association (2014)
14. Gales, M.J.: Maximum likelihood linear transformations for HMM-based speech recognition. Comput. Speech Lang. **12**(2), 75–98 (1998)
15. Acero, A., Deng, L., Kristjansson, T.T., Zhang, J.: HMM adaptation using vector taylor series for noisy speech recognition. In: INTERSPEECH, pp. 869–872 (2000)

16. Motlicek, P., Garner, P.N., Kim, N., Cho, J.: Accent adaptation using subspace gaussian mixture models. In: 2013 IEEE International Conference on Acoustics, Speech and Signal Processing (ICASSP), pp. 7170–7174. IEEE (2013)
17. Tamura, M., Masuko, T., Tokuda, K., Kobayashi, T.: Adaptation of pitch and spectrum for HMM-based speech synthesis using MLLR. In: Proceedings of 2001 IEEE International Conference on Acoustics, Speech, and Signal Processing, (ICASSP01), vol. 2, pp. 805–808. IEEE (2001)
18. Liang, H., Dines, J.: An analysis of language mismatch in HMM state mapping-based cross-lingual speaker adaptation. Technical report, Idiap (2010)
19. Llisterri, J., Mariño, J.B.: Spanish adaptation of sampa and automatic phonetic transcription. Reporte técnico del ESPRIT PROJECT, vol. 6819 (1993)
20. Caballero, M., Moreno, A., Nogueiras, A.: Data driven multidialectal phone set for spanish dialects. In: INTERSPEECH. Citeseer (2004)
21. Elra catalogue: Emotional speech synthesis database. http://catalog.elra.info. Accessed 30 Nov 2014
22. HTS: HMM speech synthesis system. http://hts.sp.nitech.ac.jp/. Accessed 20 Jan 2015
23. Yan, Q., Vaseghi, S., Rentzos, D., Ho, C.-H.: Analysis by synthesis of acoustic correlates of british, australian and american accents. In: Proceedings of IEEE International Conference on Acoustics, Speech, and Signal Processing, (ICASSP 2004), vol. 1, p. I637. IEEE (2004)

Author Index

Printed in the United States
By Bookmasters